Anonymous (The Monk)

Mechanism for the eyes and ja

(Characteristics and layout determined by radiography)

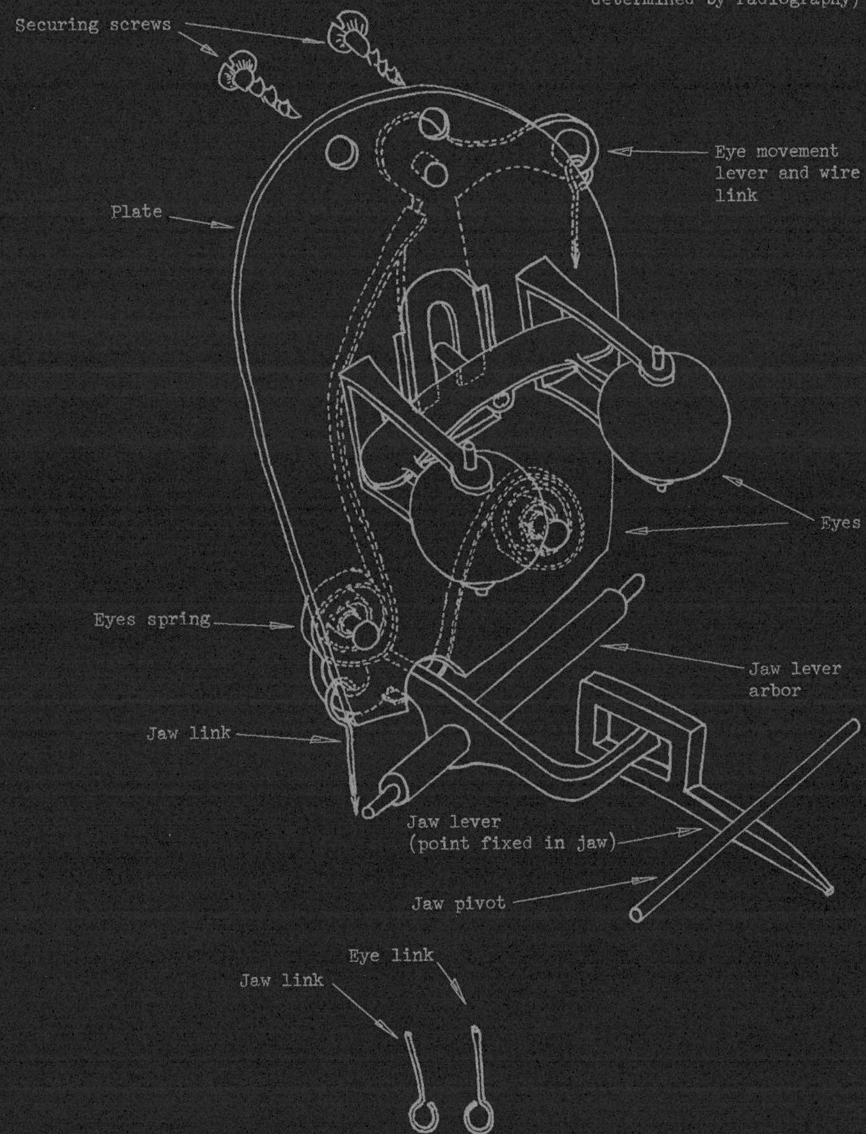

MIRACLES AND MACHINES

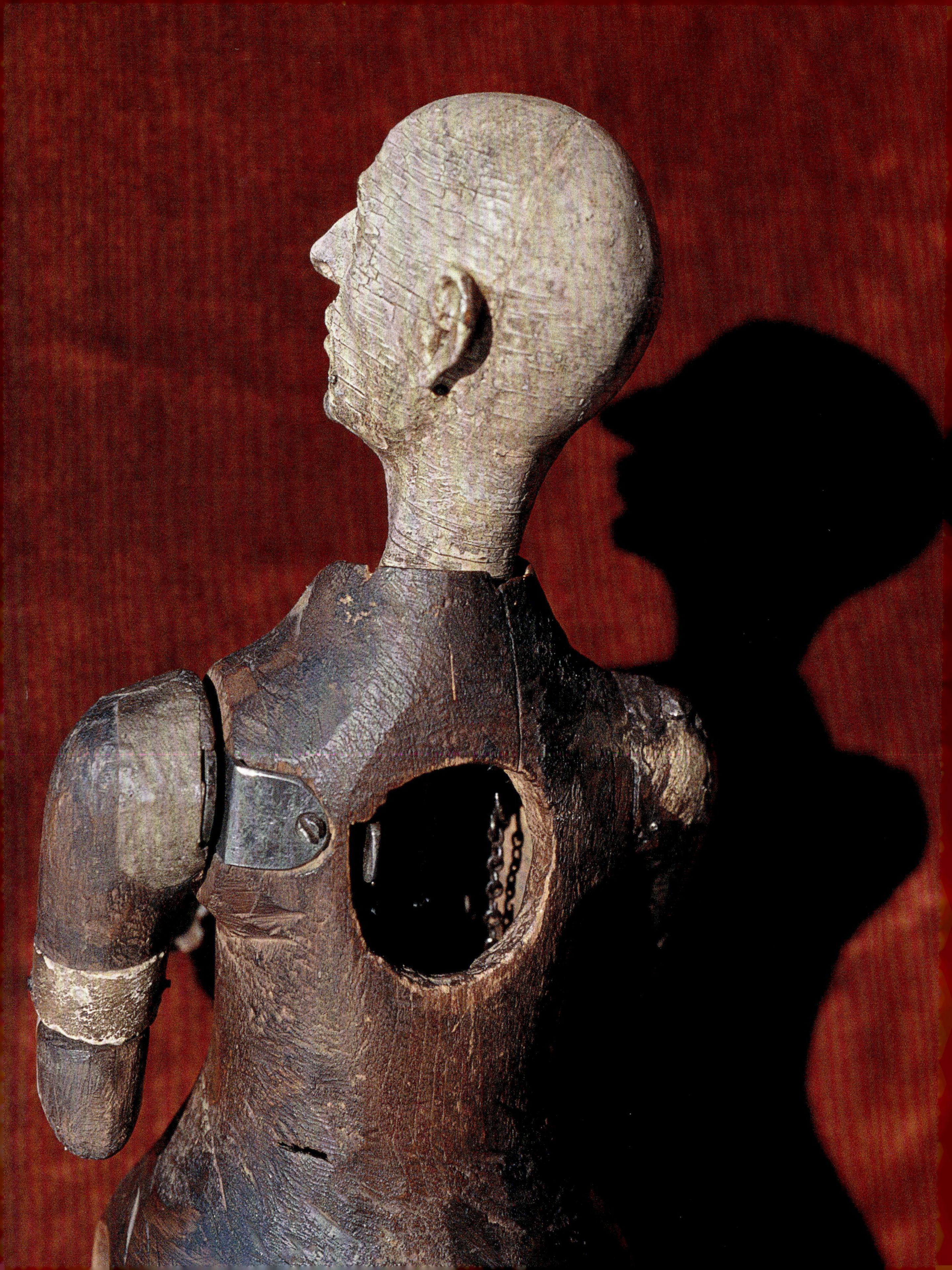

MIRACLES AND MACHINES

A SIXTEENTH-CENTURY AUTOMATON AND ITS LEGEND

Elizabeth King
and W. David Todd

WITH PHOTOGRAPHS BY
Rosamond Purcell

GETTY PUBLICATIONS | LOS ANGELES

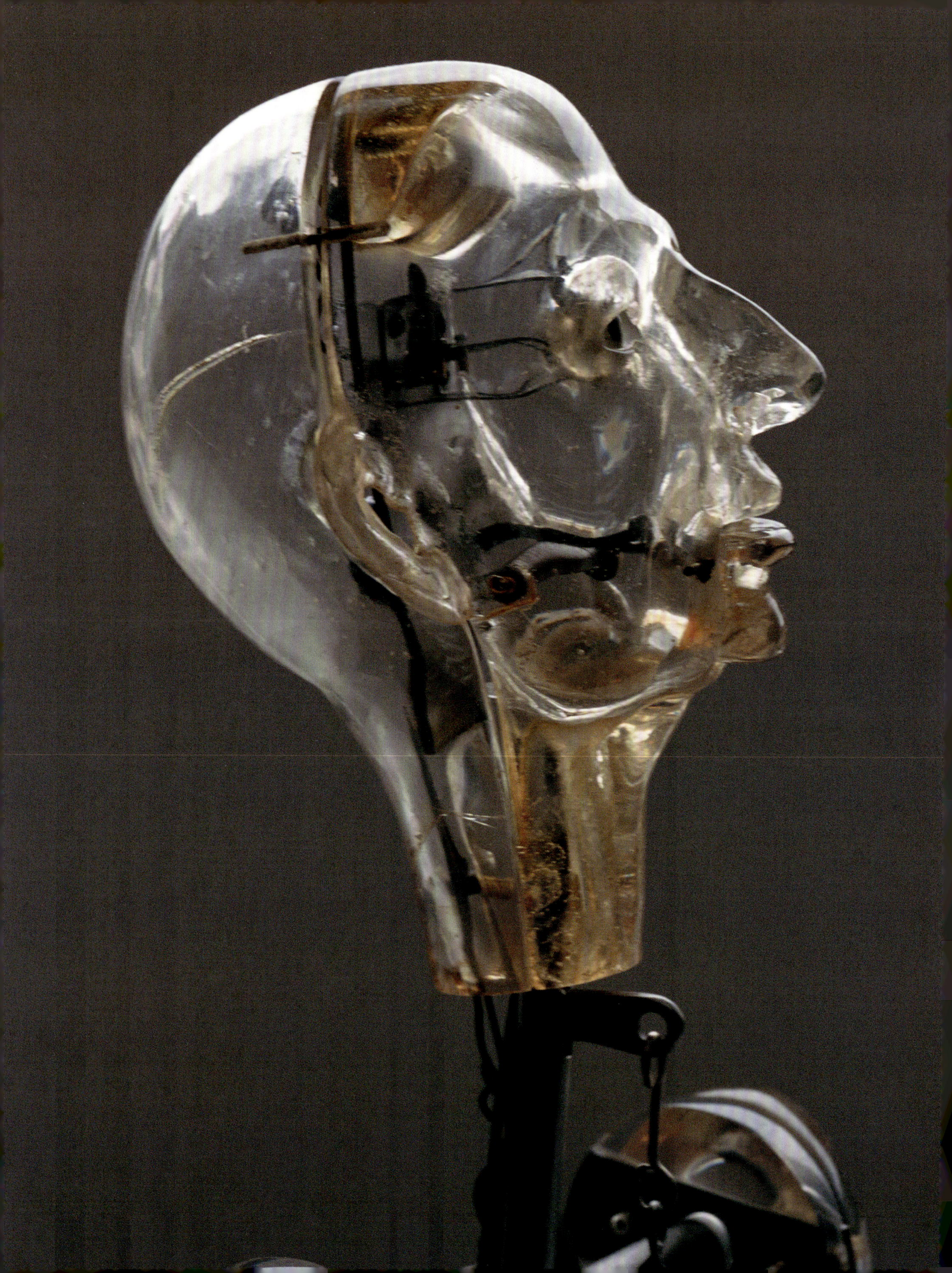

Contents

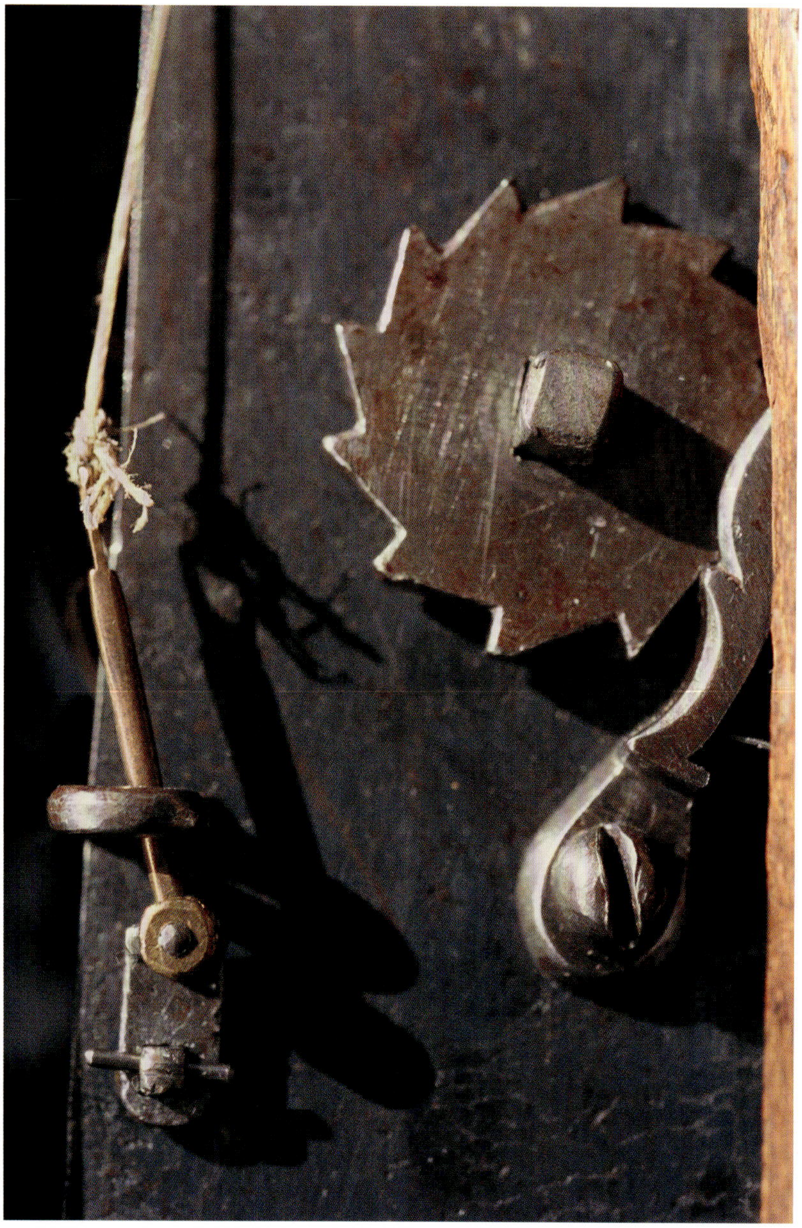

To Otto Mayr, Silvio Bedini, Carlene Stephens, and Servus Gieben
and to José Antonio García-Diego and Adelheid von Herz,
all of whom escorted the monk out of oblivion

QVI. SIM. SCIES. SI. PAR. OPVS. FACERE. CONABERIS

"You Will Know Who I Am If You Try to Make This"

—Juanelo Turriano's Latin inscription on his great astrarium, recorded by Ambrosio de Morales, *Las antigüedades de las ciudades de España*

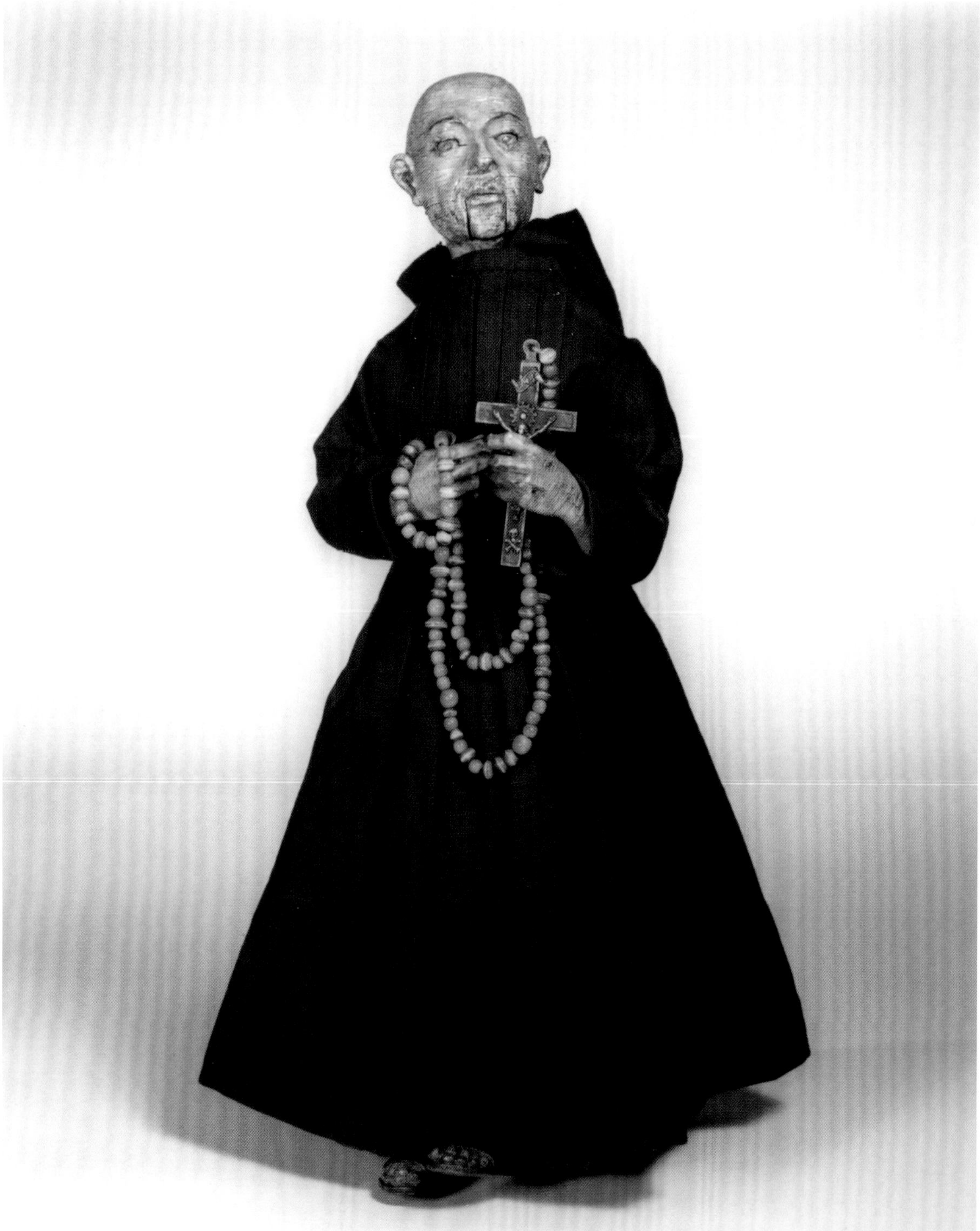

Introduction

In 1977 the Smithsonian Institution purchased a small automaton made in Europe in the second half of the sixteenth century. A diminutive figure of a friar in Franciscan tunic and sandals, the automaton possesses a jointed wooden body animated by a concealed drive mechanism of hand-forged iron (**FIGS. I.1, I.2**). Set in motion, the figure performs for spectators on any tabletop, without a mediating plinth or stage. "The monk," as he is popularly known, walks and turns along a nonrepeating path, striking his chest with one hand, raising and flourishing a small wooden cross with the other. As he walks, his head and his eyes pivot to gaze at the cross, then the onlookers, then the cross. His mouth opens and closes as in speech. From time to time he brings the cross to his lips and kisses it.

The monk's wooden hands, head, and feet are delicately carved and painted. The paint, now cracked with age, was clearly once aglow with life; a faint flush still lingers on cheek and lip. The eyeballs are made of iron, hammered round and painted. The figure wears a tunic made of real cloth. An X-ray of the head reveals the internal machinery that guides the motions of the eyes and mouth. At sixteen inches in height, the machine weighs just under five pounds. This is what we now call an android—an automaton of a human figure that moves by its own accord. In good working order after some 450 years, the monk comes to life in animated defiance of the wounds of time.

FIGURE I.1.
Automaton of a Friar, mid-sixteenth century. Spain or South Germany. Wood, fabric, iron clockwork. Height: 40.64 cm (16 in.). On arrival at the Smithsonian, 1977. Curators immediately recognized its importance to the histories of technology and artificial life.

The automaton arrived with a story. Legend claims that the figure is a votive representation of Diego de Alcalá, a Franciscan lay brother whose holy corpse was agent to the miraculous cure of the young prince Don Carlos, heir to the Spanish throne, who lay near death from a head wound in 1562. The prince's father, Philip II of Spain, petitioned Rome to recognize the miracle, and Pope Sixtus V made Diego a saint in 1588. Juanelo Turriano, royal clockmaker to the king, and to the king's father, Emperor Charles V, must have fashioned the automaton, the story goes, to reenact the cure with state-of-the-art technology. The wizardry of Juanelo, known for his automata, was celebrated in a 1575 biography. Prince, cure, and saint: this much is history. More mysterious are the documents connecting the automaton itself to the drama in the Spanish court.[1]

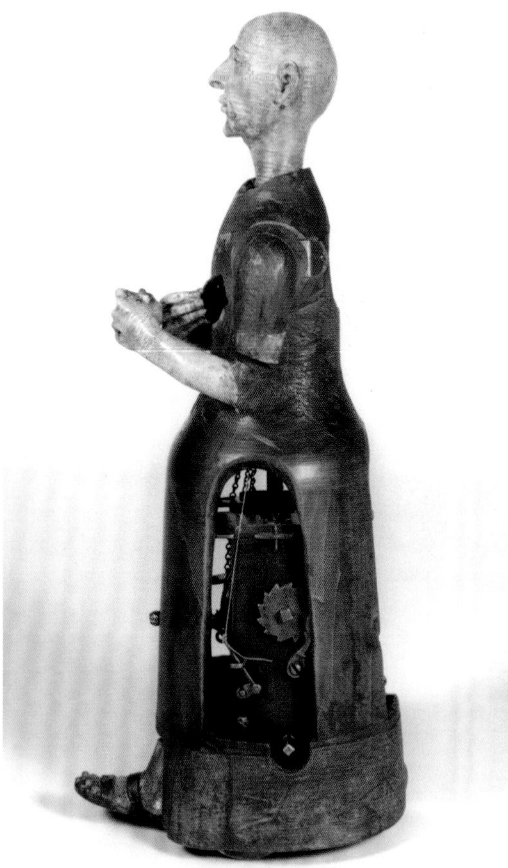

FIGURE I.2.
Side view of the automaton without the tunic, affording a glimpse of the internal clockwork.

The historian Otto Mayr, then curator at the Smithsonian's National Museum of History and Technology, wrote to his acquisition board before the purchase, "For the history of technology, this automaton is important as one of the earliest program-controlled self-acting machines that have survived to the present." Referring to makers of several great Enlightenment mechanical androids and their connection to the origins of computing, he noted that "it is a direct predecessor of the eighteenth-century automata of Vaucanson and Jaquet-Droz, of the Jacquard loom and Babbage's computer."[2] Mayr equally understood its significance to philosophy and to the literature of artificial life, for few of the famed automata attributed to inventors and thinkers from Archimedes to Descartes have materialized beyond legend. Fate is reversed in the case of the monk: it arrived intact from the tumult of history with no letter of introduction. We find no identifying mark, no inscription on the clockwork.

Every great and surviving automaton from history engenders legends. Mechanism starts them in motion; stories keep them running. The history of legends of artificial beings is the subject of an essay that Mayr himself wrote several years before encountering the monk. Makers of automata from antiquity to the Renaissance, he wrote, engaged a technology "in which rational, magical, and mystical elements intermingled."[3] Do legends spring from the object itself or from gaps in our knowledge about the object, its origins, its uses? The performance confronts even the twenty-first-century robot-savvy viewer in a demanding and urgent way. What did an onlooker in the sixteenth century see and believe, witnessing the monk in action?

Let us imagine the monk set in motion by a trusted steward, in a privileged setting. The mainspring has been wound ahead of time, out of sight of the assembled spectators, and the key withdrawn. The steward steps forward holding the figure upright in both hands and places it on a table, with a feint that discreetly releases a hidden stop-work lever. Slowly the monk comes to life. He turns his head to single out one among the company. Left foot stepping forth from under the cassock hem, then right foot, the monk advances in the direction of his gaze, raising the cross and rosary before him as he walks. His eyes move: turning his head, he looks to the raised cross and then back to his subject. His mouth opens, then closes, affording a glimpse of teeth and interior. He bends his right arm, and with the beaked fingers of his hand, he strikes his chest. The small blow is audible. Now he lowers his head as he walks, and with elbow and shoulder in synchronized motion he brings the cross to his lips and kisses it. Twenty seconds into the act, he has taken eight steps, beaten his chest three times, kissed the cross, and traveled a distance of nineteen inches. At what seems like the last moment—doubtless the guest in his sights has backed away from the table—he executes a turn to his right and begins a new advance. But as he turns, he keeps his eyes on that person for two seconds longer, then looks away. He will make seven such turns if the mainspring is fully wound, navigating the table and fixing spectators, one by one, with an iron gaze. The uninterrupted repetition gives the campaign its force.

This book narrates the search for the origins of the monk and the sources of the legend. We examine a set of intersecting histories, sites, and witnesses, for the monk until now has evaded classification. As a prototype, this early figure represents a synthesis of sculpture and mechanism in one object. The monk's carved and painted wooden body displays, in miniature, the virtuoso polychrome sculptural tradition at its height in early modern Europe. A dissection of the concealed iron clockwork firmly locates its fabrication in the early years of the golden age of clockmaking in the same period. How were these separate worlds combined in the fabrication of a *moving* sculpture? The question cannot be broached without acknowledging the monk's character as a Catholic icon. Though this is one of the best-preserved examples of an automaton that predates and anticipates the famous mechanical androids of the Enlightenment, the monk and his story must not be confined to a study of technology alone. Art, religion, natural science, medicine, folklore, and popular culture all play a part in the performance of this charismatic object from the past. Against the backdrop of Reformation and Counter-Reformation Europe, who indeed commissioned the monk, and for whom was the performance intended? Have other such walking automata survived to the present? The paradox of a rational object performing a mystical act: What can it reveal of the birth of the machine in a still-fervent age of faith?

Seen today, the monk in performance brings us face to face with our own ambivalent relationship to artificial life: the human/machine interface, the mind/body riddle, the illusion of agency, the history and psychology of belief, and our passionate arguments about what a robot can and cannot do.

We are a clockmaker (David Todd) and a sculptor (Elizabeth King), putting our two worlds together to study a 450-year-old clockwork sculpture. As artisans ourselves, we write from the perspective of workshop and studio, from the knowledge of our hands. David has been a practicing clockmaker since the late 1960s. Trained in his native England, he is an expert on sixteenth- to eighteenth-century English, European, and American clockwork. He apprenticed under the third-generation clockmaker Cyril Boniface, who preserved the family commandment that a clockmaker ought to be able to make a clock without using modern power tools. To conserve a sixteenth-century clock, the precise skills of the original maker were to be mastered first. From 1978 to 2006, David was museum specialist in Timekeeping at the Smithsonian Institution's National Museum of History and Technology (now the National Museum of American History), where he worked under Otto Mayr, Silvio Bedini, and Carlene Stephens. One of his first tasks at the Smithsonian was to examine, clean, and document the newly acquired automaton monk. He cared for it and kept it in running condition for the next thirty years. Elizabeth is a sculptor whose works in wood, bronze, and porcelain reflect her interest in early puppets and clockwork automata. She first encountered the monk in 1989 on a field trip organized by the artist Christopher Furman, then a student in the sculpture department at Virginia Commonwealth University where Elizabeth was teaching **(FIG. I.3)**. Twenty sculpture students crowded into David Todd's small workshop in the museum basement, where he murmured the legend of the monk's origins while demonstrating the methods of the sixteenth-century clockmaker who made its iron parts. A few years later Elizabeth began a search for the source of the curious story. Steeped in the history of figurative sculpture and drawn to literature's age-old legends of artificial figures that come to life, she was in the midst of an artist fellowship in 1996 at the Mary Ingraham Bunting Institute (now the Radcliffe Institute of Advanced Study at Harvard) and found she had access to a great library.

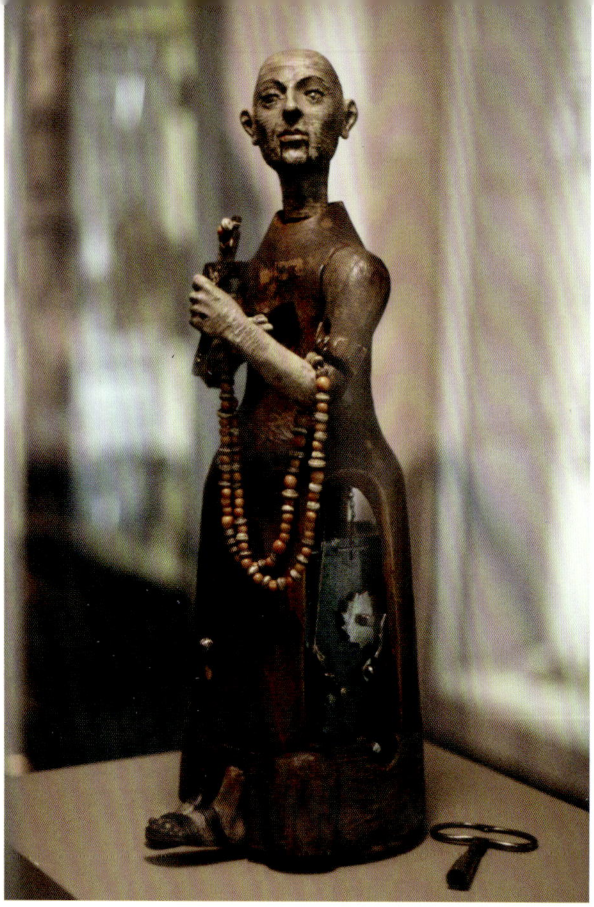

and David together compare them in part 3 and propose that this group of self-propelled walking figures constitutes a class of its own in the history of artificial life. Part 4 assembles witnesses, then and now, in pursuit of the origins of these automata. Sixteenth-century chronicles name Juanelo Turriano but also the Nuremberg clockmaker Jakob Bulmann among makers of such automata. Juanelo was a royal court clockmaker in Habsburg Spain; Bulmann, a guild clockmaker in South Germany. We sketch the differences between these two settings: David describes a typical clockmaker's workshop in Nuremberg, while Elizabeth locates descriptions of automata in the inventory of the famous *Kunstkammer* in Prague assembled under Rudolf II, cousin to Don Carlos. In tracking the cascade of references to these two clockmakers by scholars and horologists across time and into the present, Elizabeth discovers the constellation of sources for the monk's legend and lays them out in chapter 24. In part 5 Elizabeth asks how a spectator in the sixteenth century might have viewed a small moving religious icon in the context of devotional sculpture before and during the Counter-Reformation. At last, we both return to the bedside of the prince, to see why the legend behind a machine that prays has gained a life of its own.

Our book is itself a clockwork: many separate parts are assembled to describe an object that is a sculpture, a machine, an icon, and a messenger. We circle back again and again to the physical object that is the monk in our search for its origins. The search itself is left visible, for the chronology of our discoveries gained its own drama. Our two voices are distinct. In part 1 Elizabeth lays out the strands of the legend: prince, saint, clockmaker, automaton. Part 2 is David's work: the object that is the monk, what it is and how it works. By what authority can we establish a date for its manufacture, and is it fully original to that date? (We believe that only the monk's robe, cross, and rosary have been replaced in the intervening centuries.) Surely no single artisan could be adept at fashioning both sculpture and clockwork: How was the automaton made? Seven surviving automata similar to the monk are known at present in collections in Europe, several of them once attributed to Juanelo Turriano. Elizabeth

Our study is brought to life by the photographs of Rosamond Purcell, whose color plates appear throughout this book. Her closely observed portrait series of the monk was made at the Smithsonian's National Museum of American History on December 6, 2002, during the early years when we first began our collaboration. When we asked if she might come to the museum with her camera, her only requirement was to be allowed to photograph the monk in natural light. On the appointed day, the Director's Conference Room, its stately fifth-floor windows facing south onto the Mall, was free. David brought the monk upstairs from his workshop and posed it on the massive table, and Rosamond set to work in the clear winter light. Her focus on the mute material of the monk, the marks of the passage of time on the wooden body, the delicate and speaking cracks in the paint—miraculously unspoiled by previous "restoration"—tells a visceral story of age and fate. The photographs answer the monk's mystery with the immediate news of its sheer physical presence.

FIGURE I.3.
The monk on display in 1993 in what was then the Hall of Timekeeping, National Museum of American History, Smithsonian Institution, Washington, DC.

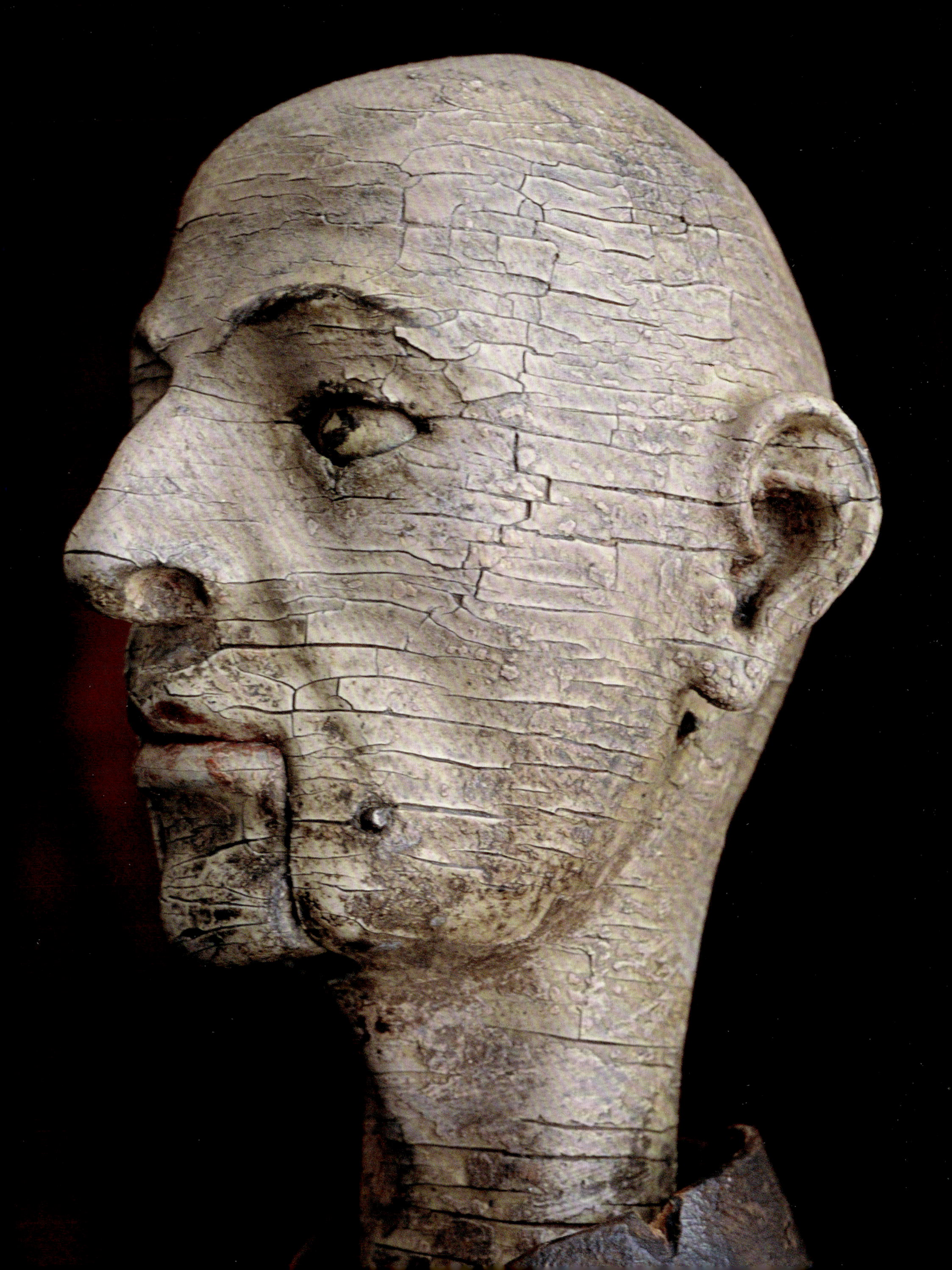

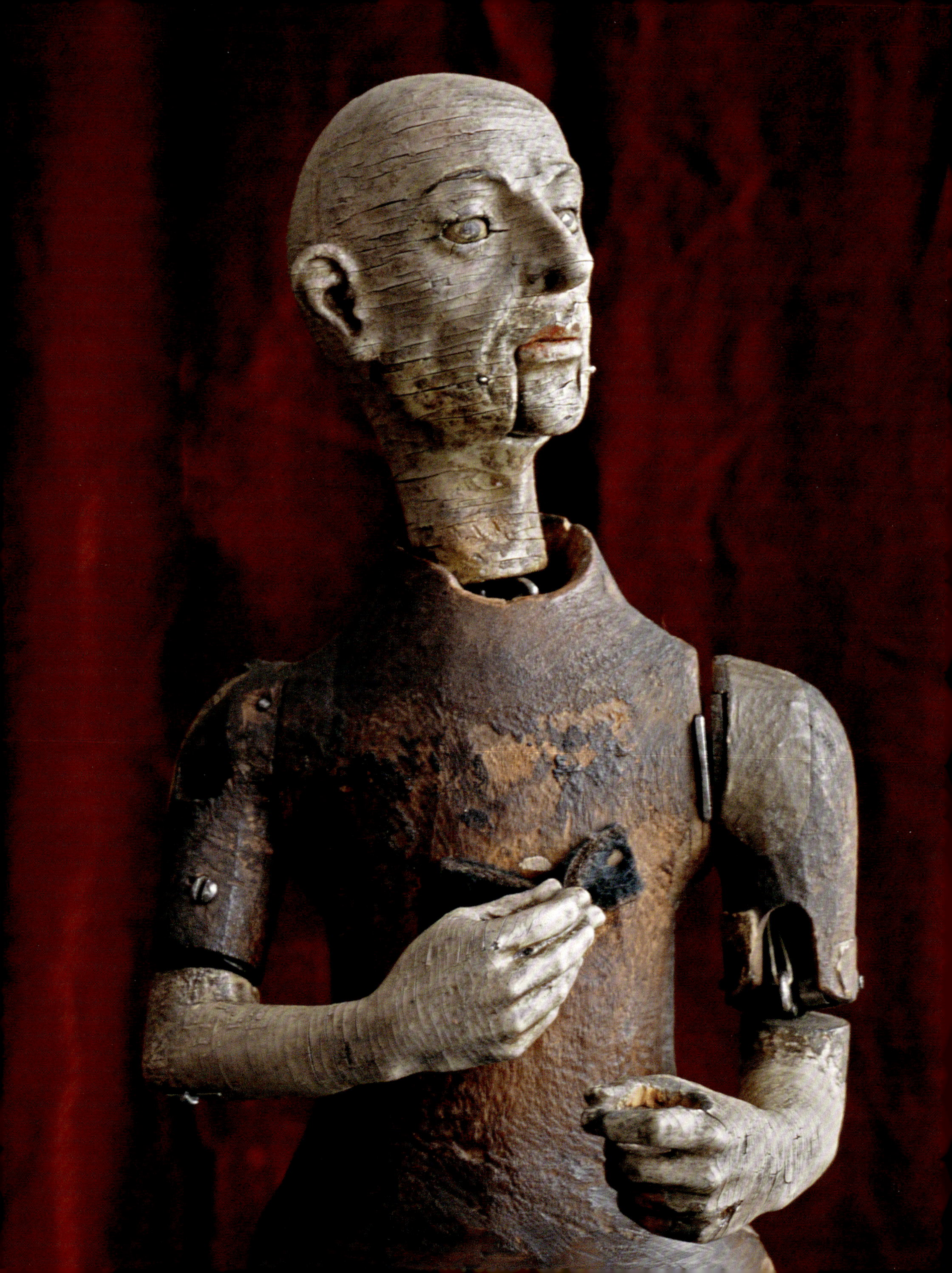

PART I

A Legend

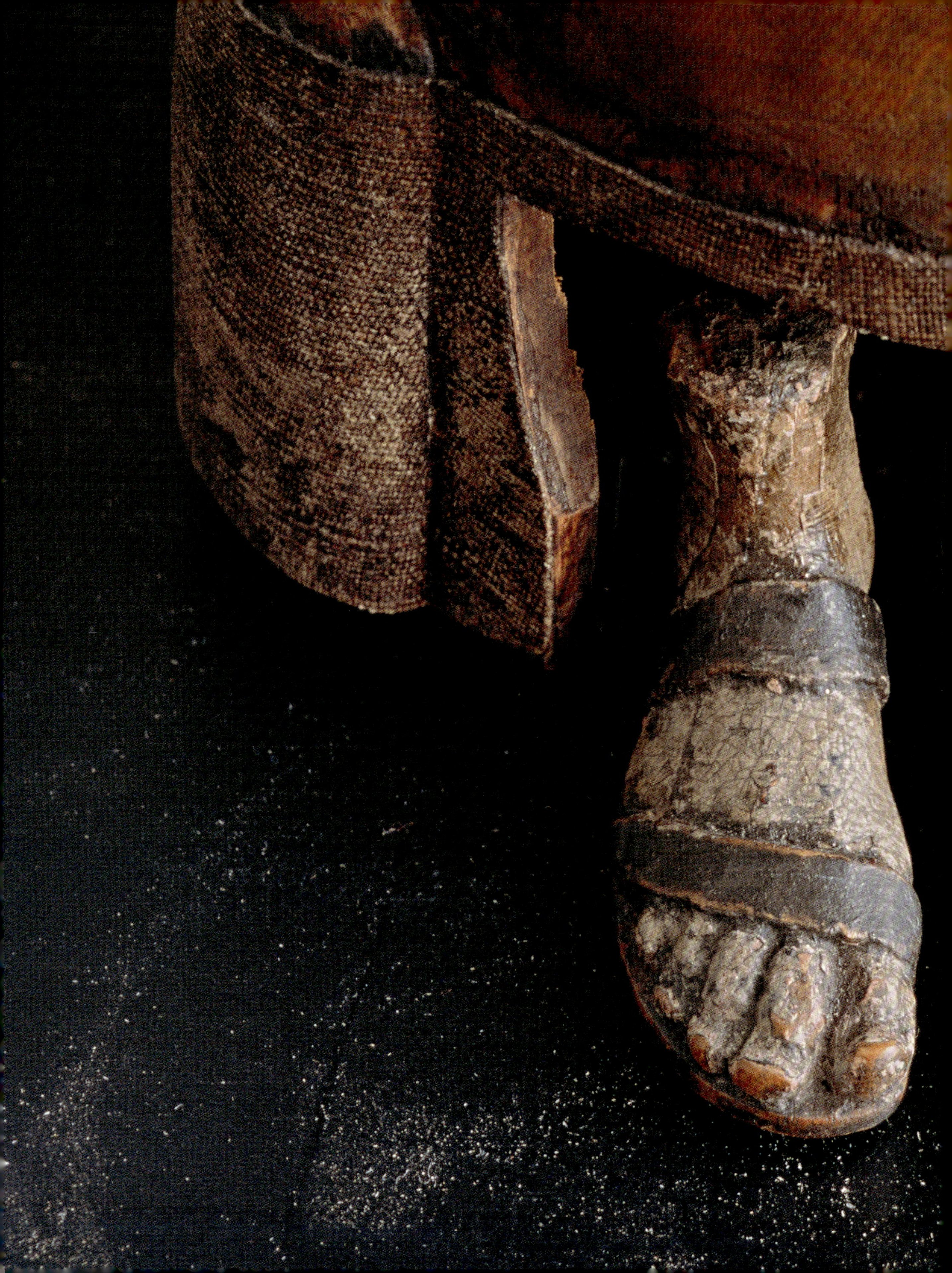

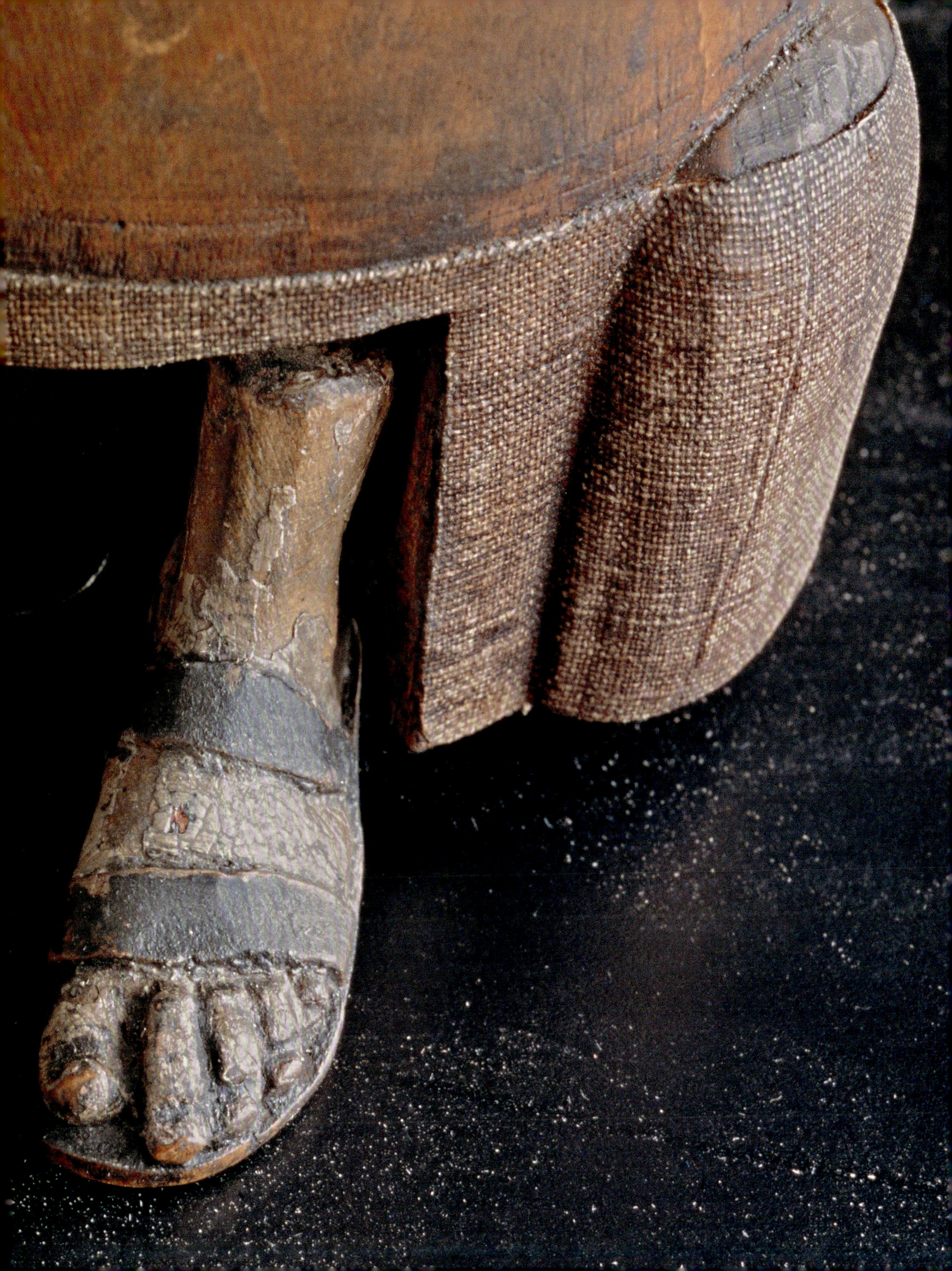

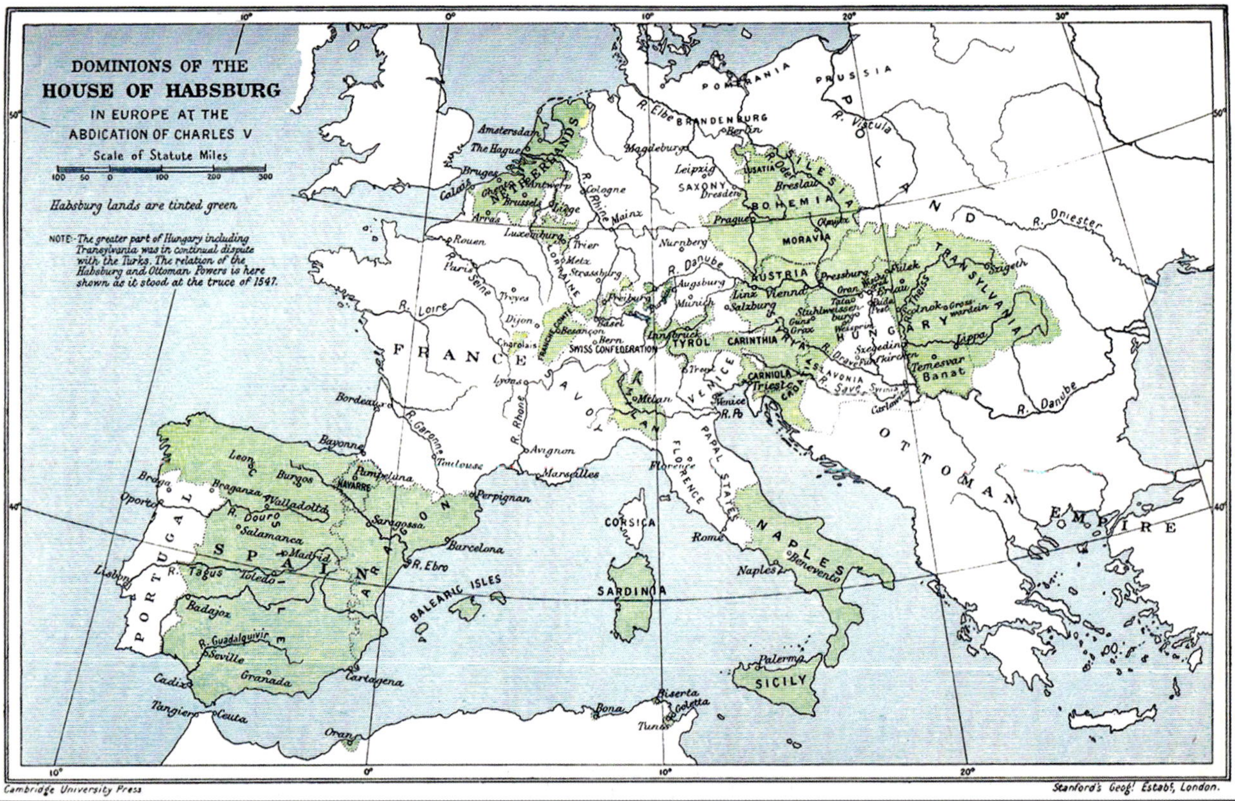

Cambridge University Press

Stanford's Geog! Estab! London.

1 To Set the Stage

In the reign of Charles V (as he is now best known) from 1516 to 1556, the House of Habsburg, a ruling dynasty in Central Europe originating in the thirteenth century, attained its broadest geographic reach (**FIG. 1.1**). Before his abdication and the division of his territories in 1556, Charles was Holy Roman emperor and monarch of the lands and cultures of Austria, Hungary, Bohemia, South Germany, Burgundy and the Netherlands, the Italian kingdoms of Naples and Sicily and the duchy of Milan, and Spain and the Spanish conquests around the world, including Mexico and Peru. The period of his rule spanned the late Renaissance and the flowering of the classical humanism of Erasmus of Rotterdam, followed by the devastating religious wars of the Protestant Reformation and the Council of Trent that launched the Counter-Reformation of the Catholics. Almost continually at war with France and with the Ottoman sultanate of Suleiman the Magnificent far to the east, Charles was equally engaged by battles within the empire: Spanish revolts, papal intrigues, German peasant uprisings. Aggressive, strategic, diplomatic, and revered, he believed he could unite all Christendom into a world-defining imperium equal to the golden age of Augustus. He was a powerful patron of arts and letters, astronomy and clockwork, and a lover of feast and pageant. The works by Titian and Albrecht Dürer that he commissioned can be seen today in the collection of the Museo del Prado in Madrid.

FIGURE 1.1.
The House of Habsburg, Europe, 1547.

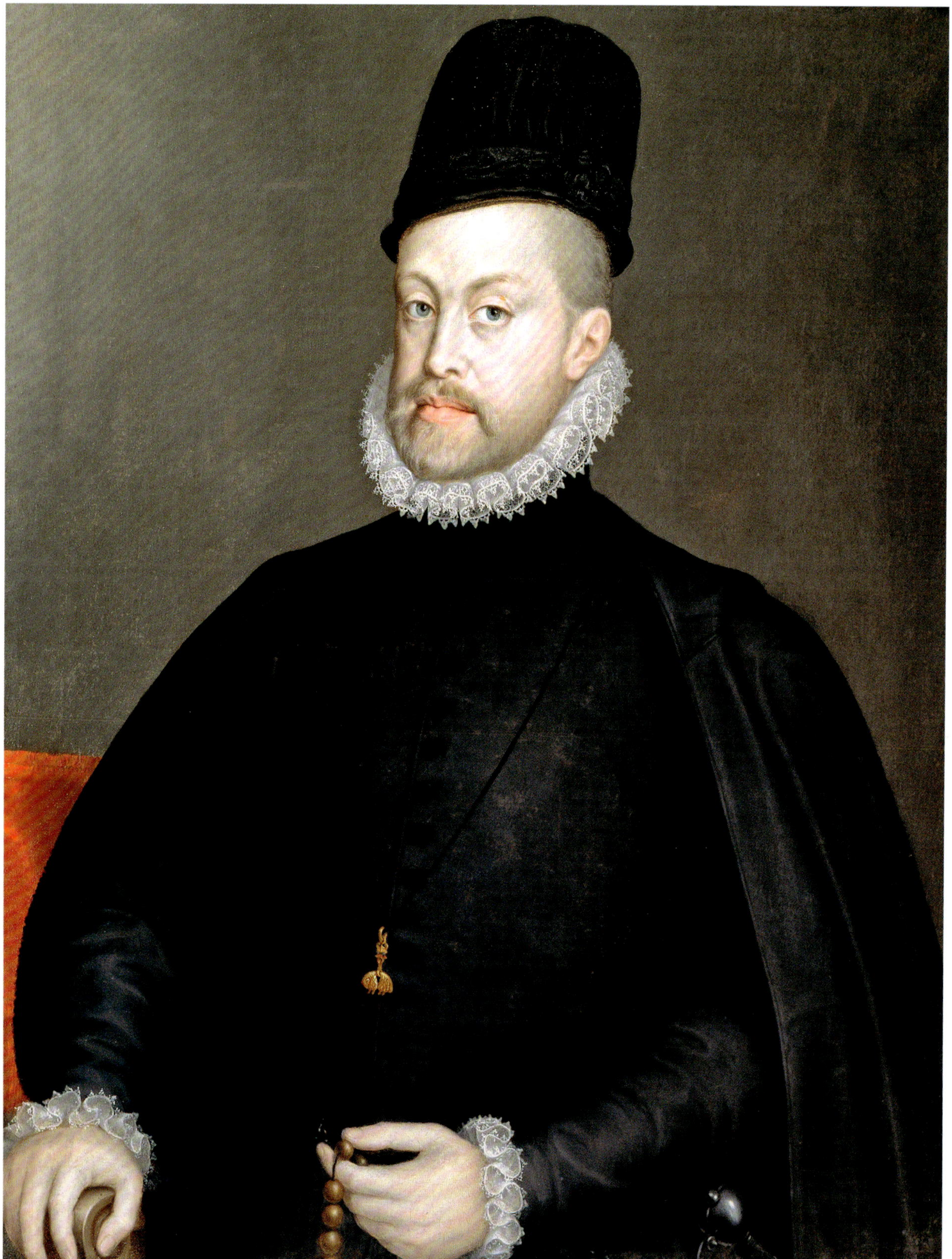

In 1556, at the age of fifty-six, Charles stunned Europe by relinquishing his crown, dividing his realms between his son Philip II and his brother Ferdinand I, and retiring to the remote Hieronymite monastery of San Yuste in Spain, where he died within two years, surrounded by his paintings, his books, and his clocks. In the now-split Habsburg line, Ferdinand I (and after him, in quick succession, Ferdinand's son Maximilian II and then grandson Rudolf II) assumed the Holy Roman imperial rule of domains in Central Europe.

Philip II (1527–1598) inherited the Netherlands, the Burgundian Franche-Comté, the Italian kingdoms, and the great kingdom of his native Spain—together with Spain's colonies in the Americas and the Caribbean, founded under the crown of Castile in the late fifteenth century and dramatically, if brutally, expanded over the half century preceding his ascension to the throne (**FIG. 1.2**). He was briefly king consort of England through his second marriage, to the reigning queen, Mary Tudor. Following Mary's death, he formed an alliance with France through his third wife, Elisabeth of Valois, although good relations with neither England nor France would last. When he secured Portugal in 1580, and thus the Portuguese claims in Africa, the Americas, India, Indonesia, and China, his authority extended across an empire larger than any known in history. In contrast to Charles's militant mobility, Philip ruled his empire almost exclusively from Spain, a poised and stationary monarch who conducted his global affairs through his viceroys and generals. Spain's vast enterprise of colonization, religious conversion, and plunder in the western hemisphere was a major source of the wealth that sustained his reign. Under the forty-year rule of this "most Catholic king," as he has often been called, the Counter-Reformation attained its most repressive form. Spain's Inquisition was already well established by Philip's time, implemented in 1478 by Philip's great-grandparents Ferdinand and Isabella, who launched the persecution and dispersal of the great populations of Muslims, Jews, and Berbers that had defined the Iberian Peninsula as one of the most culturally diverse regions in the world.

FIGURE 1.2.
Sofonisba Anguisola, *Philip II*, 1573. Oil on canvas. 88 × 72 cm (34⅝ × 28⁵⁄₁₆ in.). Museo Nacional del Prado, Madrid. In the mid-sixteenth century, Phillip's authority extended throughout western Europe, Africa, Asia, and the Americas, but he was unable to prevent misfortune from befalling his son and heir.

Yet Philip's reign inaugurated what has been called Spain's artistic golden age: Juan de Herrera, Miguel de Cervantes, and El Greco flourished within or in proximity to his court. In his massive San Lorenzo del Escorial—palace, monastery, museum, library, mausoleum—Philip installed his father's great collections of art and added his own: new works by Titian; the sculpture of Leone and Pompeo Leoni and Benvenuto Cellini; paintings by Antonis Mor, Alonso Sánchez Coello, and the great El Greco. Among the Netherlandish masterpieces he inherited or sought out was Rogier van der Weyden's incomparable *Descent from the Cross* and what is now the world's finest collection of the works of Hieronymus Bosch.

Hugh Trevor-Roper, in his 1976 book, *Princes and Artists*, ponders the strange attraction that drew Philip to Bosch: "Why did he wish to contemplate, in his private rooms, those bizarre canvases filled with strange symbolism of monstrous figures emerging from broken eggshells, incomplete, fantastic forms, weird half-animal machinery and mechanised animals, and gleeful devils toasting naked bodies in lurid, fuliginous flames?" Trevor-Roper can do no more than defer to Philip's librarian at the Escorial, the Hieronymite monk José de Sigüenza,

> who, of all men, best understood his royal master's mind. There are some people, says Sigüenza, who are so foolish as to think that Bosch was a heretic; which is absurd, for if that were so, the king would never have had his pictures near him. No: Bosch is a devout and orthodox satirist of our sins and follies: what distinguishes him from other painters is that while they seek to paint men as they outwardly appear, "he has the courage to paint them as they really are."[1]

Who was Philip's firstborn son, Don Carlos? Bedeviled from birth by poor health, he was suspected of having inherited the unsound mind of his great-grandmother Juana la Loca (Joanna the Mad). The visionary theater of Bosch, pageant of saints and demons, wounded bodies and uncanny creatures, may capture the faith and fervor of Philip's worldview, but the king's effort to preserve "outward appearance" could not prevent the young prince's fall from grace.

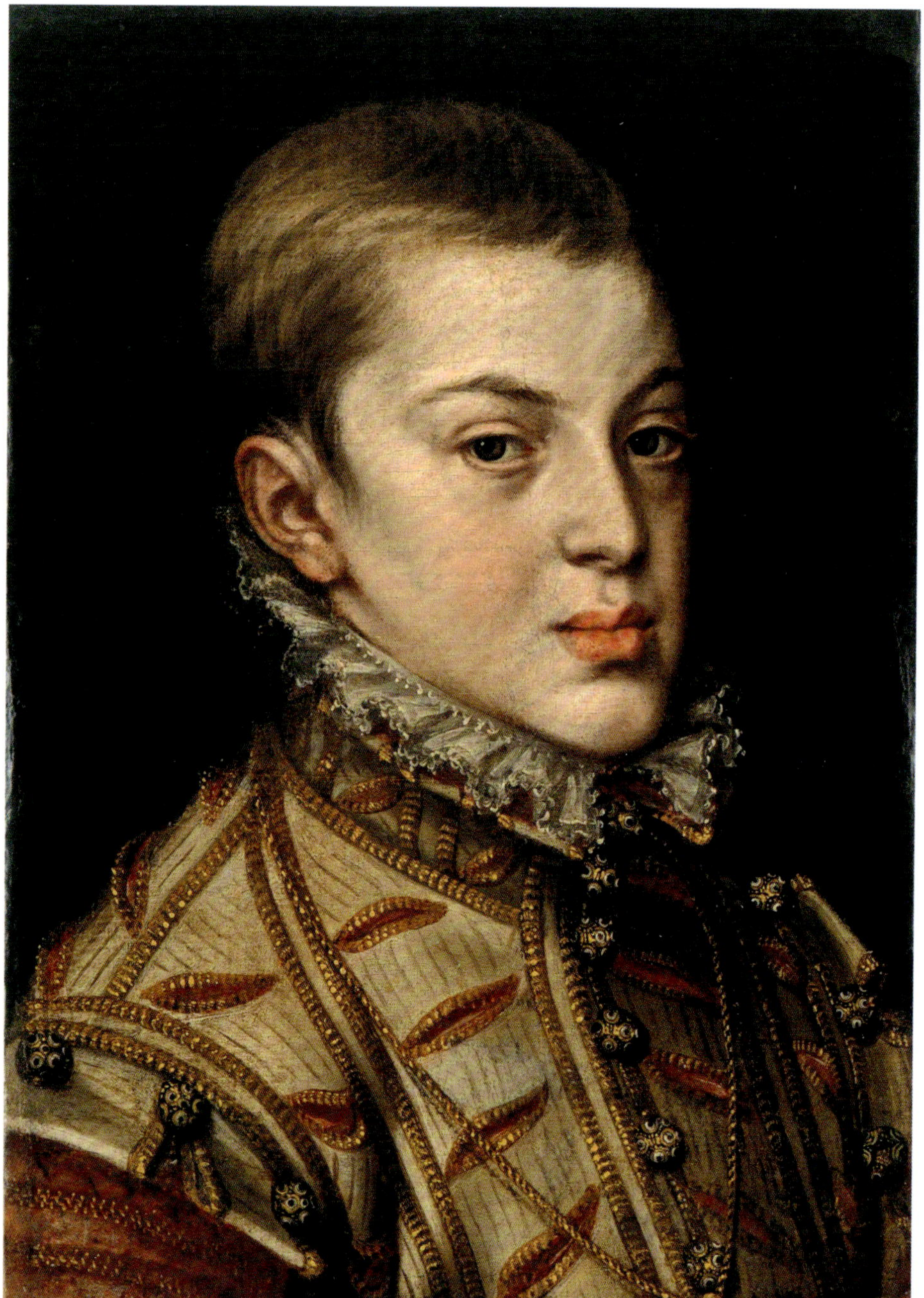

2 A Miracle and a Vow

Don Carlos is perhaps best known today as the romantic hero of Giuseppe Verdi's opera *Don Carlo*, first performed in Paris in 1867. Verdi set to music Friedrich Schiller's eighteenth-century play about a noble and defiant crown prince of Spain, destroyed by his jealous father, King Philip, in a tale of empire, doomed love, and Inquisition. The historical record reveals a real-life sixteenth-century Spanish prince, king, and court of a very different order, although even today contention clings to the short and strange life of Don Carlos. Heir to the Spanish throne, the prince was born misshapen and remained childlike in frame and mind into adulthood. Unfit for roles at court, he grew dangerously volatile; by the time he was twenty-two, Philip found no choice but to confine him. His death soon after gave Spain's enemies, especially those in the Netherlands, a pretext to claim the prince a martyred rebel. So orchestrated was the propaganda that historians, repeating it, turned fabrication into whole cloth. In the battle for command of any public narrative, such is the power of a good story.[1] But an earlier event in the life of the unfortunate prince, with repercussions just as momentous for the Spanish crown, launched a different story, the one we tell now (**FIG. 2.1.**).

FIGURE 2.1.
Alonso Sánchez Coello, *El príncipe don Carlos*, 1564–65. Oil on canvas. 43 × 31.2 cm (16 15/16 × 12 1/4 in.). Coleção da Câmara Municipal, on deposit at the Museu Nacional de Soares dos Reis, Porto, Portugal. The heir to the Spanish throne suffered a dramatic accident at the age of seventeen.

Born in Valladolid in 1545, Don Carlos was the first of eight children by four successive wives of Philip II. His mother, Maria of Portugal, age seventeen, died soon after his birth. The prince was himself seventeen years old in 1562 when, on Sunday, April 19, just after noon, rushing from his room on an ill-fated indiscretion, he fell down a flight of little-used stairs in his royal lodgings in the university town of Alcalá de Henares and struck his head against a closed door in the passageway below.[2] The historian William H. Prescott, in his 1874 *History of the Reign of Philip the Second*, describes what happened to Don Carlos after the fall.

> He was taken up senseless, and removed to his chamber, where his physicians were instantly summoned, and the necessary remedies applied. At first, it seemed only a simple contusion on the head, and the applications of the doctors had the desired effect. But soon the symptoms became more alarming. Fever set in. He was attacked by erysipelas; his head swelled to an enormous size; he became totally blind; and this was followed by delirium. It now appeared that the skull was fractured.... Meanwhile, the greatest alarm spread through the country at the prospect of losing the heir-apparent. Processions were everywhere made to the churches, prayers were put up, pilgrimages were vowed, and the discipline was unsparingly administered by the fanatical multitude, who hoped by self-inflicted penance to avert the wrath of Heaven from the land.[3]

All modern accounts of this injury describe a grief-stricken king rushing to Alcalá from Madrid with his councils of state and the best physicians in the land, among them, the Flemish anatomist Andreas Vesalius.[4] One detailed recent study, published in 1995 by the historian L. J. Andrew Villalon, takes us through day by appalling day of the ordeal, quoting from the accounts of the prince's surgeon.[5] On the afternoon of May 9, twenty-one days after the fall, in the aftermath of a halted attempt by physicians to trepan the patient's skull, the townspeople of Alcalá gathered at the Church of Saint Francis. "With Franciscan friars in the lead, they marched toward the palace, carrying with them the remains of a fifteenth-century member of the order, Diego de Alcalá, for whom they had long hoped to win sainthood."[6] Prescott tells the story a little differently, having the king himself and members of his court fetching "the mouldering remains of the good father, still sweet to the nostrils, as we are told." Laying the corpse on the prince's bed, they removed the cloth that wrapped its head and placed it on Carlos's forehead.[7] Witnesses describe the prince, eyes swollen shut, rousing himself from delirium to feel with his hands the exposed face of the corpse and then touch his own eyes and mouth.

Yet by evening death seemed so imminent that the king took his closest advisers and departed before midnight, riding to the Hieronymite monastery outside Madrid to pray and await the final news. But that night the prince slept peacefully for the first time in weeks. The next day he began a sudden, extraordinary recovery. He regained his sight a week later, his fever disappeared soon after, and within a month he was on his feet.

In the aftermath of this near- resurrection, the question arose as to how the cure was gained. Some praised the efforts of Vesalius, although a few among the ten attending physicians who labored for two months at the prince's bedside felt they deserved some credit. But attention came to focus on fray Diego de Alcalá, with a groundswell of feeling that here was the agent of a miracle. We can look back today and see two miracles, the second being the patient's survival of the physicians' purging and bleeding and the daily probing of the wound with nonsterile instruments. ("They went on placing upon the exposed portion of the skull a powder made of iris and birthwort, and on the lips of the wound a mixture of turpentine and egg yolk. Following this, they cleaned the wound with honey of roses and applied a plaster of betony.")[8] The most dramatic evidence came from Don Carlos himself. On the night of May 9, following the intercession of the corpse, he was heard speaking in his sleep, as if addressing the beatus, saying the words, "Basta, basta, padre, aparta el cordon" (lit., "Stop, stop, father, take away the cord"). On waking the next morning, he reported a vision: "This night the holy fray Diego appeared, standing at my bed, with a reed cross in his hand. I asked him if he was Saint Francis for it seemed to me that he was; he answered he was not Saint Francis but he had trust in God that I would be healed."[9] Certain that the figure had been Brother Diego himself, Don Carlos and King Philip vowed to bring this miracle before the pope. In Counter-Reformation Spain, even Andreas Vesalius himself would risk accusations of heresy were he to contest such testimony to divine intervention.

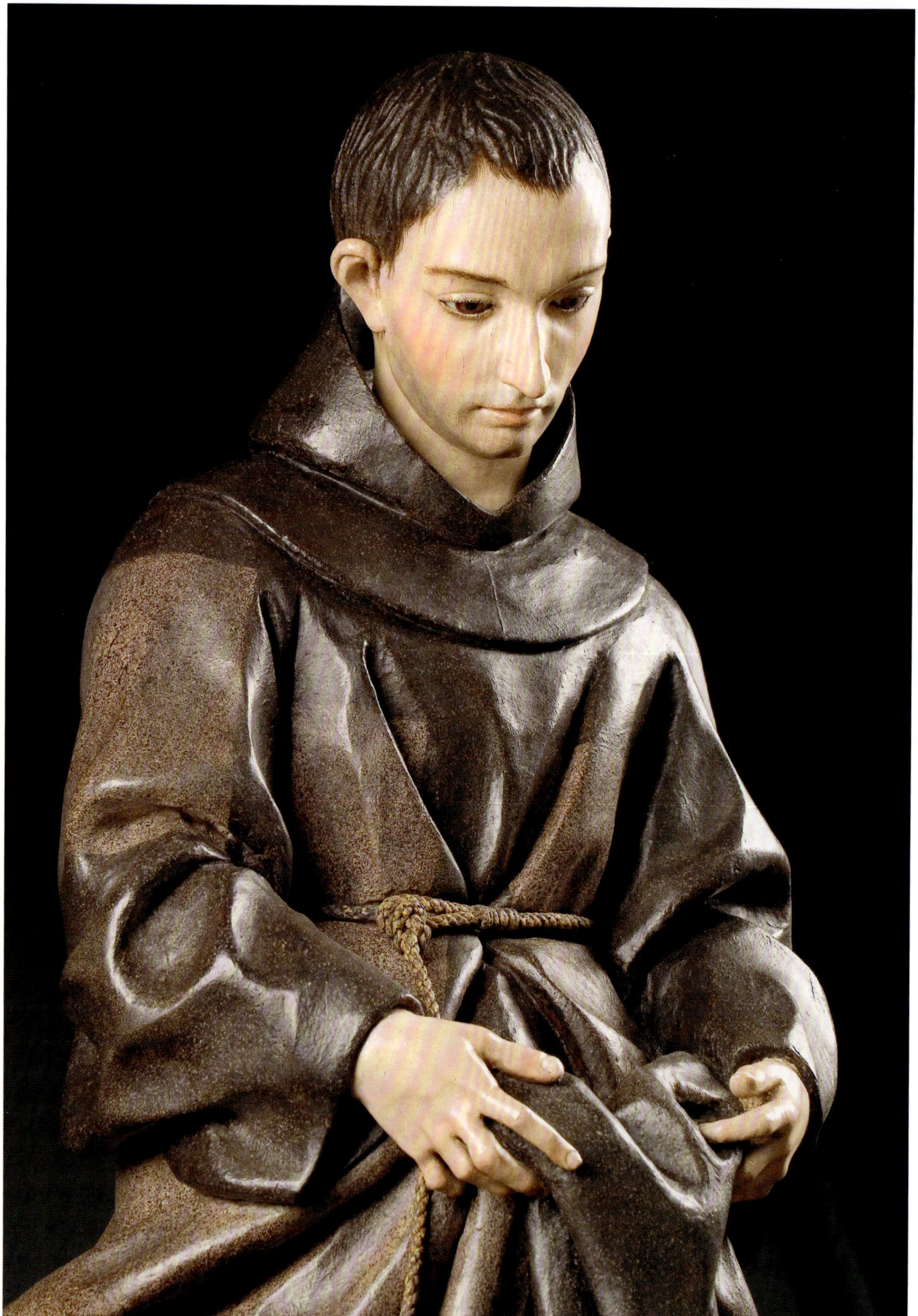

3 Diego de Alcalá

A dead body brings another body, on the brink, back to life. Don Carlos's cure was not the first wonder performed by this thaumaturgical corpse. Through the hundred years before its appearance on the royal stage, Diego's preserved body had gained a reputation throughout Castile for its divine agency. Diego had lived a life of uninterrupted poverty, famous for the asceticism of his practice, but his death in 1463 was only the beginning of his holy career. Born around 1400 in the small village of San Nicolás del Puerto, in Andalucía, Diego had become first an anchorite hermit, then a Franciscan brother, serving missions in the Canary Islands and then caring for the sick in Rome, to which city he had walked on foot from Castile to attend the Jubilee year ceremonies in 1450. In the final decade of his life he performed the duties of gatekeeper for the monastery of Santa María de Jesús in the town of Alcalá. Alonso Cano's seventeenth-century sculpture shows him holding a clutch of flowers in the folds of his robe **(FIG. 3.1)**. Confronted in the act of stealing monastery bread to give to the poor, legend goes, he opened his robe at his superiors' command and out tumbled not the illicit loaf but a fragrant bouquet. He remained illiterate all his life, and this prevented him from advancing beyond a lay affiliation with the order. After his death, as his reputation spread, numerous attempts were made to bring him to papal notice, but efforts lost traction in the declining reign of King Henry IV and the rise of Ferdinand and Isabella.

FIGURE 3.1.
Alonso Cano, *San Diego de Alcalá*, 1652–57 (detail). Polychromed wood, glass eyes, hemp cord. 60.5 × 23.7 × 23.5 cm (23 13/16 × 9 5/16 × 9 1/4 in.). Instituto Gómez-Moreno, Fundación Rodríguez Acosta, Granada. Following Diego's death in 1463, his pristine corpse gained a reputation for divine agency.

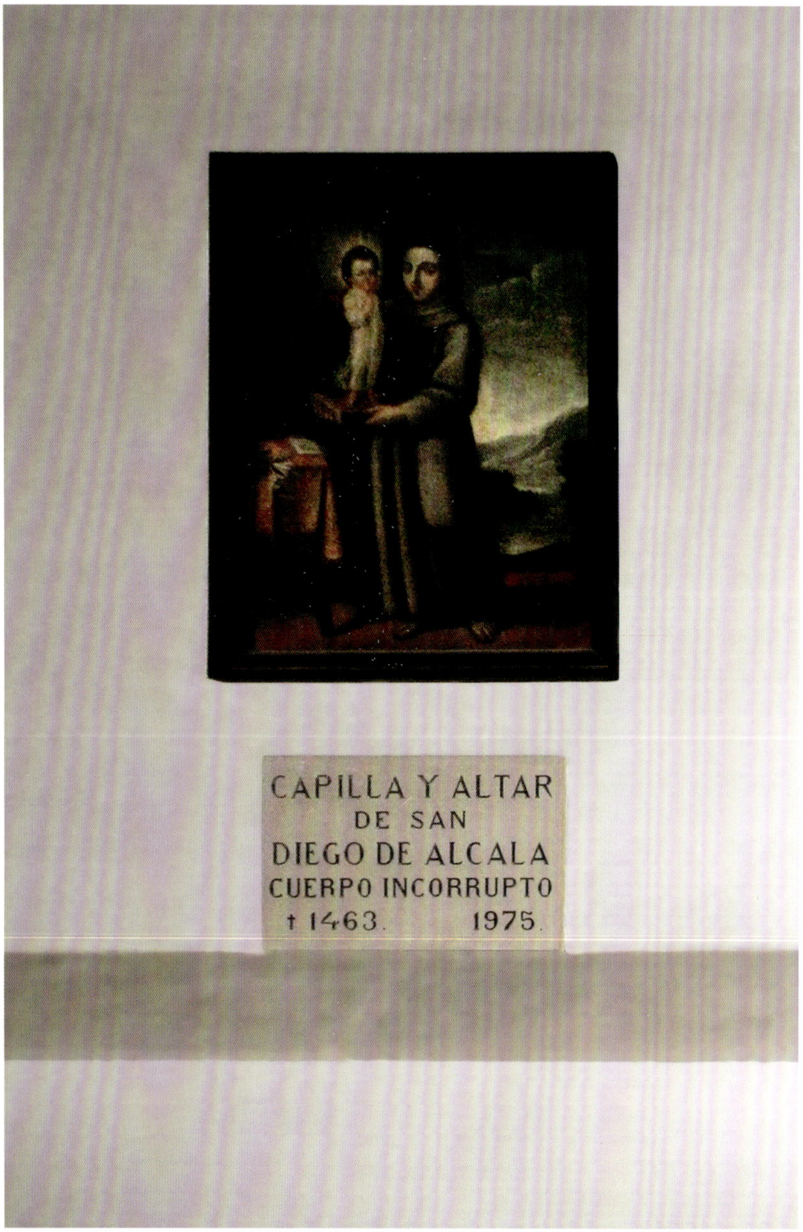
CAPILLA Y ALTAR
DE SAN
DIEGO DE ALCALA
CUERPO INCORRUPTO
† 1463. 1975.

A manuscript in the archives of the Escorial gives historians a scribe's copy of the "miracle book" of Diego de Alcalá. The original book (now lost), produced between 1463 and 1467, in the four years following Diego's death, survived at least into the next century, for Don Carlos himself ordered the copy to be made.[1] It transcribes the depositions of 159 witnesses to miracles attributed to fray Diego, observed at his deathbed and afterward in the chapel where his body was preserved. The testimony, overseen by Alonso Carrillo, the powerful archbishop of Toledo, and fray Juan de Peñalver, the monastery guardian, had been recorded in formal language in the presence of a notary. We learn that Diego died from an ulceration of his left arm and that at the moment of death he overcame his paralysis to lift a small wooden cross in both hands, breathing a prayer in Latin, "a language he had never been heard to use." That night the monastery cook, fray Pedro de Maturaba, holding vigil over the body in the church, "saw such a great light surrounding the corpse of Brother Diego that it appeared lighter than the sun and rendered the chapel as light as day." The guardian, Peñalver, was so distraught following the burial in the small monastery cemetery that on the third night, as he said in his own deposition, he sent a young brother out to unearth the corpse. Working in the dark, the brother accidentally severed the corpse's left hand with his shovel, "at which moment, he later told the guardian, he had felt the monastery walls tremble…[and] he finished the digging with only his hands." The severed hand of the dead Diego would thereafter be kept separately in the sacristy.[2]

The corpse, moved to the altar of a side chapel in the church, lay on display for six months. So perfect was the absence of any mark or scent of death that one visitor in the first month tried to take a pulse.[3] And now began a steady procession of ailing pilgrims, come to touch the holy flesh of an illiterate Franciscan lay brother.

It took King Philip twenty-six years of respectful petitions to four consecutive popes to secure recognition of the miracle of Don Carlos's cure. The petition process was arduous: commissions of bishops and ambassadors were named to draw up the *proçeso*, witnesses were assembled and their testimony was notarized, the life of Diego was researched for a *vitae sanctorum*, and the miracle book of the previous century was transcribed and extended. Diego's case was unusual. Solitary, unordained, he left no written creed, he kept no possessions, he invited no notice. Saint making as a practice was under heavy Reformation criticism. In the succession of popes between 1566 and 1585, petition after royal petition fell into the fault lines of Spanish/Roman politics. Each time, the *proçeso* had to be reinstated anew. At last a Franciscan pope, no doubt aware that Philip's great armada had recently left port in Lisbon to make war on Protestant England, allowed the full canonization to unfold. The armada, of course, would soon endure a crushing defeat. But on July 2, 1588, more than a century after his death, Diego de Alcalá de Henares was canonized by Pope Sixtus V as the Roman Catholic Church's first Counter-Reformation saint—and the first to be named in Rome in sixty-five years.[4] The king finally gained his saint—watertight proof of the divine status of the dynasty—but he had long since lost his firstborn heir, Don Carlos having died in 1568, only six years after his famous ordeal. In his will, the prince had implored his father to pursue the campaign for Diego's canonization.[5]

San Diego's symbols, humble ones, are a small reed or wooden cross and a rosary. His Latin name, Didaci Complutensis, is derived from the ancient Roman town of Complutum, later conquered by the Moors, who renamed it Al-Qal'ah al-Nahr. He is commemorated in paintings by Francisco de Zurbarán, Bartolomé Estéban Murillo, and Annibale Carracci and in sculptures by Gregorio Fernández, Alonso Cano, Pedro de Mena, and Luisa Roldán. His remains are still in Alcalá de Henares, although the body was moved in 1975 to the Iglesia Magistral, a twelfth-century cathedral newly rebuilt following its destruction in the Spanish Civil War. A side chapel is devoted to the display of San Diego's small tomb (**FIG. 3.2**). Over it, an inscription on the wall reads:

CAPILLA Y ALTAR
DE SAN
DIEGO DE ALCALÁ
CUERPO INCORRUPTO
† 1463 1975

A mission in his name was established in America in 1769. It later became the city of San Diego.

FIGURE 3.2.
Chapel and tomb of San Diego, Iglesia Magistral, Alcalá de Henares.

4 An Automaton by Juanelo Turriano

The canonization documents that Philip's court amassed are intact today in the Vatican library; they are the source of the close historical detail scholars have of the story of Don Carlos and San Diego.[1] The king had charged his court annalist, Ambrosio de Morales (1513–1591)—Córdoban priest and professor of rhetoric at the University of Alcalá de Henares—to act as *procurador* to oversee the compilation of documents to be submitted to Rome. The massive dossier included Morales's recorded testimony of eighty-three witnesses to events surrounding the miracle cure: the opening of the tomb and the transfer of the corpse, and Don Carlos's waking description of his midnight vision of the friar, followed by his sudden and swift recovery. Morales was further entrusted with the research and writing of Diego's *vitae sanctorum*, providing Catholic history with the first biography of San Diego.[2]

As it happens, Ambrosio de Morales was the author of another encyclopedic oeuvre for his king. Titled *Las antigüedades de las ciudades de España* and published in Alcalá de Henares in 1575, it is a partial history of the cities and geographic features of the Iberian Peninsula, presented as a tour, with particular emphasis on Spain's religious institutions, relics, and libraries (**FIG. 4.1**).[3] Chapter 22 is devoted to the city of Toledo, and here we find a detailed introduction to the works of Juanelo Turriano, Philip's fabled engineer, whom Morales knew as a friend. In eloquent Old Spanish, Morales chronicled Juanelo's inventions large and small—among them, his famous clocks—relating how he made them and what he had to say about them. Among the smaller works is an automaton.

> Juanelo as a diversion also wanted to create anew the ancient statues which moved and, on that account, were called automata by the Greeks. He made a lady more than one *tercia* [about eleven inches] high who, placed on a table, dances all over it to the sound of a drum which she meanwhile beats herself, and goes round in circles, returning to where she started. Though it is a toy and fit for mirth, it is nevertheless a great proof of his high intelligence.[4] (**FIG. 4.2.**)

"La dama que tañe y dança," reads Morales's paragraph title: "The lady that plays and dances."

Juanelo Turriano was born in Italy sometime around 1500 to parents of humble means. He was first known as Gianello Torriano, or della Torre, after the great medieval tower, il Torrazzo, landmark of his native Cremona. Little is certain about his education or apprenticeship.[5] He was characterized by contemporaries as a prodigy, and for the first decades of his working life he practiced in Milan, where he restored and built some of the most advanced clocks and astronomical models of his time. During this period—scholars disagree on precisely when—he attracted the notice of Charles V, whose love of clocks was well known, and he was ultimately engaged as imperial court clockmaker in the later years of the emperor's reign, working in Brussels, Madrid, and finally Toledo. Following the emperor's death in 1558, Juanelo served Philip II as court engineer. He died in Toledo in 1585 (**FIG. 4.3.**).

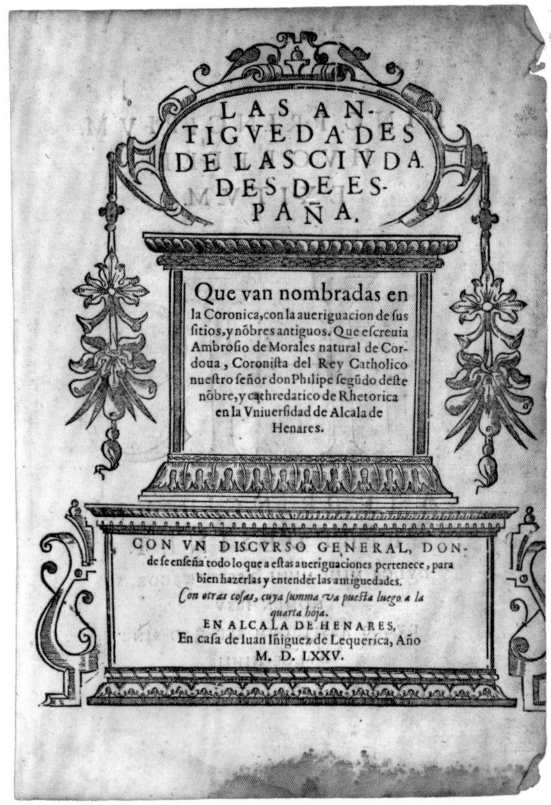

LAS ANTIGVEDADES DE LAS CIVDADES DE ESPAÑA.

Que van nombradas en la Coronica, con la aueriguacion de sus sitios, y nõbres antiguos. Que escreuia Ambrosio de Morales natural de Cordoua, Coronista del Rey Catholico nuestro señor don Philipe segũdo deste nõbre, y cathredatico de Rhetorica en la Vniuersidad de Alcala de Henares.

CON VN DISCVRSO GENERAL, DONde se enseña todo lo que a estas aueriguaciones pertenece, para bien hazerlas y entender las antiguedades.

Con otras cosas, cuya summa va puesta luego a la quarta hoja.

EN ALCALA DE HENARES,
En casa de Iuan Iñiguez de Lequerica, Año
M. D. LXXV.

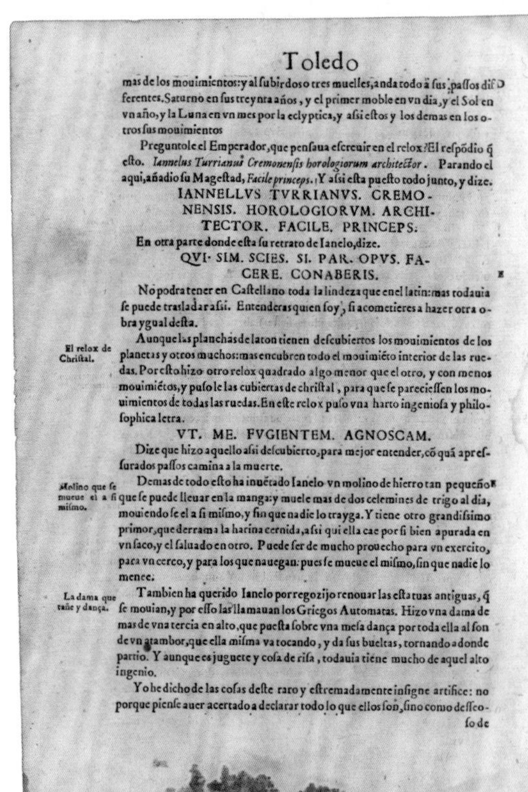

Toledo

mas de los mouimientos: y al subir dos o tres muelles, anda todo a sus passos diferentes. Saturno en sus treynta años, y el primer moble en vn dia, y el Sol en vn año, y la Luna en vn mes por la eclyptica, y assi estos y los demas en los otros sus mouimientos

Preguntole el Emperador, que pensaua escreuir en el relox? El respõdio q̃ esto. *Iannelus Turrianus Cremonensis horologiorum architector*. Parando el aqui, añadio su Magestad, *Facile princeps*. Y assi esta puesto todo junto, y dize.

IANNELLVS TVRRIANVS. CREMONENSIS. HOROLOGIORVM. ARCHITECTOR. FACILE. PRINCEPS.

En otra parte donde esta su retrato de Ianelo, dize.

QVI· SIM. SCIES. SI. PAR. OPVS. FACERE. CONABERIS.

No podra tener en Castellano toda la lindeza que en el latin: mas todauia se puede trasladar assi. Entenderas quien soy, si acometieres a hazer otra obra ygual desta.

El relox de Christal.

Aunque las planchas de laton tienen descubiertos los mouimientos de los planetas y otros muchos: mas encubren todo el mouimiẽto interior de las ruedas. Por esto hizo otro relox quadrado algo menor que el otro, y con menos mouimiẽtos, y puso le las cubiertas de christal, para que se pareciessen los mouimientos de todas las ruedas. En este relox puso vna harto ingeniosa y philosophica letra.

VT. ME. FVGIENTEM. AGNOSCAM.

Dize que hizo aquello assi descubierto, para mejor entender, cõ quã apressurados passos camina a la muerte.

Molino que se mueue el a si mismo.

Demas de todo esto ha inuẽtado Ianelo vn molino de hierro tan pequeño que se puede lleuar en la manga: y muele mas de dos celemines de trigo al dia, mouiendo se el a si mismo, y sin que nadie lo trayga. Y tiene otro grandissimo primor, que derrama la harina cernida, assi qui ella cae por si bien apurada en vn saco, y el saluado en otro. Puede ser de mucho prouecho para vn exercito, para vn cerco, y para los que nauegan: pues se mueue el mismo, sin que nadie lo menee.

La dama que tañe y dança.

Tambien ha querido Ianelo por regozijo renouar las estatuas antiguas, q̃ se mouian, y por esso las llamauan los Griegos Automatas. Hizo vna dama de mas de vna tercia en alto, que puesta sobre vna mesa dança por toda ella al son de vn atambor, que ella misma va tocando, y da sus bueltas, tornando a donde partio. Y aunque es juguete y cosa de risa, todauia tiene mucho de aquel alto ingenio.

Yo he dicho de las cosas deste raro y estremadamente insigne artifice: no porque piense auer acertado a declarar todo lo que ellos son, sino como desseoso de

FIGURES 4.1 and **4.2.**
Ambrosio de Morales, *Las antigüedades de las ciudades de España* (Madrid, 1575), title page and page 93v, identifying Juanelo's automaton, "La dama que tañe y dança"—The lady that plays and dances.

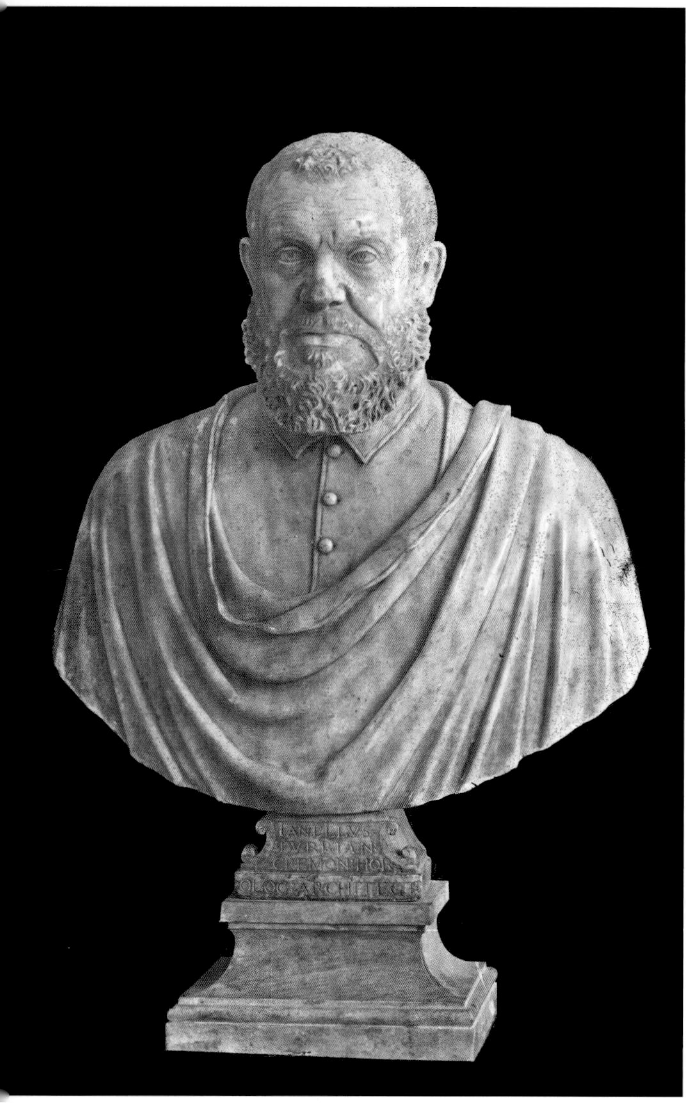

FIGURE 4.3.
Unknown artist, *Bust of Juanelo Turriano*, ca. 1560. Carrara marble. 85.5 × 60 × 31 cm (33⅝ × 23⅝ × 12¼ in.). Museo de Santa Cruz, Toledo. A photograph extraordinary in its own right, made between 1860 and 1886 by Jean Laurent.

Ambrosio de Morales was not the first contemporary of Juanelo Turriano to write about him, nor was he the only one to mention an automaton. As vivid as Morales's account of Juanelo's reputation in the Spanish court is, earlier writers from Juanelo's native Italy had already proclaimed him a new Archimedes. Bernardo Sacco (1497–1579), nobleman and fellow Lombard, tells the story of Juanelo's introduction to the court.

> During the reign over the regions beyond Padua of Joannes Galeacius Vicecomes [Visconti], it is reported that there was completed a clock, showing not only the hours, but also the stars [i.e., the planets], and the courses of the Sun and the Moon. The author of this work is not known. The clock was set up in the stronghold, or castello, of Pavia, where, after the death of the Prince, a work as marvellous as this lay despised, and even some of its circles [dials?] were removed. In the very next century, in the 1500's, about the 29th year, when Charles the Fifth received the imperial crown at Bologna, that clock, incomplete and marred with rust, was brought from its place to the aforesaid Emperor, who, having inspected and admired the machine, ordered the repair of so great a work by workmen called from all parts. While they worked in vain on the task of reconstruction, there arrived a certain Joannes of Cremona, called Ianellus, of misshapen appearance, but bright of mind, who when he examined the work said he could repair the machine, but this was of no use to anyone as the iron parts were brittle with rust and eroded away, unless a new instrument, similar to the old one and of the same proportions, be put together. Having set about this work, he completed the task by daily labour, imitating, emulating and equalling the earlier device. When the Emperor wanted to take it to Spain, he took Master Ianellus at the same time.[6]

The mathematician Gerolamo Cardano (1501–1576) and the poet-bishop Marco Girolamo Vida (1485?–1566) had also published accounts of Juanelo's clocks and astraria, Vida coining the Greek word *microcosm* to describe the greatest of them.[7] In 1585 Antonio Campi (or Campo, ca. 1524–1587), painter, architect, historian, and fellow Cremonese, published a multivolume history of his native domain and proclaimed, "But of all the craftsmen from our city, there is none more illustrious than Lionello Torriano, a lowborn man, but gifted by God with such a sublime intelligence that he astounded the world, and was reputed by everyone to be a miracle of nature." Among the wonders of his craft, Campi writes, "Lionello had fashioned (like a new Archytas) birds that not only beat their wings, but also sang, to the amazement of everyone, as if they were real."[8] Such lifelike birds appear again in accounts by later writers (with ever greater rhetorical flourish), deployed in larger theaters of automata.

> For often, when the Cloth was taken away after dinner, [Turrianus] brought upon the board little armed figures of Horse and Foot, some beating Drums, others sounding Trumpets, and divers of them charging one another with their Pikes. Sometimes he sent wooden sparrows out of his chamber into the Emperours Dining-room, that would flie round, and back again; the Superiour of the Monastery, who came in by accident, suspecting him for a Conjurer.

The quotation above is from a history of the Low Country wars printed in 1647 by the Italian Jesuit historian and poet Famianus Strada (1572–1649), translated from Latin into English in 1650.[9] Juanelo Turriano's fame had carried him into the dawn of the scientific revolution (**FIG. 4.4.**).

When the infirm Charles abdicated in 1556 and retired to the remote monastery at San Yuste in Extremadura, Spain, he took a small group of retainers with him into seclusion, Juanelo among them. Strada relates Juanelo's evening diversions for the ailing emperor, surrounded by his clocks, "whose wheels he governed with more ease, than Fortunes wheel."[10] Strada's source for the anecdote may be Campi, but other references to Juanelo in his text bear a strong resemblance to Morales's phrasing. The passage has been much quoted. Here it is paraphrased by Sir William Stirling-Maxwell (1818–1878), in *The Cloister Life of the Emperor Charles V* (1852), where it joins forces with Morales's description of the dancing lady:

> [Torriano] also constructed...the figure of a lady who danced on the table to the sounds of her own tambourine. Other puppets were also attributed to him, minute men and horses which fought, and pranced, and blew tiny trumpets, and birds which flew about the room as if alive; toys which, at first, scared the prior and his monks out of their wits, and for awhile gained the artificer the dangerous fame of a wizard.[11]

Stirling-Maxwell credits Morales and Strada in his notes, and in citing them both he builds an ever stronger case for Juanelo's automata. By the time our own contemporary historian Silvio Bedini narrates the scene, it is tradition speaking.

> During the sojourn of Charles V at San Yuste, Torriano devoted himself to averting the Emperor's moods of depression by creating little automata for his diversion. Tradition relates that Torriano's little figures often appeared on the dinner table after the Emperor's meal in the form of armed soldiers which marched about, rode horseback, beat drums, blew trumpets, and engaged in battle with lances. At another time Torriano is said to have released little birds carved of wood which flew about the room, out of the windows and returned, to the great disapproval of the Father Superior, who considered them to be works of the devil.[12]

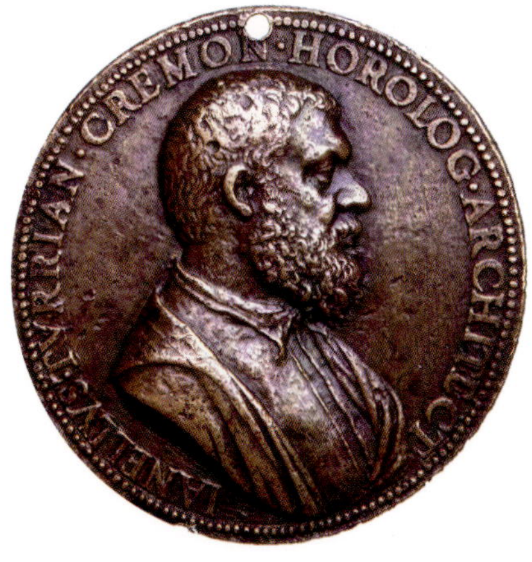

FIGURE 4.4.
Attributed to Jacopo Nizzola da Trezzo or Leone Leoni, *Juanelo Turriano*, ca. 1555. Bronze medal. Diam.: 80.5 mm (3 3⁄16 in.). Museo de la Fundación Lázaro Galdiano, Madrid. "Ese buey de forma humana," Leone described Juanelo: "An ox in the shape of a human."

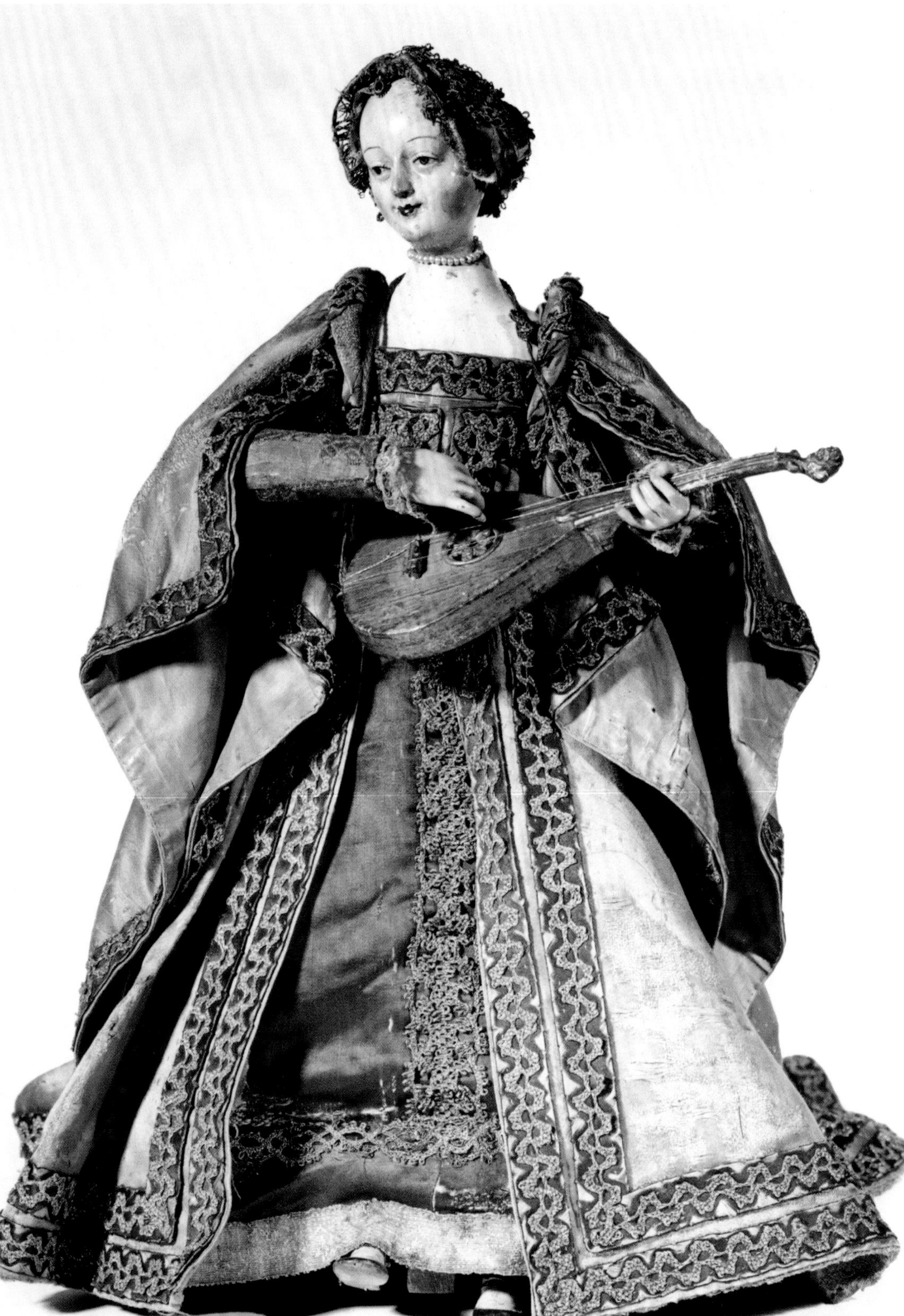

5 The Attribution of the Monk

Around 1922 an automaton turned up in Austria that so closely resembled Ambrosio de Morales's account of the dancing lady that the horological historian Ernst von Bassermann-Jordan (1876–1932) cited the passage to suggest an attribution to Master Juanelo. Bassermann-Jordan, who had seen this figure in a private collection, described it in his comprehensive volume, *Alte Uhren und ihre Meister* (Historical clocks and their masters), published in 1926.[1] The figure plays a cittern (an instrument similar to a lute) rather than a drum, but her dancing steps and circular tour and her small size certainly fit the slipper. The material and design of her clockwork, like that of the Smithsonian monk, confirm a sixteenth-century pedigree. Bassermann-Jordan's argument is cited in turn in the great two-volume encyclopedia that remains indispensable to any present-day study of the history of automata, *Le monde des automates: Étude historique et technique*, by Alfred Chapuis and Édouard Gélis, first published in Paris in 1928. Chapter 23, "Les automates et androïdes marchants," opens with an illustration of this small masterpiece.[2] In 1934 the figure was given to Vienna's Kunsthistorisches Museum, where she can be seen today (FIG. 5.1). Juanelo Turriano's name has followed her, with the caveat that no firm provenance has been possible. His name, once connected to this dancing lady by respected

FIGURE 5.1.
Automat, Cisterspielerin (Automaton, Cittern player), second half of the sixteenth century. Spain. Painted wood, linen, silk brocade, iron clockwork. Height: 44 cm (17 5/16 in.). *Kunstkammer*, Kunsthistorisches Museum, Vienna. Similarities in the size and clockwork between this figure and the automaton monk led some scholars to speculate a common provenance.

scholars, has moved in turn to circle the Smithsonian monk. The functional similarity of the figures' iron clockworks and their comparable size was well evident.

Is it quixotic, then, in searching the chronicles of the secular and religious celebrations following Don Carlos's return to health, to hope for mention of a clockwork automaton, a votive offering of thanks, fabricated with the latest technology? Long before the mysterious monk arrived at the Smithsonian in 1977, Bassermann-Jordan had cited Morales's "dama que tañe y dança" to pose a connection between the automaton in Vienna, Juanelo Turriano, and the Spanish court. We find Juanelo's name connected with yet another automaton, a musical saint, in the Iparművészeti Múzeum (Museum of Applied Arts) in Budapest, purchased in 1915. Both museums now emphasize that the attribution is hypothetical. The attribution comes down to us from the history of technology: Ambrosio de Morales's testimony and other sixteenth-century firsthand records of Juanelo Turriano's clocks are the source for a cascade of references to Juanelo's legendary automata. But how striking to learn that Morales, the first historian to systematically assemble the fragments of San Diego's life, was the same historian to observe and record the accomplishments of Juanelo Turriano. Eyewitness to events from both the theological and technological spheres of the Spanish court, he is the very man we should like to consult in the search for the origins and legend of the Smithsonian monk. As tantalizing as this is, we have no further words from Morales on any other automaton by Juanelo's hand.

A startling clue appears in a monograph titled *Los relojes y autómatas de Juanelo Turriano* (The clocks and automata of Juanelo Turriano), published in a small subscribers' edition in Madrid in 1982. The author is José A. García-Diego (1919–1994), a civil engineer who became interested in the history of science and technology in Spain.[3] His book is devoted to piecing together the fragmented record of Juanelo's life and works, in an effort to recuperate the legacy of an innovator whose rightful place in the history of technology has been overlooked. In an inventory of surviving or lost or alleged machines from Juanelo's hand, García-Diego includes photographs of two small automata. Not surprisingly, one is the dancing lady with the cittern in Vienna, and he notes the stream of references to Juanelo Turriano, running backward through Bedini, Bassermann-Jordan, Strada, and so on, to Morales. The other is the Smithsonian monk. "This second automaton represents a friar," he writes. "In contrast with the musical woman, I would say that its personality is, for me at least, considerably unpleasant."[4] As it turns out, he has had some previous experience with this figure, having seen it in Geneva in 1975. The following year, he tells us, it was offered (unsuccessfully) for sale in two auctions, one in Zurich and one in Geneva. The auction catalog from the Zurich sale speculates a South German origin for the automaton;[5] the Geneva catalog argues for Spain. In García-Diego's words:

> But in the cited catalog of Geneva a hypothesis was given about when the religious figure could have been fabricated, of which I think I should record mention. It is supported in an announcement by Father Servius Gieben [*sic*], director of the Instituto Histórico de los Capuchinos de Roma. According to him, it would represent San Diego de Alcalá, a Franciscan who died in 1463. That would permit connecting it to the Spanish court, since the episode was known of how his mummy [*momia*] was placed in the bed of Prince Don Carlos, wounded in the head in a supposedly fatal fashion, having fallen downstairs (1562); which had the effect of an immediate cure and later, upon petition of Philip II, the canonization of the friar. Father Servius continues and I copy: "It is in this climate of religious exaltation—processions, public prayers, pilgrimages for obtaining a cure for the prince—that the fabrication of the automaton found itself as a kind of votive offering.... The date should be 1562 or a little later, and the author Juanelo Turriano (or de la Torre, dead in 1585) who was chief engineer for Philip II."[6]

Who is "Father Servius"? Is he a contemporary or a historical source? García-Diego doesn't say, and the reader is left with an author quoting an auction catalog quoting a hypothesis. But two more images are reproduced in his book. Taken from the same auction catalog and apparently provided by the mysterious "historiador capuchino," a close-up photograph of the head of the automaton is paired with an engraving said to have been made in 1588: a portrait of San Diego de Alcalá (**FIG. 5.2**).

Dans une communication du 28 août 1975, José A. García Diego estime que l'automate ici reproduit « peut être attribué à Juanelo Turriano qui assurément le fit quand il accompagnait l'empereur Charles Quint dans sa retraite de Yuste. Ceci en raison de la machine et de la ressemblance avec l'automate de Vienne qui est considéré par Bedini comme le seul fait par Juanelo Turriano ».

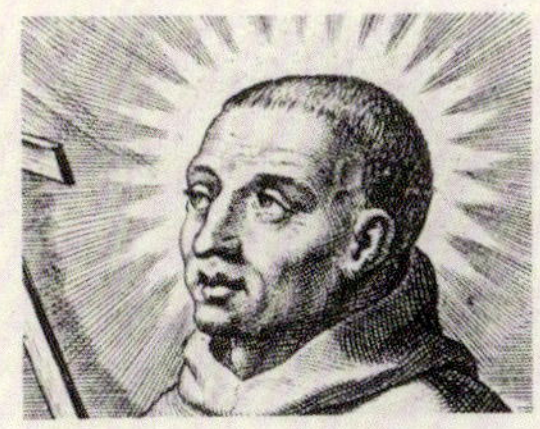

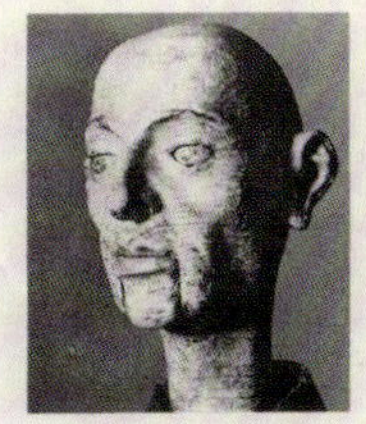

Des recherches entreprises ensuite à l'Institut Historique des Capucins de Rome, concernant l'identité du personnage représenté, ont permis de cerner de plus près le problème. D'après des gravures du XVIe siècle, l'une datant de 1588, année de la canonisation de Saint Diego d'Alcala, il s'agirait d'une représentation de ce saint, frère convers de l'ordre de Saint François d'Assise, mort en 1463. Ceci nous rapproche de nouveau de la cour espagnole, mais quelques années plus tard, en 1562, quatre ans après la mort de Charles Quint. Nous extrairons ici d'une communication du 16 novembre 1975 la conclusion à laquelle est parvenue le Père Servus Gieben, directeur de l'Institut Historique des Capucins de Rome:

« En avril 1562, le prince Don Carlos (1545-1568), fils de Philippe II, roi d'Espagne, se trouvait au palais d'Alcala, gravement malade. En tombant dans l'escalier (19 avril 1562) il s'était fait à la tête une plaie que les chirurgiens jugèrent mortelle. Tentant l'extrême, on apporta dans la chambre du prince le corps resté intact du Saint et à l'instant il fut parfaitement guéri. Philippe II, roi d'Espagne, père de Don Carlos, va solliciter immédiatement, par reconnaissance, la canonisation du Serviteur de Dieu. Sixte V le mit au nombre des Saints en 1568.

« C'est dans ce climat d'exaltation religieuse — on faisait des processions, des prières publiques, des pélerinages pour obtenir la guérison du prince — qu'on doit localiser la fabrication de l'automate comme une espèce d'ex-voto et — pourquoi non ? — comme une exhortation au jeune prince à une vie plus sérieuse (il était très capricieux). La date devrait être 1562 ou peu après, et l'auteur Juanelo Turriano (ou de la Torre, mort en 1585) qui était « ingeniero mayor » de Philippe II ».

Cet automate aurait donc appartenu au malheureux prince Don Carlos qui,

84

FIGURE 5.2.
A page from the 1976 auction catalog *Très importante collection de tableaux espagnols du XIIIe au XVIIIe siècle… un automate de Juanelo Turriano ingénieur de Charles Quint.* An engraved portrait of San Diego de Alcalá appears next to a photograph of the automaton. The catalog text suggests that they are images of the same saint. Even the lighting and shadows are matched.

6 A Letter

Elizabeth had encountered José A. García-Diego's book on Juanelo Turriano in 1996. Upon reading it, she dropped everything in order to track down the quoted auction catalog. She found a copy in the library of Connecticut College. The auction had been held at l'Hotel Richemond in Geneva on June 21, 1976, and the catalog is titled straightforwardly by the list of items to be sold: *Très importante collection de tableaux espagnols du XIIIe au XVIIIe siècle, un chef-d'œuvre de Martin van Heemsckert [*sic*], un automate de Juanelo Turriano ingénieur de Charles Quint.*[1] The entry for the monk identifies it as "un automate représentant Saint Diégo d'Alcala." Juanelo Turriano is introduced via the seventeenth-century authority of Famianus Strada, transcribed in full. Next García-Diego himself, an expert on Juanelo Turriano, is quoted from a communication dated August 28, 1975, offering the opinion that the automaton was made for the emperor at San Yuste. The images of the two heads, automaton and engraving, are presented and identified:

Research subsequently undertaken at the Institut Historique des Capucins de Rome, concerning the identity of the individual represented by the monk, permits us to come closer to the problem. According to engravings from the sixteenth century, one dating from 1588, the year of the canonization of Saint Diego d'Alcalá, the figure would appear to be a representation of the saint, a lay brother of the order of Saint Francis of Assisi, who died in 1463. Here we encounter again the Spanish court, but several years later, in 1562, four years after the death of Charles V. We excerpt here a communication from November 16, 1975, and the conclusion reached by Father Servus Gieben, director of the Institut Historique des Capucins de Rome.[2]

And the opinion of Fr. Servus Gieben follows (the name "Servus" here spelled without the *i*), containing the passage on Don Carlos's miraculous cure, which is quoted by García-Diego in his subsequent book. So Elizabeth learned at least the date 1975 for the work of the "historiador capuchino." Was his research commissioned by the broker? But the catalog text proceeds briskly to its conclusion:

The automaton would have thus belonged to the unhappy prince Don Carlos, who after having plotted against his father, was imprisoned in the Alcázar in Madrid, where he died on July 25, 1568....

The authorship of Juanelo Turriano seems to be without any doubt and, as for its intent, one sees that the automaton well resembles Saint Diego de Alcalá. The hypothesis of Father Gieben seems to us perfectly coherent. Regarding the automaton itself, which has come down to us marvelously well preserved, is it necessary to emphasize the principal importance of an object nearly unique, living testimony, dare one say, of the technological preoccupations of the Renaissance and of the inventive genius of the men of that era?[3]

The monk was not sold at auction that day; the Smithsonian's purchase took place the following year as a direct transaction through a private dealer in the same city.

Among the documents in the Smithsonian accession records for the monk is a letter to Otto Mayr dated April 4, 1977, from a Messr Georges Sedlmajer, in Geneva, paraphrasing much of the catalog copy. "With this letter," Sedlmajer writes, "I send you the complete dossier that I have assembled on this automate. The only letter which is missing is of Father Servus Gieben of the Institut in Rome, photocopy of which I can send if you wish." Mayr must have asked for it, for indeed the photocopy is included in the museum's records. Written in French and dated November 16, 1975, it is the precise "communication" quoted by the Geneva catalog and later by García-Diego. "Monsieur Sedlmajer," this important letter begins,

Our research to identify the personage represented by your interesting automaton has brought us to the following conclusion.

After searching in vain among the priests of the period, the iconography of the image, namely, a Franciscan brother with the attributes of the cross and the rosary, sent us to Saint Didace, or Diégo d'Alcalá. This is a lay brother of the order of Saint Francis of Assisi, who died on November 12, 1463, and was canonized in 1588. As a lay brother he does not wear the full tonsure, but presents instead a shaved head. His attributes are precisely the cross, often made of reed, and the rosary. Is it possible that your automaton originally carried both attributes separately, rather than together as now? In other words: not a cross *of* the rosary, but a cross *and* a rosary?

Regarding the resemblance between the antique engravings of San Diego (believed to be somewhat idealized, of course) and the figure of your automaton, I wouldn't overly insist on it. You will find enclosed with this letter two photocopies of engravings, one from 1588 and the other a little later, for making a comparison. For my part, the likeness is more than sufficient.

But I have further clear grounds for bringing this lay Franciscan saint into a royal Spanish milieu, and precisely in the year 1562. They are as follows:

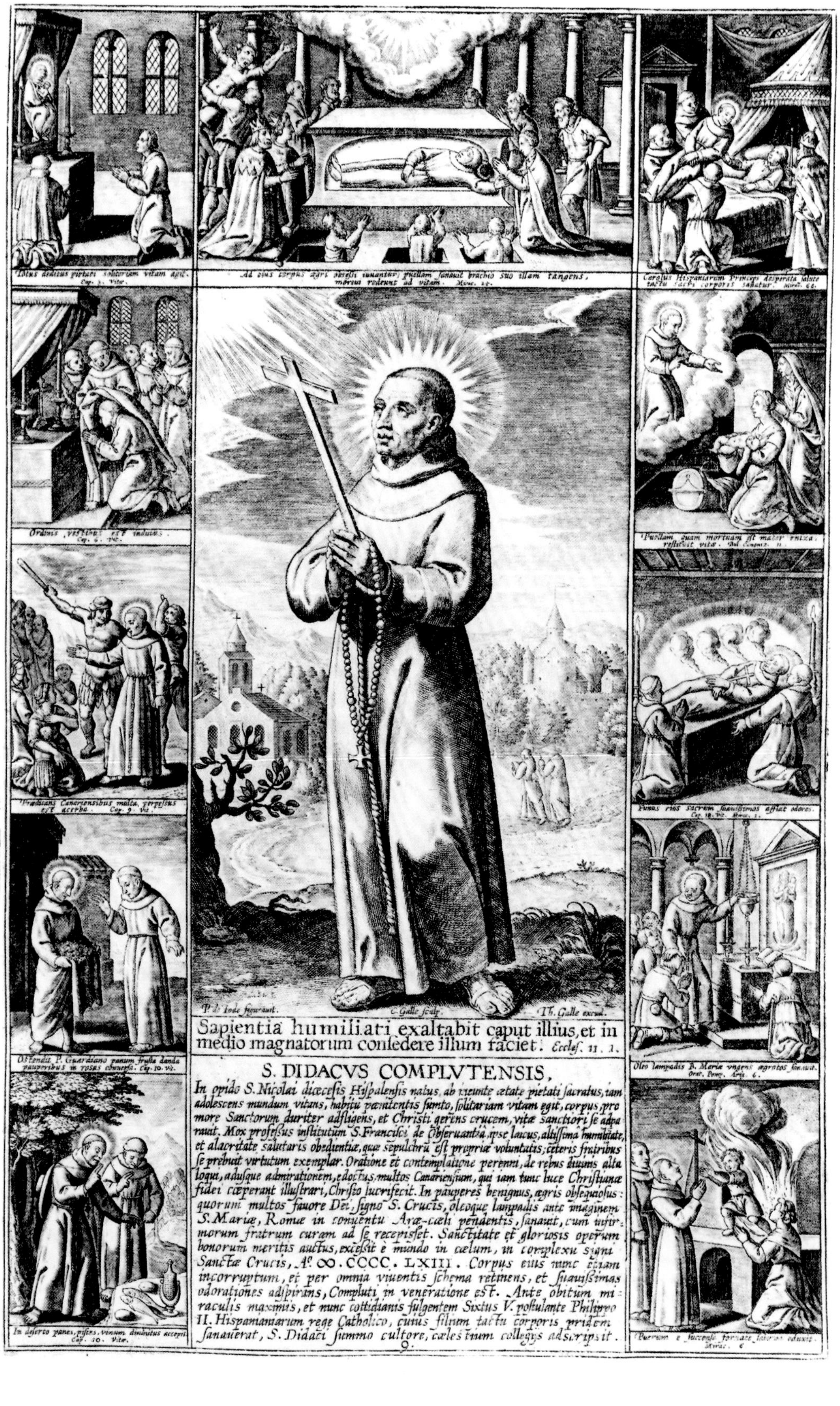
P. de Iode figurauit.
C. Galle sculp.
Th. Galle excud.
Sapientia humiliati exaltabit caput illius, et in medio magnatorum consedere illum faciet. Eccles. 11. 1.
S. DIDACVS COMPLVTENSIS,
In opido S. Nicolai diœcesis Hispalensis natus, ab ineunte ætate pietati sacratus, iam adolescens mundum vitans, habitu pœnitentis sumto, solitariam vitam egit, corpus, pro more Sanctorum duriter adfligens, et Christi gerens crucem, vitæ sanctiori se adparauit. Mox professus institutum S. Francisci de Obseruantia, ipse laicus, altissima humilitate, et alacritate salutaris obedientiæ, quæ sepulchrũ est propriæ voluntatis; ceteris fratribus se prebuit virtutum exemplar. Oratione et contemplatione perenni, de rebus diuinis alta loqui, adusque admirationem, edoctus, multos Canariensium, qui iam tunc luce Christianæ fidei cœperant illustrari, Christo lucrifecit. In pauperes benignus, ægris obsequiosus: quorum multos fauore Dei, signo S. Crucis, oleoque lampadis ante imaginem S. Mariæ, Romæ in conuentu Aræ-cœli pendentis, sanauit, cum infirmorum fratrum curam ad se recepisset. Sanctitate et gloriosis operum bonorum meritis auctus, excessit è mundo in cælum, in complexu signi Sanctæ Crucis, A°. ∞. CCCC. LXIII. Corpus eius nunc etiam incorruptum, et per omnia viuentis schema retinens, et suauissimas odorationes adspirans, Compluti in veneratione est. Ante obitum miraculis maximis, et nunc cottidianis fulgentem Sixtus V. postulante Philippo II. Hispaniarum rege Catholico, cuius filium tactu corporis pridem sanauerat, S. Didaci summo cultore, cælestium collegijs adscripsit.
9.

In April of 1562, the prince Don Carlos (1545-1568), son of Philip II, King of Spain, was in the palace at Alcalá, seriously ill. In falling down a flight of stairs (19 April 1562) he had suffered a wound to the head that the surgeons judged fatal. In desperation, they brought to the prince's room the intact remains of the body of the saint, and at once he was completely cured.

Philip II, king of Spain, father of Don Carlos, in gratitude immediately set out to seek the canonization of a Servant of God. Sixtus V placed him in the company of Saints in 1588. (On the miracle of the cure, see Fr. Lucio María Núñez, "Documentos sobre la curación del príncipe D. Carlos y la canonización de San Diego de Alcalá," in the review *Archivo Ibero-Americano* 2 [1914] 424-446.)

It is in this climate of religious exaltation—processions, public prayers, pilgrimages for obtaining the prince's cure—that one should situate the fabrication of the automaton, as a kind of ex-voto and—why not?—as an exhortation to the young prince to a more serious life (he was very capricious). The date should be 1562 or a little later, and the author Juanelo Turriano (or de la Torre, died 1585), who was chief engineer to Philip II.

It would be very interesting to check these dates in the manuscript "The Twenty-One Books of Devices and Machines" of Juanelo Turriano to be found in the National Library of Madrid.

As you can understand, we are always very interested in all other documentation that you succeed in obtaining on this interesting piece of Franciscan art.

I beg you to accept my best regards,

Servus Gieben
Directeur de l'Institut Historique des Capucins[4]

The photocopies of the engravings, one by Cornelius Galle (**FIG. 6.1**) and one by Johann Sadeler, both show the figure of San Diego, with cross and rosary, surrounded by a cycle of images of miracles attributed to the saint. Each engraving includes in its cycle a panel depicting Don Carlos's sickroom, complete with a kneeling king and a group of three Franciscan priests lifting a composed and haloed *momia* onto the bed.

FIGURE 6.1.
Cornelius Galle, *Diego de Alcalá (vita)*, first published 1614. Engraving. Sheet: 25.6 × 16.4 cm. (10 × 6½ in.). Courtesy of Museo Francescano, Istituto Storico dei Cappuccini, Rome. The central image of the saint is ringed by a cycle of depictions of miracles attributed to him, including, at upper right, the arrival of his miraculously preserved corpse in the sickroom of the young Don Carlos.

7 A Monk's Opinion

The Geneva auction catalog presents a genuine effort to construct an authoritative provenance for an object emerging from obscurity onto the world antiquities market—in hindsight, one could say groomed (and rightly so) for ascent into one of the world's major museum collections. A story all its own, with a striking similarity to the delayed recognition of San Diego himself, its connection to this fugitive automaton has come about not just through sheer repetition but, perhaps more compelling, because the automaton demands an explanation equal to its charisma. If the proposed link to the historical record might be too good to be true, it is nonetheless not implausible.

One interesting thing has changed since the sale of the monk to the Smithsonian in 1977. When García-Diego first encountered the automaton monk in Switzerland in 1975 he attributed it to Juanelo, and his authority was cited the following year in the Geneva auction catalog. But six years later, when he published his book on Juanelo Turriano, he wrote that he had changed his mind, concluding that only the dancing lady could still be a candidate for the claim (acknowledging his former opinion in an endnote). Then, in 1986, he published a second edition of his book, this one translated into English under the title, *Juanelo Turriano, Charles V's Clockmaker: The Man and His Legend.*[1] And here he withdrew the lady, too, from Juanelo's hand, as well as the word *automata* from his book's title. As he appealed to the reader, "Correcting one's mistakes seems to me the right thing."

Intrigued as he clearly was with these figures, García-Diego based his conclusions on the belief that such objects were only toys, presumably beneath the genius of Juanelo, the dignity of the emperor, and the tolerance of the king.[2]

In 1980 the Smithsonian included the monk in an ambitious show produced in partnership with the Bavarian National Museum in Munich. Curated by Otto Mayr and Klaus Maurice, *The Clockwork Universe: German Clocks and Automata, 1550–1650* presented a comprehensive selection of works from the golden age of German clock-making. From extravagant tower clocks and planetary clocks to mechanical terrestrial globes and from gilt bronze and enameled automata to common domestic clocks of the period, the show brought together masterworks from collections throughout Europe and the United States. In the landmark catalog published for the exhibition, the monk, one of the earliest pieces in the show, is labeled as follows:

> Automaton figure of a monk
> South Germany or Spain, c. 1560
> Figure: head of poplar wood; head and limbs rendered naturalistically; modern habit
> Movement: iron; height: 39 cm (15⅜ in.)
>
> Movement has a wooden fusee. The monk is programmed to move in a square approximately 2 feet wide. Although his feet step out from beneath his robe, he is actually rolling on wheels. As he proceeds he strikes his right arm against his chest and moves the left arm up and down. He is constantly turning his head, nodding, moving his mouth, and rolling his eyes. The mechanism is pinned for the most part; only the pillars and one cam are secured by means of screws.
>
> The ascription to Spain is based upon the realistic conception and modelling of the monk's head and upon the datum that Emperor Charles V's clockmaker, Juanelo Turriano, had also made comparable mechanisms for automaton figures. Yet similar moving figures were also made by the Nuremberg clockmakers. Jakob Bulmann (master in 1497) made "figures of men and women that walked around and struck their measure upon lutes and kettledrums." In Emperor Rudolf II's collection of art treasures there were a number of spring-driven clothed dolls that beat upon drums and walked about.[3]

South Germany or Spain? Amid the cascade of narratives supporting a Spanish origin, Servus Gieben's hypothesis takes wing like a meme, for its fit with a likeness to San Diego. Could the greater weight of the hieratic tale of how and why this little machine came to be made rest on Gieben's 1975 letter? An automaton fabricated as an act of gratitude, a votive offering, if not precisely the fulfillment of a king's commission. Under what circumstances did Gieben come to perform his research? Was he aware of how his hypothesis would be utilized? He doesn't mention any other automata, or the chronicle of Juanelo Turriano by Ambrosio de Morales. Did he know of these connections?

In 1999, twenty-four years after his letter was written, Elizabeth was lucky to contact Servus Gieben himself. At the age of seventy-five, he was once again head of the Istituto Storico dei Cappuccini in Rome and director of its Museo Francescano. A scholar of medieval theology and a specialist in the lives of the saints as depicted in the history of art, he responded to her letter in fluent English (one of his six languages), writing, "I certainly will be pleased to read your paper on the 'Mechanical Monk,' of which I preserve nice photographs and curious memories."[4] Elizabeth's first question for him: Had the Geneva broker mentioned either Juanelo Turriano's name or the precedent of the Vienna automaton? No, he replied, and no link whatsoever had been made to another automaton. At most, he recalled the suggestion of a sixteenth-century origin for the piece. His opinion that the monk represented a Spanish figure was based on the physiognomy of the carved face, a judgment confirmed by his search through the Museo Francescano print archives and the match, not just with the person of San Diego, but with the saint's symbols, lay status, and gestures.

Gieben's hypothesis thus comes not from the history of technology but from the history of theology, for he arrived at his conclusion without an awareness of the attribution to Juanelo of any similar automaton figure. It is more than remarkable that two separate tracks converge on so close a hypothesis of the monk's origin.

The clockmaker himself evades us yet, for none of the machines Juanelo Turriano made with his own hands has survived in a form we can identify.[5] Ambrosio de Morales tells us Juanelo signed his greatest astronomical clock, the work of a lifetime, with an engraved inscription in Latin: "QVI. SIM. SCIES. SI. PAR. OPVS. FACERE. CONABERIS," which can be very roughly translated as, "You will know who I am if you try to make this."[6]

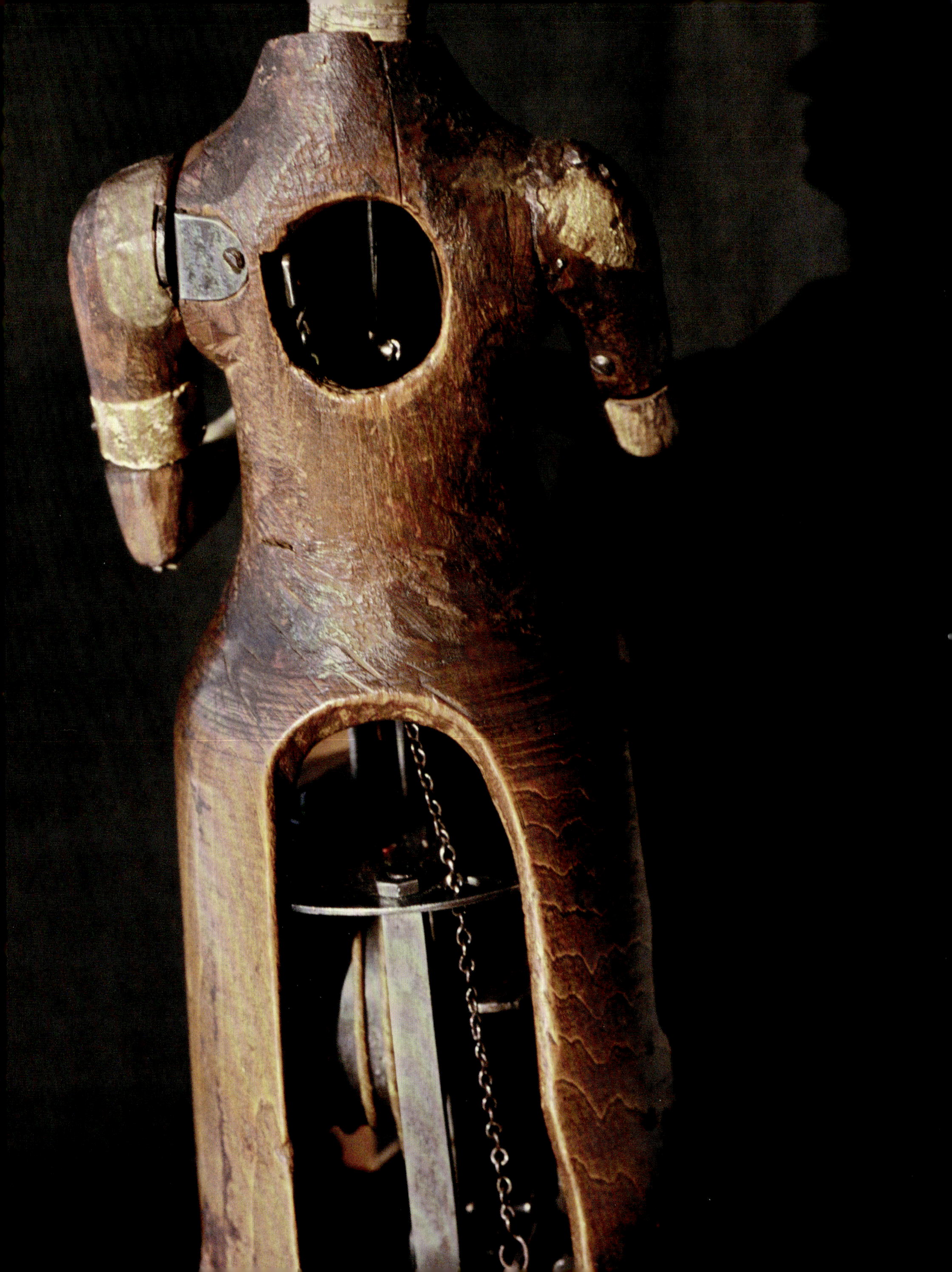

PART II
The Monk

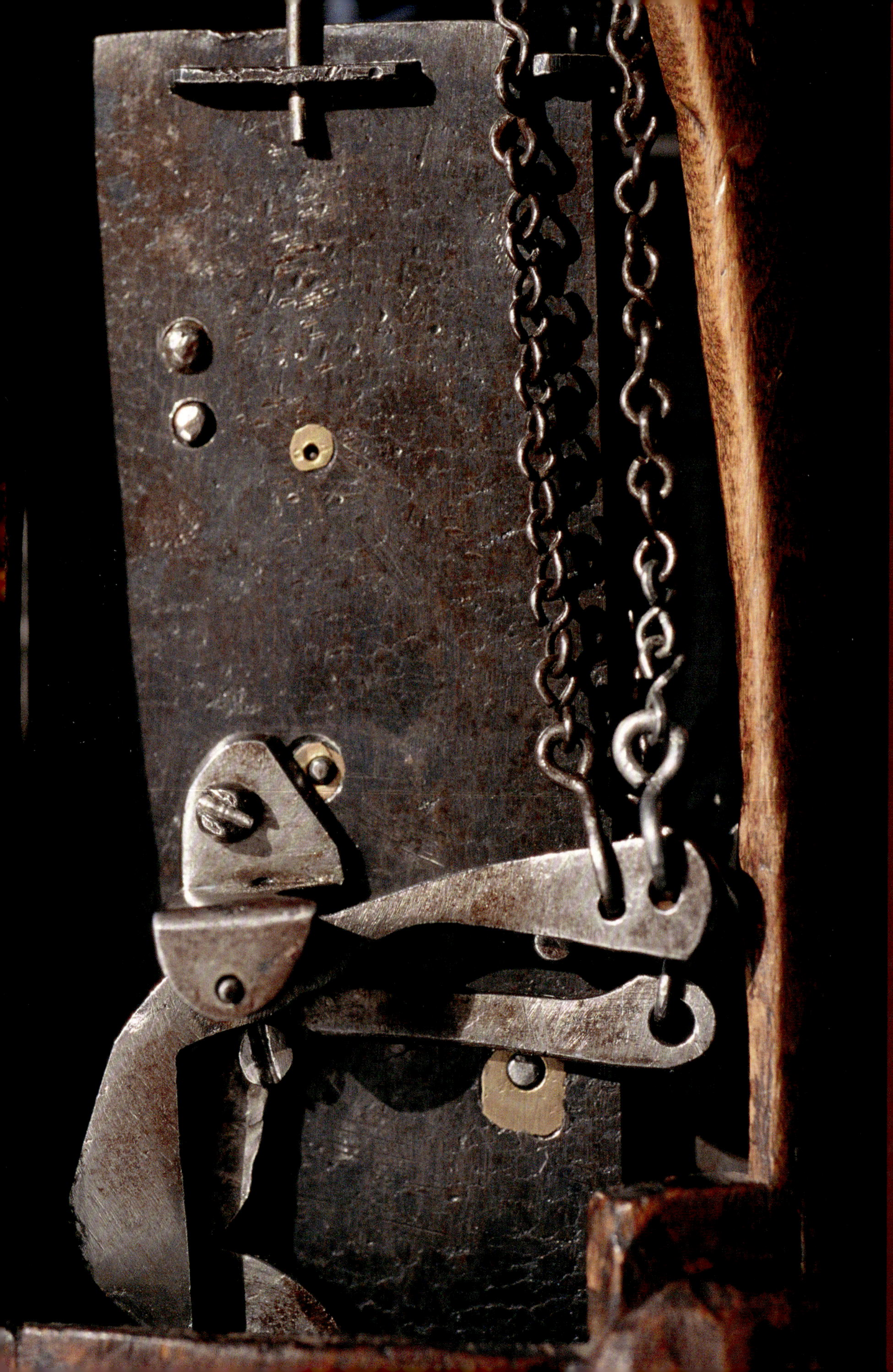

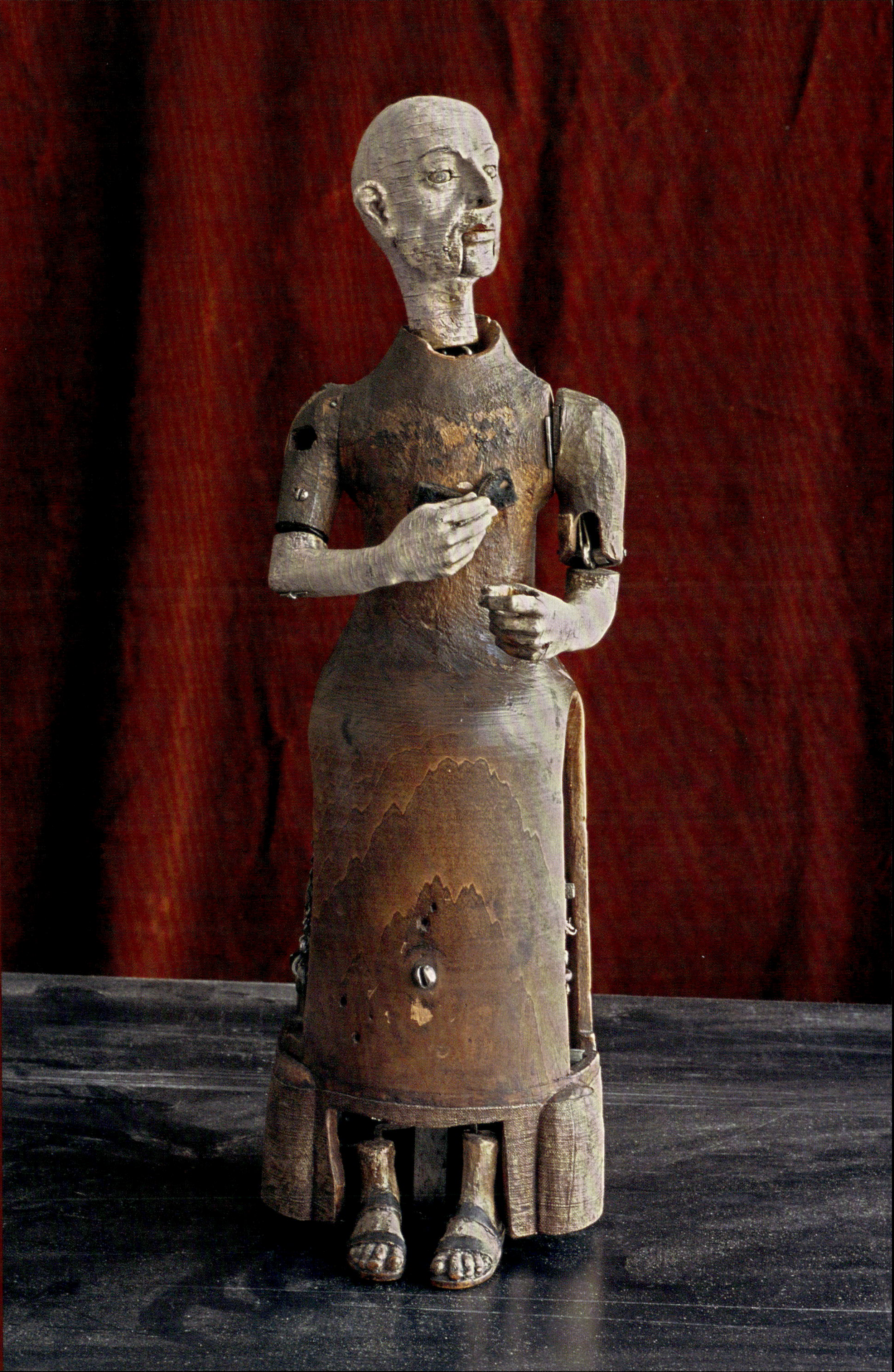

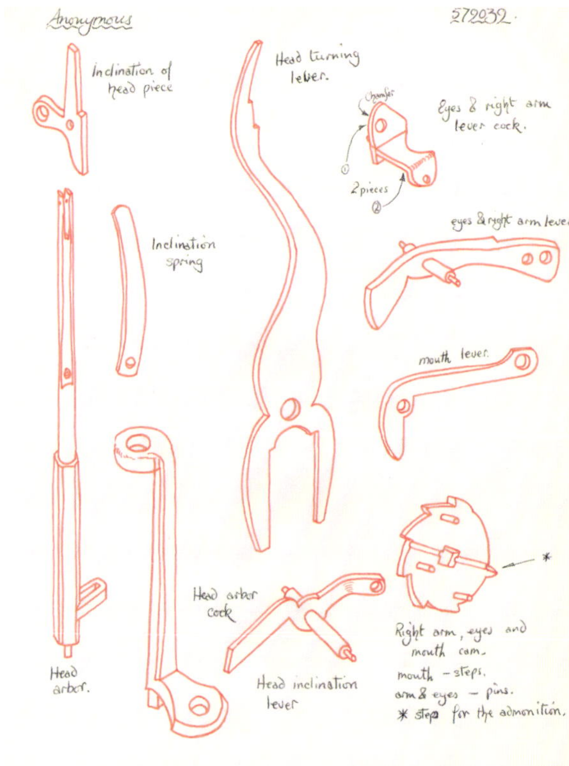

8.1.

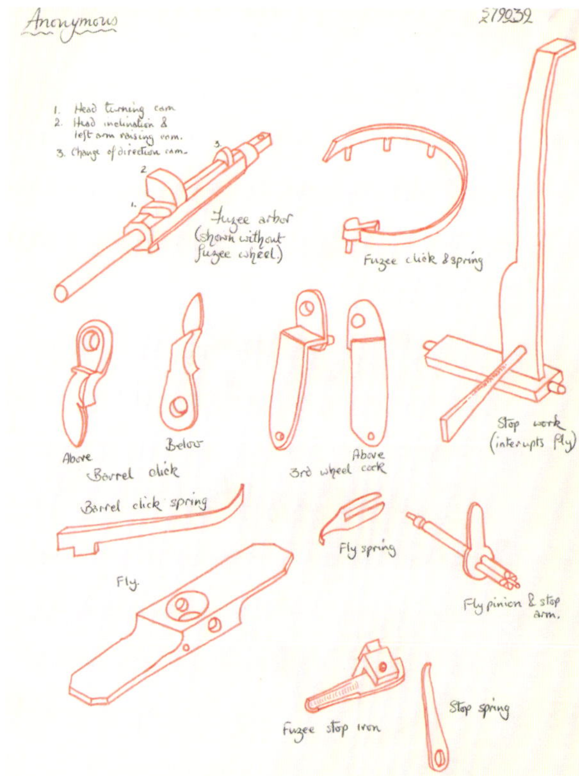

8.2.

FIGURES 8.1–8.3.
A few of David Todd's drawings of individual parts of the monk's clockwork, 1978. 8½ × 11 in. (21.6 × 27.9 cm), each sheet. The drawings are color coded: red for iron, green for brass, brown for wood.

FIGURE 8.4.
David routinely makes rubbings of each part of a clockwork, to record true scale. This is the third wheel of the monk's clockwork wheel train, excerpted from pages of rubbings of all parts of the machine, printed here at actual size. Its diameter is 1.875 inch.

8 An Immaculate Object

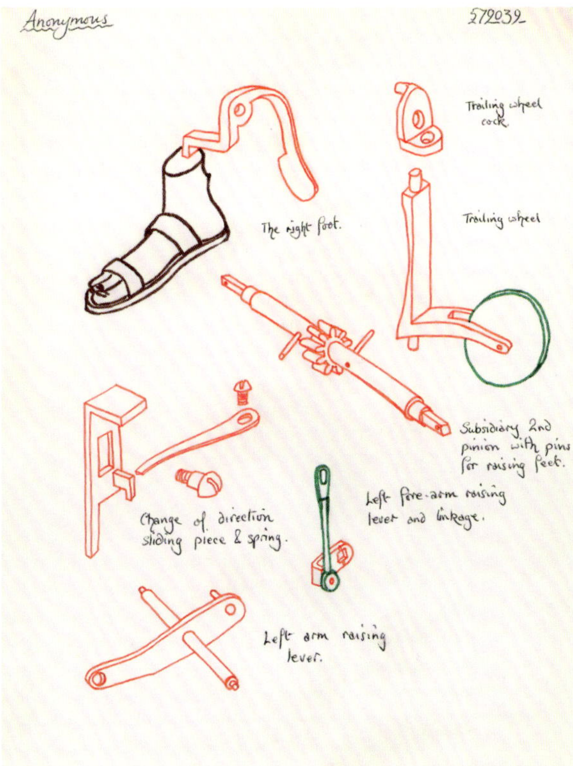

8.3.

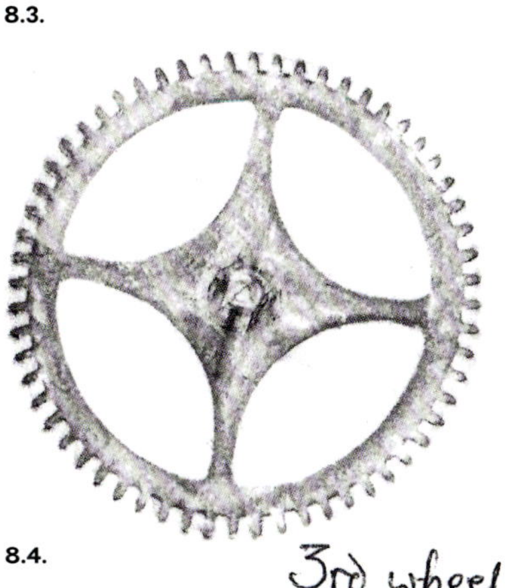

8.4.

When David Todd arrived at the Smithsonian's National Museum of History and Technology in 1978, one of his first tasks was to examine the newly acquired automaton monk. Delicately removing the habit and wooden casing shell, he dismantled the full internal mechanism in order to clean remnants of corrosion from the iron parts. In the process, he produced an index of drawings of each component (**FIGS. 8.1-8.4**). He found the clockwork to be original, except for the mainspring, some of the wire linkages, and the thin catgut line to one forearm. The mainspring had clearly been replaced, possibly in the nineteenth century, with a spring of a different type. He elected to leave these alterations in place since they were visible only on disassembly. Well aware of the rare virtue of intact original paint on all exterior carved wooden body surfaces, he preserved this untouched historical patina, allowing its full age to remain visible. To avoid damage to the fragile carving of the head, he had it X-rayed in the Smithsonian's Conservation and Analytical Laboratory in order to view the internal iron components and their design (see **FIG. 10.15**). The X-ray revealed the inside of the head in graphic detail, the iron parts of the mechanism immediately recognized by radiography. A beautiful and mysterious image in itself, a machine within a ghost, it showed all interior connections were intact and fully functional.

Throughout this work, David searched for signatures, identifying stamps, or later restorers' cleaning marks that could yield a clue to the automaton's history. He found nothing. Its provenance would have to be gleaned through comparison of its materials and design with those of kindred clockworks.[1] The most prominent distinguishing feature: the internal mechanism is made almost entirely of hammered iron, bearing the unmistakable marks of period tooling. Later mechanisms of similar design and fabrication are made of brass. In the era prior to production gear-cutting machinery, each clockwheel had its teeth cut one by one, with handsaw and file. A count of wheel and pinion teeth in the monk yields some 186 teeth, each calculated, cut, and filed to exact measure by hand and eye: the labor is immediately recognizable. Parts are assembled largely with handmade screws, wedges, and pins, each fashioned separately for its job rather than drawn from a standardized pattern. These features, along with a characteristic decorative style of workmanship of the iron, point to the second half of the sixteenth century, consistent with well-documented clock technology of the period.

The wooden parts of the figure—hands, head, and body—also display little interim repair or alteration, presenting a characteristic array of paint breaks, fading, and wear. Only the sandaled feet bear possible signs of relatively newer paint.[2] The interface between sculpture and clockwork is pristine. The sculpture is sensitively carved, and painted in the trompe l'oeil polychrome style ascendant in sixteenth-century European workshops. Paint was applied only to those parts of the wooden body visible outside the figure's intended cloth habit, making it possible to observe the boundaries between painted and unpainted wood surfaces. These boundaries, some of which betray wear and tear over time from the motion of the automaton's movements, reveal no prior paint underlayers, and we can reasonably conclude that the paint is original.

With the monk's iron clockwork disassembled—133 separate parts in all—David could map the mathematics of power source and gear train and parse the exact sequence of the movement cycle that choreographs the figure's motions. He used this knowledge to build a fully working model of the automaton in steel and cast acrylic, one-fourth larger in scale, to be installed in the museum gallery next to the original (**FIGS. 8.5-8.7**; see also view of model on p. iv.). Visitors could wind the replica themselves to view the performance. For the model, he used modern rolled steel, but he fabricated a majority of the parts of the clockwork by hand, following sixteenth-century methods, cutting each of the wheel teeth individually. The labor gave him an intimate knowledge of the orchestration of forces in this small machine. Only late in the process did he realize that if he had attempted to size his model any larger, he would not have been able to provide a spring strong enough to operate it. While the model is only one and a quarter times larger than the original, he found that the mainspring had to be almost twice as powerful. The thickness of the spring blade had to go from the

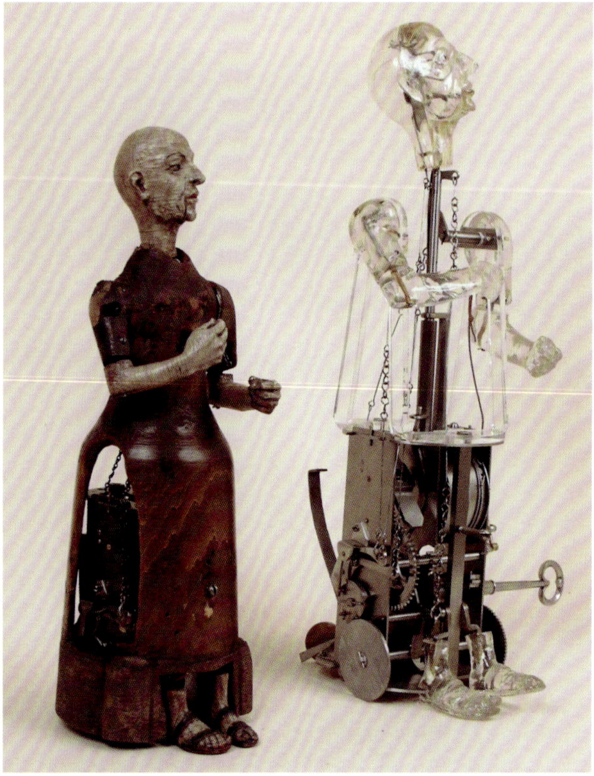

FIGURE 8.5.
Monk and model. David Todd built a working replica of the monk, complete in all its parts, made of cold-rolled steel and cast acrylic, seen here in 1980 next to the original.

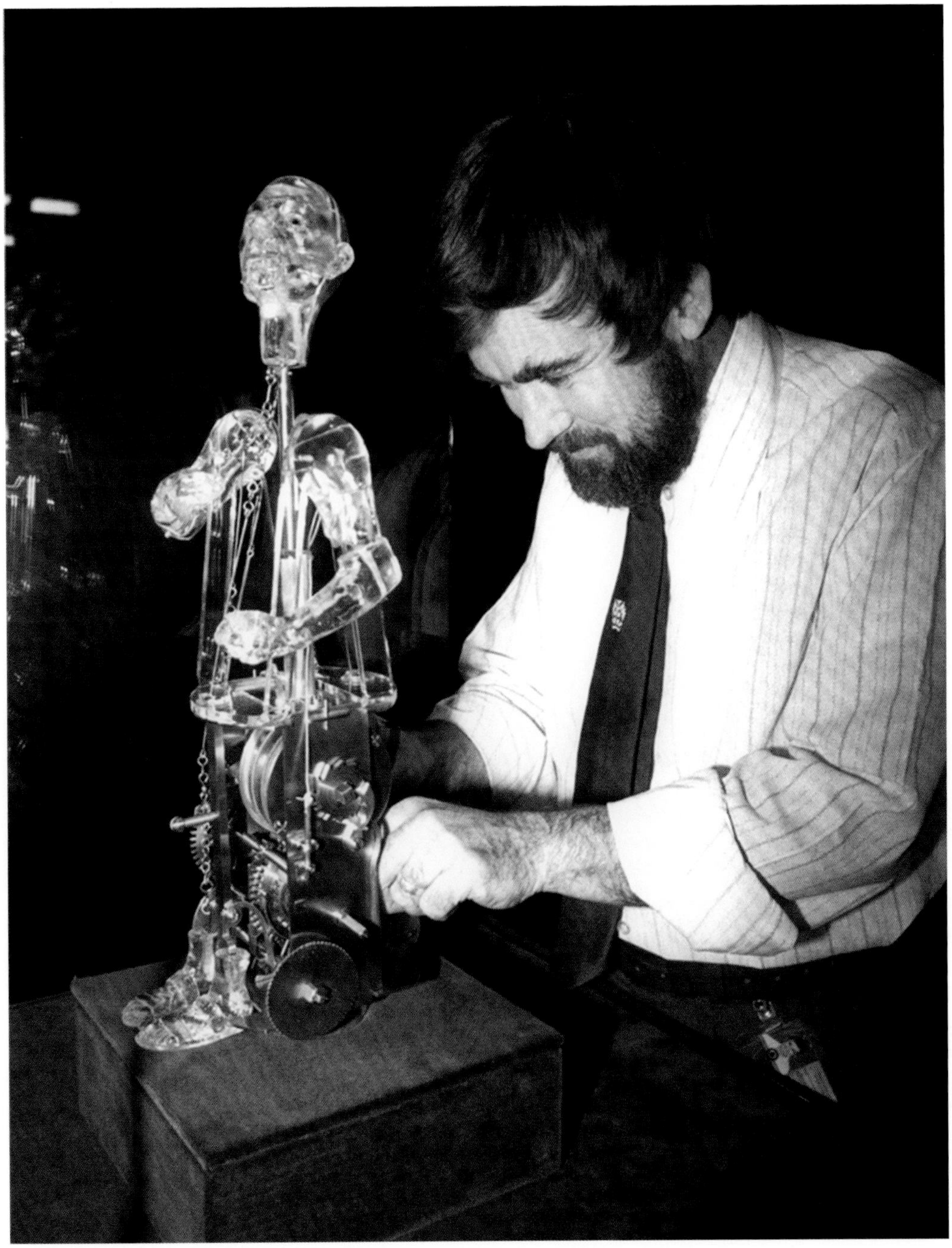

FIGURE 8.6.
David installing his model in the museum. From 1980 to 1996 the model was on view in its own vitrine next to that of the monk. Visitors could wind it themselves and view its performance—all but the walking tour. Running wheels "up," it walked in place.

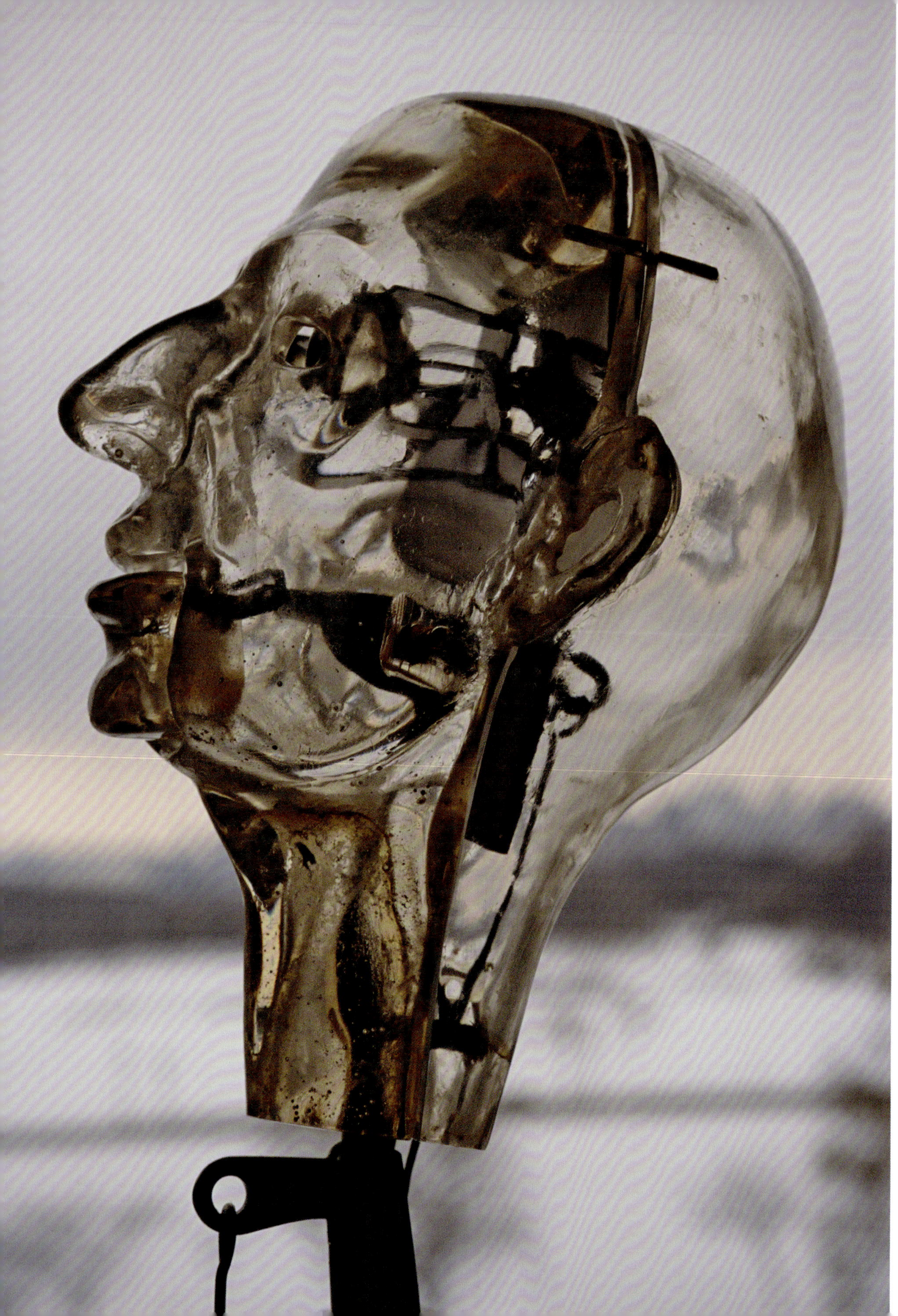

monk's 0.013 inch to the model's 0.023 inch. Even then, David found it close to impossible to find a commercially made modern spring that strong.[3] The monk, he then understood, was a sensitive compromise between the power of its mainspring and the size of the mechanism made to house it. The performance of the monk, we must remember, called for a mechanism that goes well beyond the requirements of the hands of a clock. With wooden arms and feet to be lifted in three dimensions against gravity and with the diverse actions of the head—each with its counteracting return spring—and, most important, with the task of moving the machine itself across a table, the monk needed an extremely forceful spring. The monk's size is not an accident. It is a critical calculation of matter, weight, and motion.

One day in the lab Elizabeth asked David whether he removed the mainspring entirely from the mechanism during his cleaning and study of the monk's clockwork. "Yes," he said, "I released the clickwork, let down the spring, and removed the barrel cap."

> **ELIZABETH:** And did the spring just *burst* out like an escaped bird?
>
> **DAVID:** Goodness, no. The spring must first remain contained inside the barrel wall. First, I unhooked the barrel arbor, by turning it backwards. This unhooks the arbor from the inside end of the coiled mainspring. And then I very carefully removed the spring, inch by inch. It's actually a moment for great concentration. If you lose control of a spring whilst you're taking it out—remember, the spring is still under great tension because it's contained within the barrel—it'll cut you to bits. People have actually been killed doing this. Because the *outer* end of the spring is still hooked to the barrel (the inside of the barrel wall), and the sudden explosion of the spring out of the barrel is like a kind of slingshot. It'll propel the barrel like a missile.
>
> **ELIZABETH:** People have been *killed*?
>
> **DAVID:** You take a pair of strong round-nose pliers. And you tightly grab a few center coils of the spring together with the pliers, and rotate a little as if to wind the spring. You're grasping the barrel in your other hand, with your fingers closed around the front to hold the majority of the spring safely inside. And you very carefully pull the center of the spring out of the barrel, a coil at a time, just easing the first few coils out slowly between your fingers and thumb. You've got to go slowly and concentrate because if too many coils come out at a time, that's when you lose control. After the first coils have been eased out this way, you can drop the pliers and grasp the spring right at the point where you've got it emerging from the barrel. Then, bit by bit, turning the thing like a yo-yo between your hands, always holding it tight with the one hand, and slightly winding the spring each time to take tension off the point of exit, you ease out the whole spring.
> The monk's spring, when out and at rest, is a great open spiral ribbon of steel, the spiral about 2 feet in diameter, the ribbon (or blade, as we call it) $1\frac{3}{8}$ inches wide, and the overall length about 4 or 5 feet if you straightened it out.
>
> **ELIZABETH:** All from a barrel the size of a pillbox!... And when you have the spring out—
>
> **DAVID:** —then you have a cup of tea. Because the part most fraught with danger is done.[4]

With cleaning, study, analysis, reassembly, and a few minor repairs to linkages, David at last wound that spring. The monk performed flawlessly on cue.

FIGURE 8.7.
David Todd's model of the monk, photographed by Rosamond Purcell, 2002. See also view of the model on p. iv.

9 The Monk Performs

The monk is a quintessential object of theater. The cloth tunic is the closed curtain concealing the stagecraft. Hidden under the tunic hem, behind the monk's right foot, is a small crank called the stopwork lever. Easily pressed or released by a knowledgeable handler, this lever interrupts the automaton's gear train at its point of lowest torque (the fly) and acts as a switch that can halt, or resume, the action of the fully wound clockwork. The mainspring may thus be wound ahead of time with its key, the key removed, and the automaton held motionless until the moment the performance is to begin. A practiced attendant sets the monk on the table and discreetly releases the stopwork lever. The ready gear train is set running, and the monk starts moving. In turn, the performance ends when the stopwork lever is secretly reengaged as the monk is lifted, and the stilled figure is removed from sight.

Perhaps the monk was seen only during the three minutes of the performance and spirited away before observers had time to look twice. Primed to believe that the monk moves of its own accord, spectators might wonder, "*What else* will it do?" Whoever is seen to handle or possess the monk—clockmaker, prince, king, or steward—could claim authority to the mystery. Above all, winding the monk out of sight of observers lets the operator determine exactly when, in the clockwork's movement cycle, to set the stopwork lever and thus what the monk will do first when released in front of onlookers.

Suppose you are one of a small group of visitors assembled for the occasion. You are seated at the table, by the light of a flickering oil lamp, or candle; what would you see? A silent steward appears from an adjoining room and places a small cloaked figure on the table before the company, then steps back. A compact statuette, about half an arm's height, in the form of a friar, wears a rough-spun cassock and sandals. The shaved head is erect, the face severe. In its left hand it grips a tall wooden cross.

Murmurs break out around the table as the figure moves and stirs. It glides forward very slowly, footsteps barely touching the surface of the table. Its eyes move. Its mouth opens and then closes. A few of your companions have backed away; others lean in for a closer look.

As it advances, the manikin-friar flourishes the cross, left arm hinging at the elbow to bring the cross upright like a pike. Now the right arm moves. With fingers beaked to a point, the figure strikes its own chest.

It—no, *he*—is walking in the direction of a person across the table. You see the friar in profile. As he raises the cross, his mouth opens as if to say, "See this?" A few steps more and without breaking pace, he looks down and raises the cross slowly to his mouth. With a fierce snap of his lips, he kisses it.

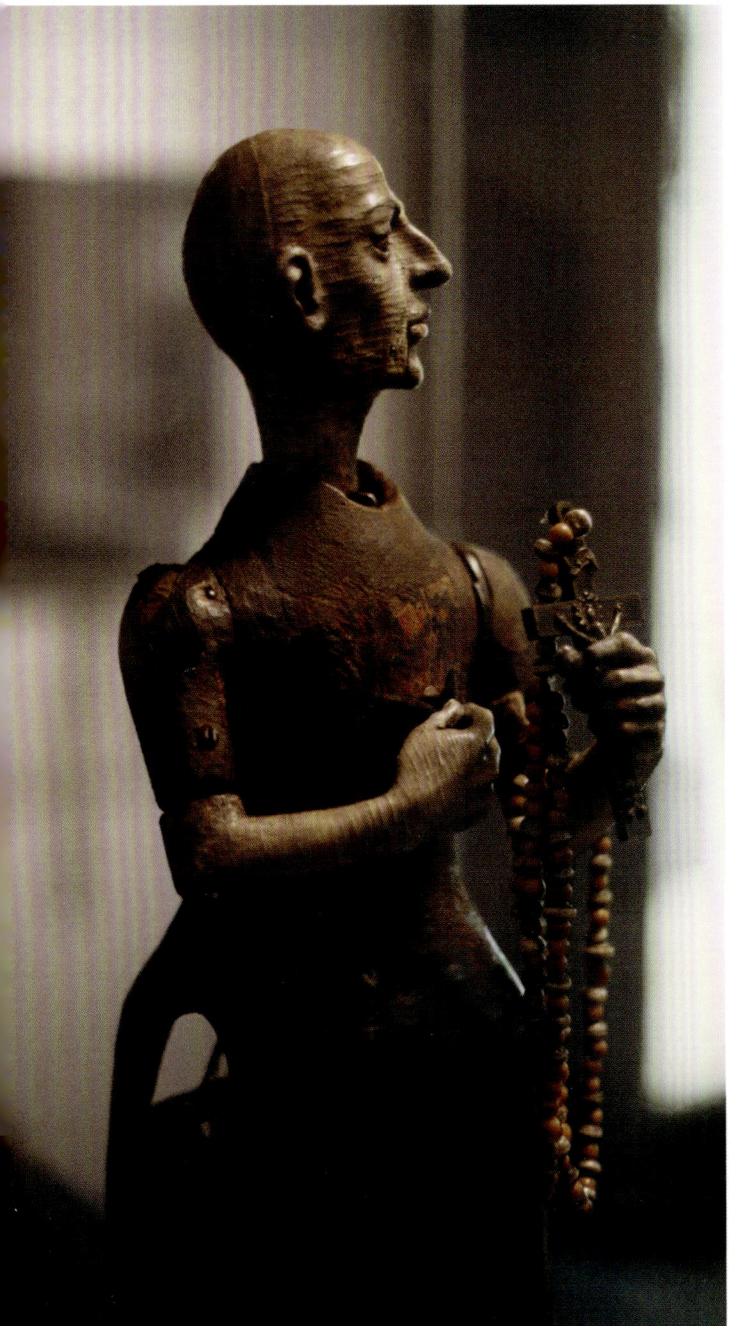

Still moving forward, he looks up and resumes brandishing the cross. Twice more he beats his chest. The impact of each beat makes his head shudder. Closing in on your dismayed neighbor, the friar comes within inches of the edge of the table and at the last moment makes a smooth walking turn to his right—and heads your way.

Now you see him head-on, and your own flight urge stirs. His eyes scan the table, then fasten on you. Everyone's eyes fasten on you. He's moving slowly, but he's doing more than you can take in. You notice his protruding ears. He glances left at the cross, then back at you. Mouth opening and closing, he hikes the cross repeatedly in front of him as he comes. Step by step, each foot lifting the cassock hem. He is like an emissary acting under orders, or under a spell. Again he brings the cross to his mouth, his eyes focus on it, and again the formidable kiss. He strikes his chest, the mea culpa, but there's no pause in his step. Just inches from you he begins a turn. You let out half a breath. But as his full body pivots on the table, feet in motion, head forward, his eyes slide left in their sockets to stay fixed on you! Then they snap forward as he strikes his chest again, and he locks his eyes on another guest across the table.

Now he moves away from you. His face is hidden but not the wide-eyed face of the person in his path. You try to picture his gaze. Each time he beats his chest his head jerks back with the impact. This time, as he brings the cross up for the kiss, you see the arm now moving at both elbow and shoulder, the cassock sleeve sliding and creasing with the motion. He lowers his head to the cross.

Five more turns and five more advances. On each turn he selects a different person, never the same person twice. You see his head turn: to the left, pause, to the right, pause. Some of his motions are repetitive; others are new with each step. Or is it that you notice something new with each step? His tour back and forth across the table keeps changing the angle of your view, in a performance too complex to parse (**FIG. 9.1**).

The steward steps forward, and the figure is gone before you quite know what you've seen.

FIGURE 9.1.
The monk "works" the table. If the automaton was whisked away at the close of the performance, spectators might be left to imagine further actions and added consequences.

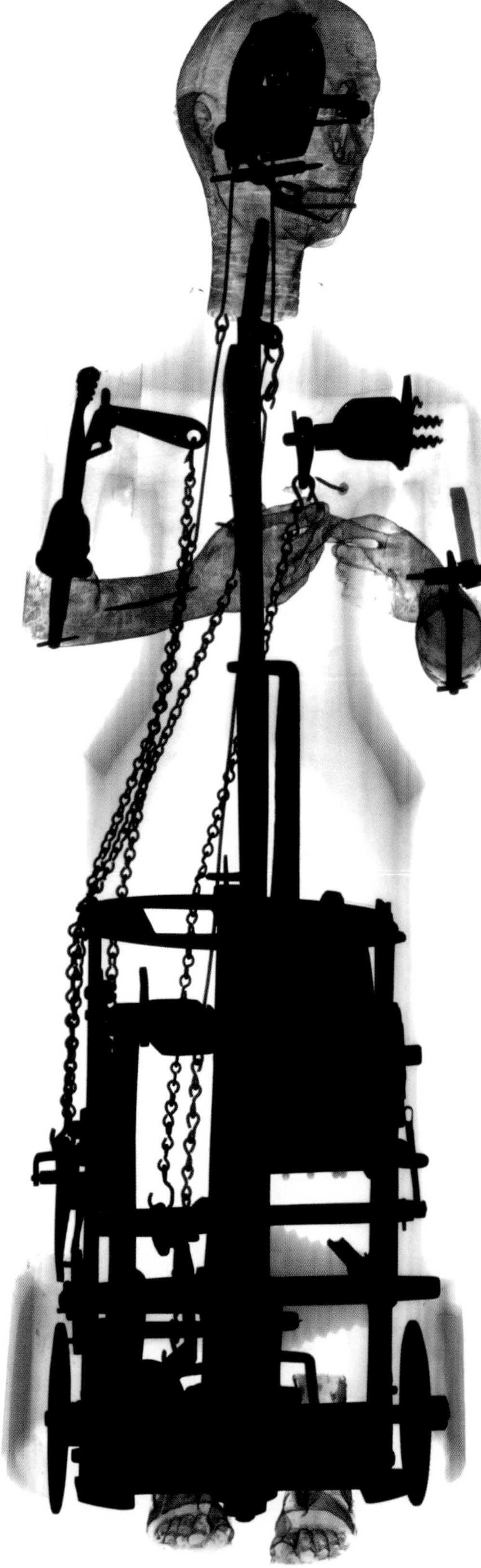

10 A Mechanical Anatomy

Understanding a few of the mechanical principles underlying the monk's performance will let us grasp the relationship between technology and art in this automaton. When we look at the clockwork that brings the monk alive, we look back in time to a world just awakening to what a machine could do and could be (**FIG. 10.1**). To art's dramatic depictions of the quick and the dead, the introduction of real-time motion illustrates how the monk might have been considered an object of magic or miracle by some who saw it.

Key to the power of the performance is the combination of separate motions and arcs as the automaton moves in three dimensions. The monk moves his head, his mouth, his eyes, his two arms, and his feet, all while advancing along the table: the onlooker cannot perceive one action in isolation from the others. Both arms are in motion when the monk makes his walking turn, and they sweep out three-dimensional gestures in space. As he turns to his right, his eyes shift left, effectively holding the prior direction of his gaze as his body pivots. His feet are stepping as he turns, each foot tracing successive arcs en route. All these actions happen simultaneously, each at its separate speed, with the overall movement quite fluid. At five pounds of iron and wood, this is a lot to ask of a coiled spring.

FIGURE 10.1.
In X-ray view, the monk's iron clockwork appears darkest; the painted head, forearms, hands, and sandaled feet show in midtones; the unpainted wood body and fusee are pale shadows. National Museum of American History and Smithsonian Museum Conservation Institute, Washington, DC.

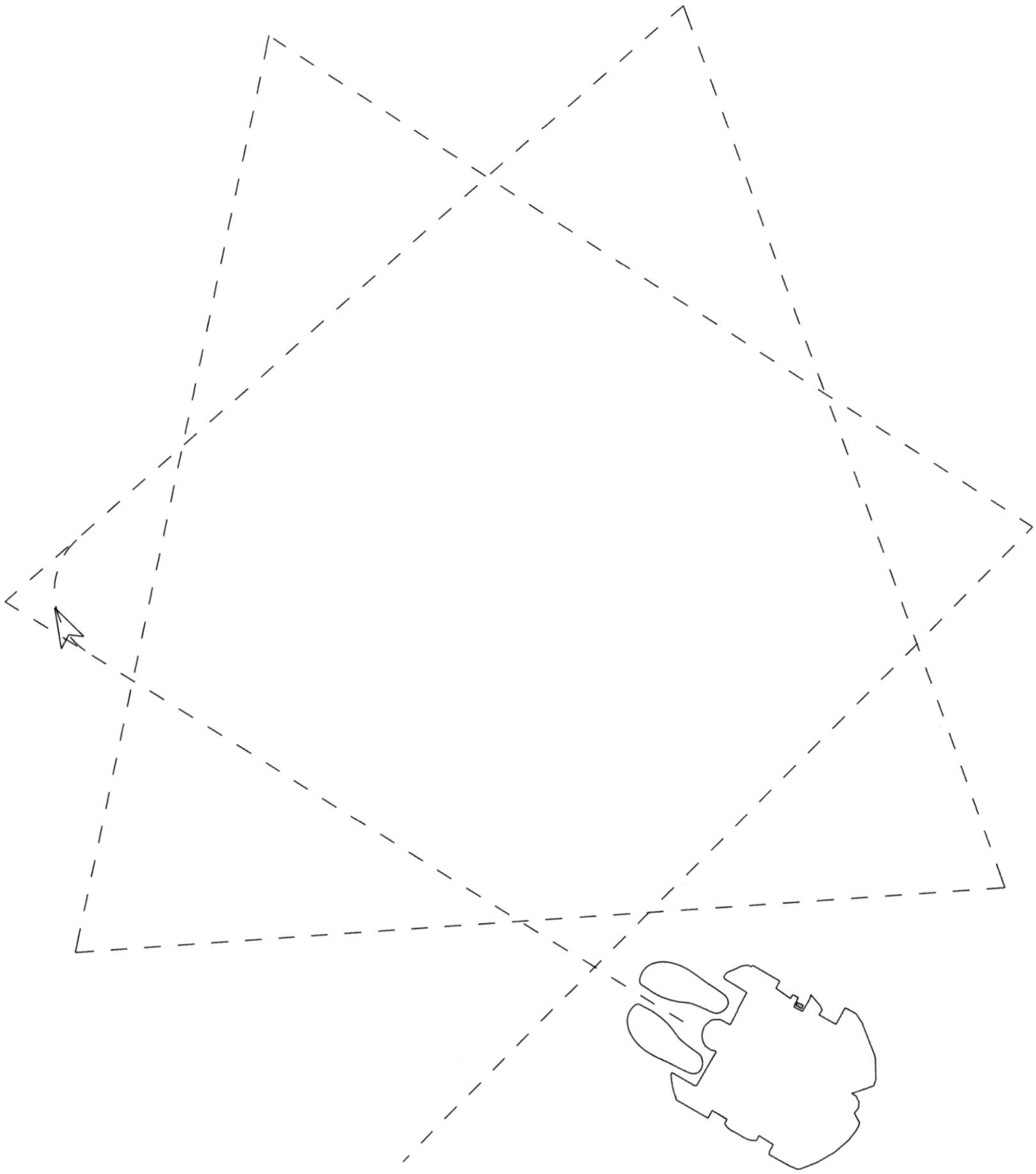

FIGURE 10.2.
The monk's walking path: a crisscross tour of the table, each "address" revealed as onlookers watch and wait. Each turn angle can vary by a few degrees.

The monk's feet were designed to step out from the hem of a cloth habit in imitation of walking, but the figure actually rolls forward on three wheels hidden beneath the habit. Two "perambulation wheels" at right and left are driven by the clockwork mainspring—the monk's power source—and one trailing wheel at the rear keeps the machine stable as it advances. The monk will "walk" a straight distance of approximately 19 inches, then turn to his right and pace out another 19 inches, turn again, and so on, tracking the path of a nonrepeating irregular polygon on a tabletop. The turn is accomplished by means of a pivot that periodically lifts the right perambulation wheel off the table surface just enough so that the left wheel rolls in an arc around the pivot. If the mainspring has been fully wound, he will complete a total of seven such lengths and turns. The angles of the turns are not exact and depend on the texture of the tabletop. Angles from 100 to 118 degrees have been measured: the variability means that the monk's new walking direction after a turn is unpredictable. Onlookers will be in some suspense about where he will head next, as he "works" the table. **FIGURE 10.2** shows a diagram of the monk's path based on video taken from above, in December 2021.[1] Including the turns, the net length of the monk's travel adds up to almost 14 feet.

The *slowness* of the monk's actions lends his character an agency not quite of this world. Only his eyes and his striking hand move abruptly—all the more disturbing in contrast to his otherwise ghostly pace. The net time of the performance can differ, due to variable traction and drag from the tabletop surface. Tests in the lab have yielded twenty-five to thirty seconds for one straight walk and three to four seconds for a turn. The monk's full performance might last no more than three and a half minutes.

On a single walking span, the monk performs a complex sequence of body motions, to be repeated on each of the seven segments of the polygonal path. Picking up the performance just after a turn, here is a summary:

- Each foot is raised and lowered four times.
- The head turns from center to left and back.
- The head bows forward once, for the kiss.
- The lower left arm bearing the cross is raised and lowered at the elbow six times.
- The upper left arm is raised at the shoulder for the kiss, then lowered.
- The lower right arm, moving at the elbow, strikes the chest three times.
- The eyes move from center to left and back again, three times.
- The mouth opens and closes six times, and performs one kiss.

Each action is precisely synchronized with all others. The kiss, for example, involves simultaneous motions of the head, mouth, eyes, and both the elbow and the shoulder of the left arm, all in coordination. Below, David touches on key parts of the monk's clockwork: the frame, the mainspring, and its ingenious fusee that is the monk's power source, the gear train, the primary cams and levers that create the performance, the speed-governing fly, and the hidden stopwork "switch" that starts or stops the action after the mainspring has been wound.

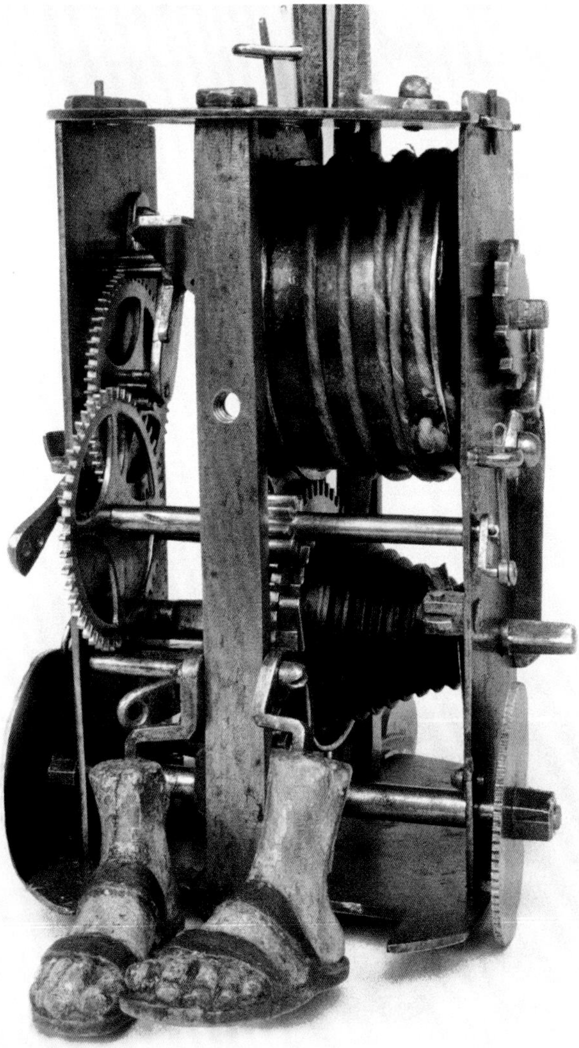

10.3A.

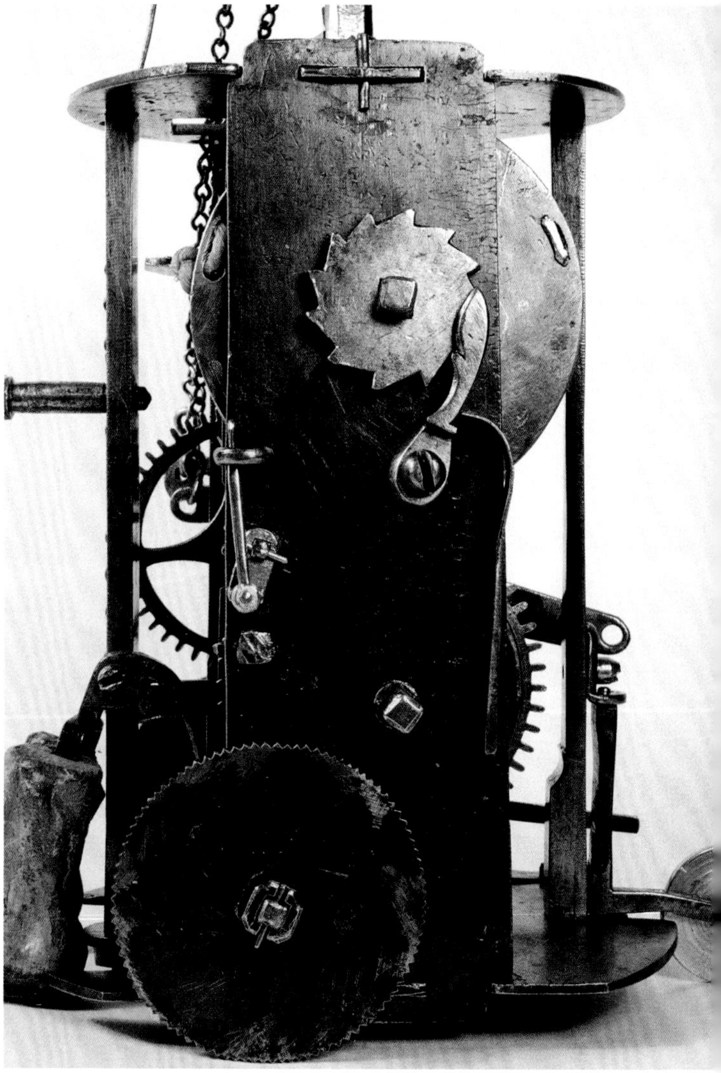

10.3B.

The Clockwork Frame and the Monk's Wooden Body Shell

In the sixteenth century the posted frame configuration of the traditional weight-driven wall clock was adapted to the newer spring-driven table clocks, and the monk's clockwork structure follows suit. Such a frame consists of the *horizontal top and bottom plates*, the vertical *pillars* (or posts) that separate them, and three vertical plates, called the *train bars*. The plates and pillars form the initial "cage" of the frame; the train bars anchor the various elements of the gear

FIGURES 10.3A–10.3D.
Four views of the monk's animating clockwork.

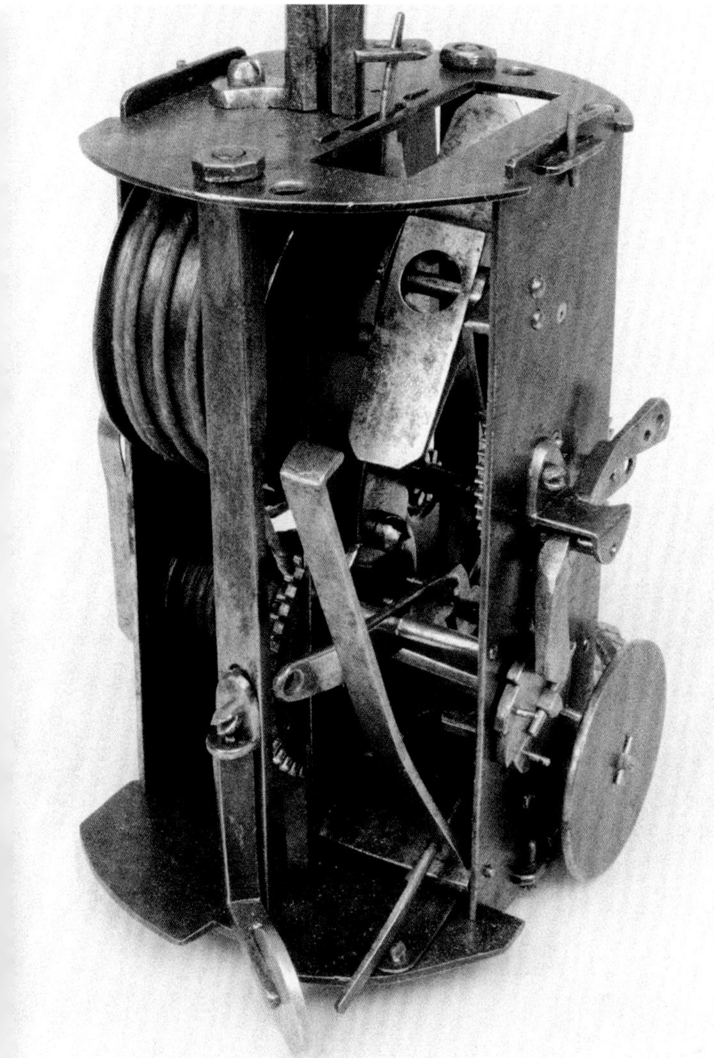

10.3C.

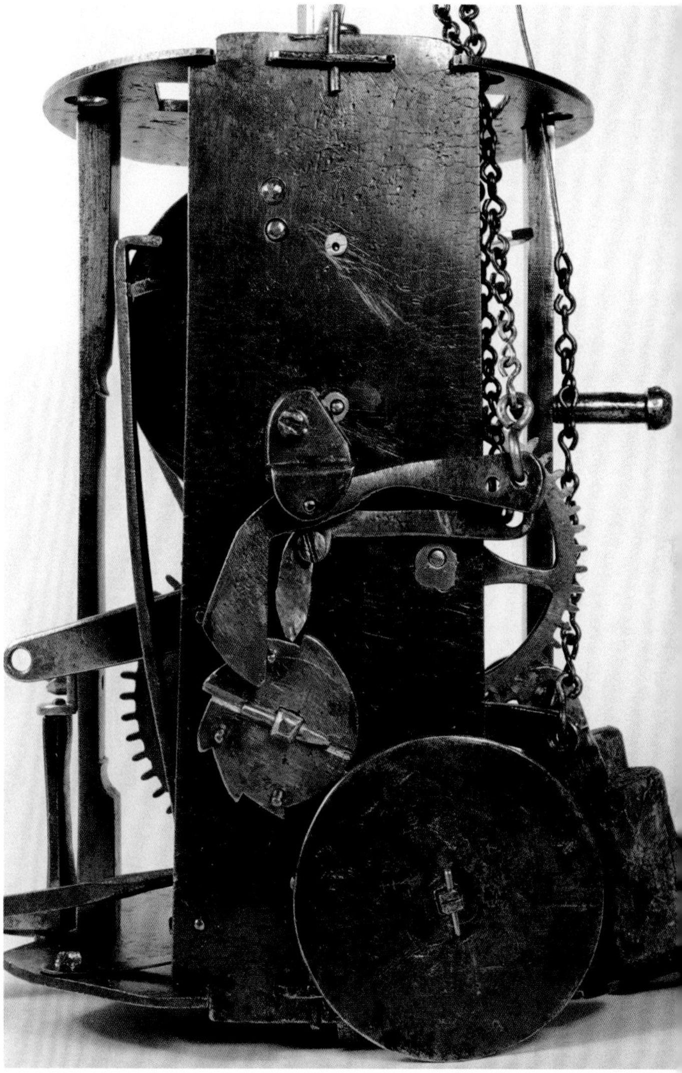

10.3D.

train **(FIGS. 10.3A-10.3D)**. Note the elegant tab-and-slot joints that secure the train bars to the plates, held fast by a tapered pin or wedge hammered through the tab. The pillars, square in cross section, are turned and threaded at each end and fastened to top and bottom plates with custom-made nuts.

The bell-shaped wooden shell of the monk's body (see introduction, **FIG. I.2**), lathe-turned and hollowed out, serves as a casing for the mechanism. Made from a single piece of wood, the shell is attached to the front pillar of the frame with, remarkably, just one substantial iron screw. Of a piece with the shell but in contrast to its lathed skirt, the roughly carved upper torso is also hollow and contains the linkages to the limbs and head. Openings at each side and at back give access to the mechanism for assembly. The wood itself is identified as poplar or beech, the museum's current estimate. No microanalysis has yet been carried out for a sure identification. The shell's thin bottom edge is reinforced with glued cloth to give it strength (this could be original or a later repair). If the less refined shell is intended to be hidden beneath the monk's robe, it nonetheless harbors the proudly fashioned clockwork.

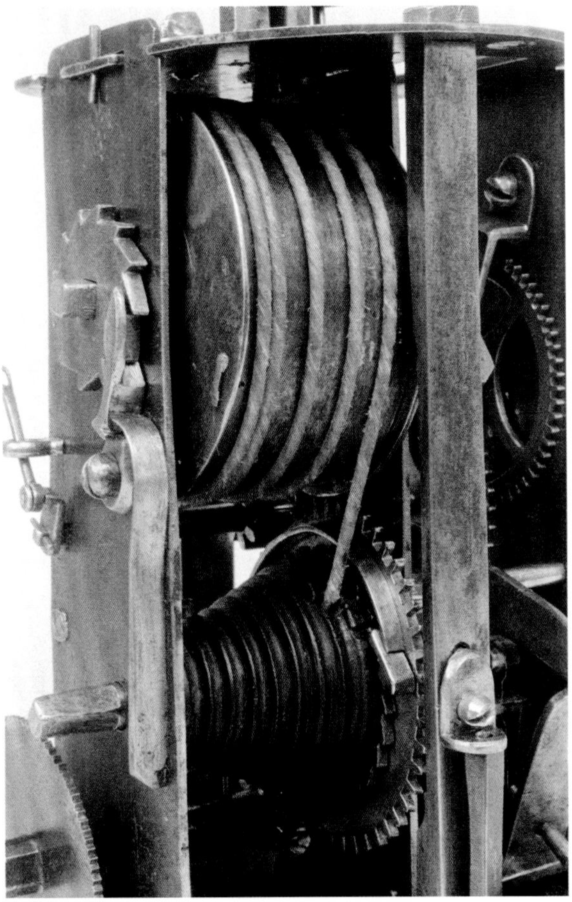

FIGURE 10.4.
Mainspring barrel and fusee.

A view into the clockwork from the rear: the mainspring barrel, with gut line wrapped around it, one end attached to the wooden fusee. The fusee winding square (to wind up the clockwork) is visible at lower left. Note the fusee clickwork at the wide end of the fusee and next to it the toothed fusee wheel, the first wheel in the gear train. The mainspring barrel arbor also has a clickwork, visible at upper left (and more clearly in **FIG. 10.3B**) to keep the barrel from spinning backward as it is wound. The curved pawl, called the "click," with its decorative shape, is pushed against the toothed ratchet wheel by the long vertical leaf spring mounted beneath it.

Mainspring and Fusee: Power Source and Perfect Torque

The monk was made only about a century and a half after the introduction of the *mainspring*—the tightly coiled ribbon of tempered steel used to drive clockwork mechanisms. The spring's greatest advantage—it released clocks from the captivity of the hanging weight and thus heralded the table clock—also allowed the introduction of freely roaming automata, portable and mobile, and the technology of true androids began. The monk's mainspring is coiled inside the *mainspring barrel*, an iron drum mounted between train bars just above the cone-shaped *fusee* **(FIG. 10.4)**. Inside the barrel, the spring's outer end is hooked to the barrel wall; its inner end is hooked to an axle, called the *barrel arbor*. If the barrel is turned while the arbor is held still, the spring inside will be tensioned, or wound. The tighter the spring is wound, of course, the greater the captured latent force.

A mainspring may render a clock portable, but a spring by itself does not provide a uniform release of power. A fully wound mainspring is extremely powerful, while one that is almost unwound has close to no torque. Of all the solutions devised to address the problem of the uneven force of a spring, the fusee is by far the most elegant. Its use became widespread after its introduction to clockwork around the middle of the fifteenth century.[2] The monk's fusee is a graduated cone made of hardwood, with a spiral groove cut into it, turning on its own arbor, parallel to and beneath the mainspring. The fusee is connected to the mainspring barrel by a thin strong catgut line.[3] One end of this line is tied into the fusee cone at its large diameter; the other end is wrapped around the barrel wall for a number of turns and then attached to it with a knot or a hook. Winding the mainspring is accomplished indirectly by winding the fusee, with a key applied to the squared end of the *fusee arbor*. As the key turns the fusee, the catgut line is drawn off the barrel and on to the fusee cone, thus turning the barrel and winding the spring inside it. As the winding continues, the line seats into the spiral groove on the fusee cone **(FIG. 10.5)**.

When the machine is then set running, the driving power of the wound-up spring is transmitted from the barrel through the gut line back to the fusee: the line is now pulled off the fusee and back on to the barrel, thus turning the fusee. The force emanating from the spring is determined at the fusee by the diameter of the cone where the line meets it. When almost fully wound and very strong, the spring pulls line from the small diameter of the fusee cone; when half unwound, the spring pulls line from the midsection of the cone; and when almost unwound, the now-weakened spring pulls line from the large diameter of the cone. To visualize this action differently, think of the fusee as a lever and the point where the line meets the cone as its fulcrum. (Recall that the longer the lever, the less force is required to move it.) With the driving force of the spring lessening as it unwinds, the fulcrum of the lever must be moved in order to keep an unvarying force delivered to the clockwork gear train. The greater the distance from the fulcrum to the fusee's axis of rotation, the less force is required from the spring. This ratio explains why the line meets the cone at its smallest diameter when the spring is fully wound (and at its strongest) but meets the cone at its largest diameter when the spring is almost unwound (and at its weakest). The curved profile of the fusee cone, as you view it from the side (see **FIG. 10.5**), is a perfect graph of the diminishing power of the mainspring.[4] The partnership of spring to fusee delivers a torque so consistent and precise that it is still used in high-end mechanical clocks today.

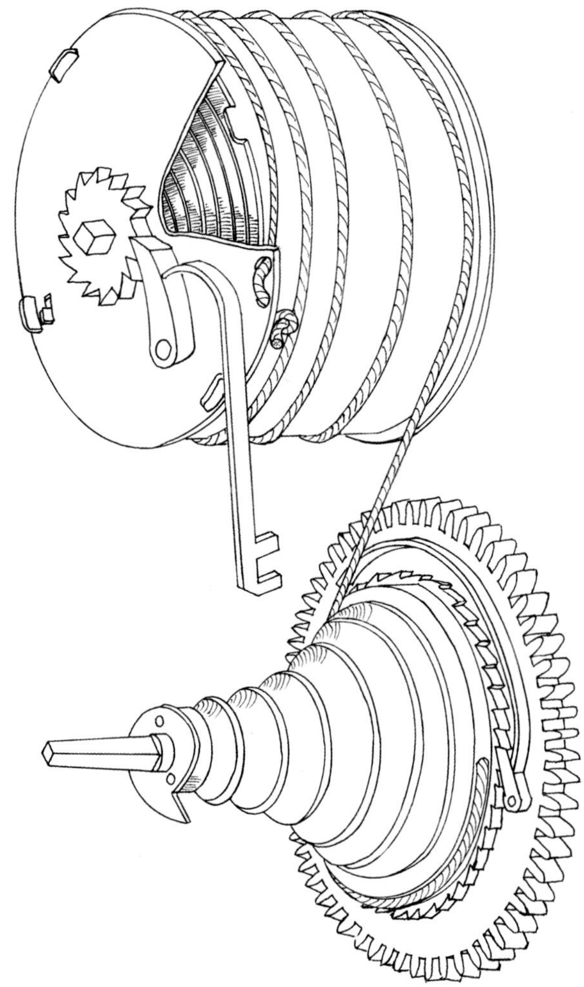

FIGURE 10.5.
Mainspring barrel and fusee diagram.

The automaton's power source. The barrel is shown with a portion of its lid cut away to reveal the partly wound mainspring inside. The outer end of the coiled mainspring blade is secured to the barrel (with strong steel tabs, one of which can be seen in the barrel cap); the inner end is hooked to the barrel arbor. The spring is wound indirectly from the fusee below: a key set on the long winding square of the fusee arbor and turned counterclockwise will draw catgut line off the barrel and on to the fusee, turning the barrel on its arbor and thus winding the spring inside it. The force of the mainspring is held in check by the fusee's clickwork. After winding, the stored power is delivered evenly and slowly, as the fusee keeps the force constant. The engaged gear train and fly comprise a workload calculated to spend the force at a specific and gradual rate.

The wood fusee cone is seated against but is not directly secured to the *fusee wheel*, the first toothed clockwheel in the gear train. Both fusee and fusee wheel are secured to the fusee arbor, but the fusee wheel and the gear train must remain stationary while the fusee is being wound. The clockmaker accomplishes this by separating the fusee arbor into two parts—one shaft turning inside another—and installs a *clickwork* between them.[5] The fusee clickwork, a ratchet and pawl, also prevents the mainspring and fusee from violently unwinding when the key is removed. The clickwork can be seen in **FIGURES 10.4** and **10.5**, its ratchet-toothed clickwheel at the wide end of the wood cone, its spring and pawl attached to the fusee wheel. The interaction of pawl and clickwheel makes the staccato sound we hear as the monk is being wound. **FIGURE 10.6A** shows the fusee and its arbor. **FIGURE 10.6B** shows just the right half of the arbor, fusee wheel removed, revealing the shaft that inserts into the wood fusee cone. This shaft remains stationary inside the fusee cone during winding, as the fusee turns around it. Both shafts then turn together when the machine begins to run.

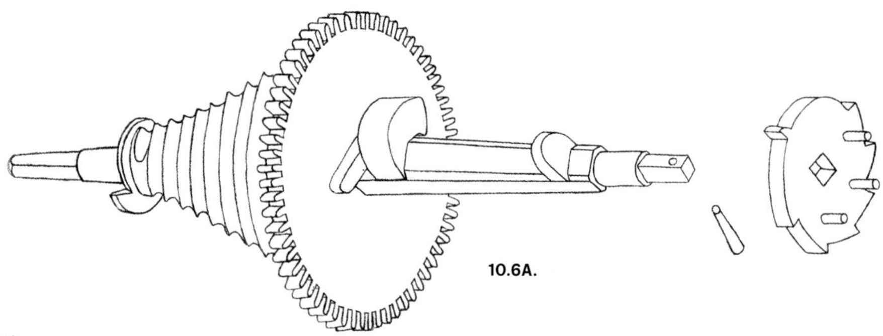

10.6A.

FIGURE 10.6A.

Fusee assembly and its cams.

The fusee on its arbor, bearing the cams that are the "brain" of the monk. Starting at the far left, we see the winding square, then the comma-shaped stopwork that prevents overwinding of the fusee line, then the wood fusee cone itself, then the first train wheel, called the fusee wheel. That wheel in turn is attached to the right half of the fusee arbor, bearing its three primary cams, and finally at far right to the multipurpose cam, which fits onto the squared right end of the fusee arbor, retained with a tapered pin. A clickwork, not visible in this view, allows the separation of the fusee for winding, then its reengagement for the gear train to run.

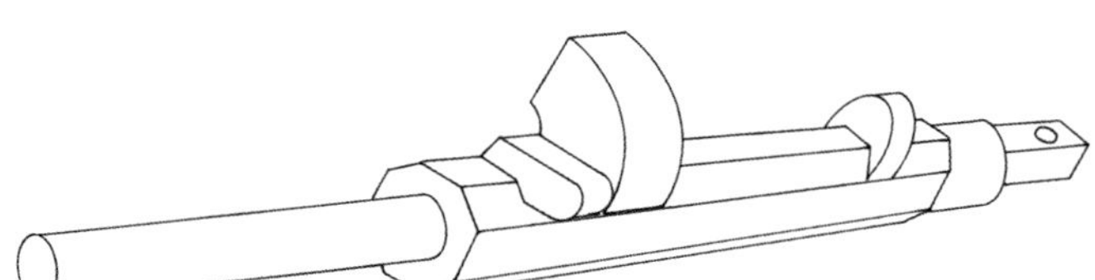

10.6B.

FIGURE 10.6B.

Fusee arbor and cams.

Fusee arbor shown without the fusee and fusee wheel. At left is the long pivot that is inserted into the wood fusee cone, permitting the winding of the fusee without engaging the gear train. Three cams are pinned or brazed onto the arbor shaft. The first cam (1) engages the serpentine lever that turns the head from side to side. Next to it, the large second cam (2) both tilts the head and raises the left arm at the shoulder. The third cam (3) initiates the monk's walking turn. The square at the far end receives the multipurpose cam.

Transmission and Regulation

In the world of clockmaking, the words *arbor*, *wheel*, and *pinion* have precise meanings. Arbor is the horologist's term for axle. A clockwheel, or wheel, refers to the metal disk with teeth cut into it, called in other trades a toothed gear. A wheel by definition has more than sixteen teeth. Any wheel with sixteen teeth or less is called a pinion. Pinion teeth are known as *leaves*. A wheel is typically riveted to its arbor and spins with it, though in some cases it may spin *around* its arbor. Pinions, in contrast, are always riveted to, or of a piece with, their arbors. A gear train (or wheel train)—a mating set of wheels and pinions—divides and distributes the mainspring's force through the stepped ratios of driving wheels to driven pinions.[6] For example, one revolution of a wheel of forty teeth will drive a ten-leaf pinion to revolve four times. Such ratios, in combination, will progressively increase the speed of a train of wheels. But speed is bought at the price of power. In any clock mechanism the high-force/slow-speed work is done closest to the power source. Actions requiring less force and/or greater speed take their power from calculated points progressively more distant from the power source as the gear train steps up. Each of the monk's actions requires a specific and different speed and force. Ratios between wheels and pinions determine what the monk does, when and how fast he does it, and how much power is available for each function.

Although the fusee equalizes the power of the mainspring, it does not control the rate—the speed—at which this power is spent. Will a given action take 5 seconds, 10 seconds, 20 seconds? How fast will the wheels turn; how much time will the monk's performance take? The net number of uses—the load—asked of the mainspring's force determines the speed at which the spring unwinds and thus the faster or slower rotation of the gear train. If the spring is asked to do too little work, the gear train will run faster. Too much, and the train will run too slowly—or finally not at all. The clockmaker takes care to underuse the mainspring force and then calculate the weight and drive of the fly to precisely top off the net load and determine the speed of the full machine behind it.

The *fly*, so called because its vanes fly around at great speed (**FIG. 10.7**), spins at the tail of the gear train. It functions as an air brake: the resistance of the air on its vanes keeps the speed of its rotation in check. It is the governor of the machine, providing a calculated drag on the full system to keep it running at a specific rate. The beauty of the fly is the way that it operates "backward" through the train of wheels to govern the rate at which the train turns. In summary: while the fusee equalizes the force of the mainspring, the rate at which that force is spent is regulated by the gear train and its fly. Finally and serendipitously, because the fly spins with such minuscule torque on the wing, it can easily be interrupted, to halt the entire machine. It would take a firm grasp, and torn flesh, to stop the machine at the fusee. At the fly it can be done by one finger. This is the perfect place to introduce a "switch"—the fly stopwork lever—that the operator of the monk can discreetly press or release, to bring the monk to life or to bring the performance to a halt.

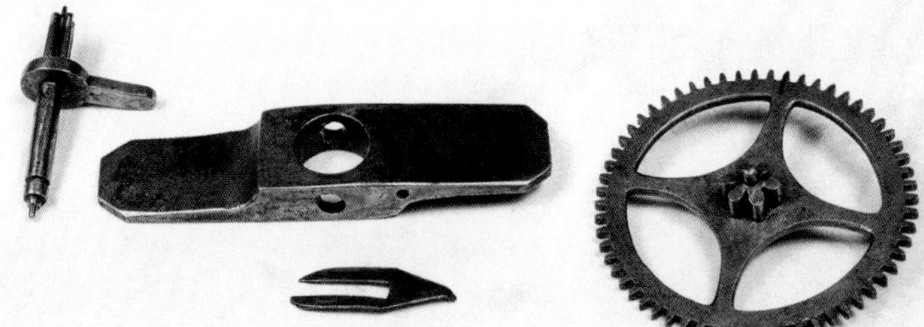

FIGURE 10.7.
Fly components, laid out.

The large and heavy fly, in the center, regulates the speed of the monk's performance by wind resistance. On the left is the fly arbor with its locking lever and four-leaf pinion (top end) and groove (bottom end) for its spring. Beneath the fly is the forked spring that keeps the fly located and friction-tight on its arbor and also acts as a shock absorber when the gear train's movement is suddenly halted by the operator (via the fly stopwork lever). The small hole in the edge of the fly receives the bent and pointed end of the spring and retains it in position. At right is the third wheel, the last wheel in the train, with its sixty teeth and its pinion of six leaves. This wheel drives the fly pinion. **FIGURE 10.3A** shows the third wheel, the fly pinion, and the fly in place. **FIGURES 10.3C** and **10.10** show the fly from above.

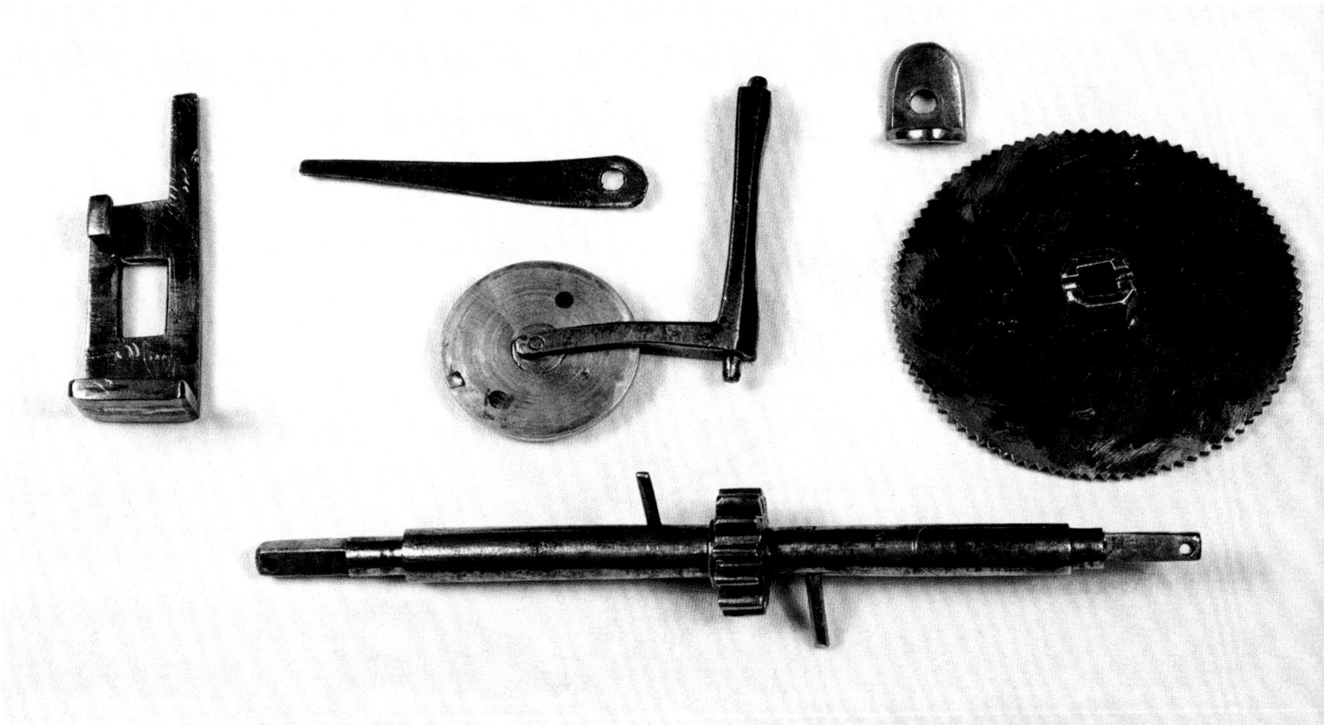

The Cam: Mechanical Memory

The fusee and mainspring have been wound, the stopwork lever is released, the fusee arbor and fusee wheel start turning, and the theater begins. The fusee wheel drives not one but two separate pinions (both glimpsed in FIG. 10.4: one at seven o'clock and the other at eleven o'clock on the wheel). The lower pinion, the *perambulation pinion*, transmits power to the monk's two *perambulation wheels* secured at each end of that pinion's arbor, and the automaton advances on the tabletop. This is the heaviest work, done closest to the power source. The ratio of fusee wheel to perambulation pinion yields four revolutions of the pinion (and its arbor) for every one revolution of the fusee arbor. Two iron pins secured in the perambulation arbor initiate the actions of the monk's feet (**FIGS. 10.8A, 10.8B**).

The upper pinion driven by the fusee wheel, the *second pinion*, has two roles: it initiates the complex motions of the left forearm, and it drives the full remainder of the gear train.

FIGURE 10.8A. (above)
Perambulation components, laid out.

CLOCKWISE FROM LEFT: turning pivot (seen here upside down), turning pivot leaf spring, trailing wheel in its bracket, trailing wheel bracket mount, serrated perambulation wheel, and perambulation arbor bearing its pinion. Note the perambulation arbor's squared extensions at each end to receive the pinned-on perambulation wheels. The two spikes lift the feet alternately. The arbor pinion is driven by the fusee wheel, the first wheel in the gear train.

FIGURE 10.8B.
The feet.

The delicately carved wooden feet are mounted on U-shaped iron levers that pivot at the front pillar. Each foot is lifted by a spike on the perambulation arbor that pushes against the lever's inner-facing flattened blade.

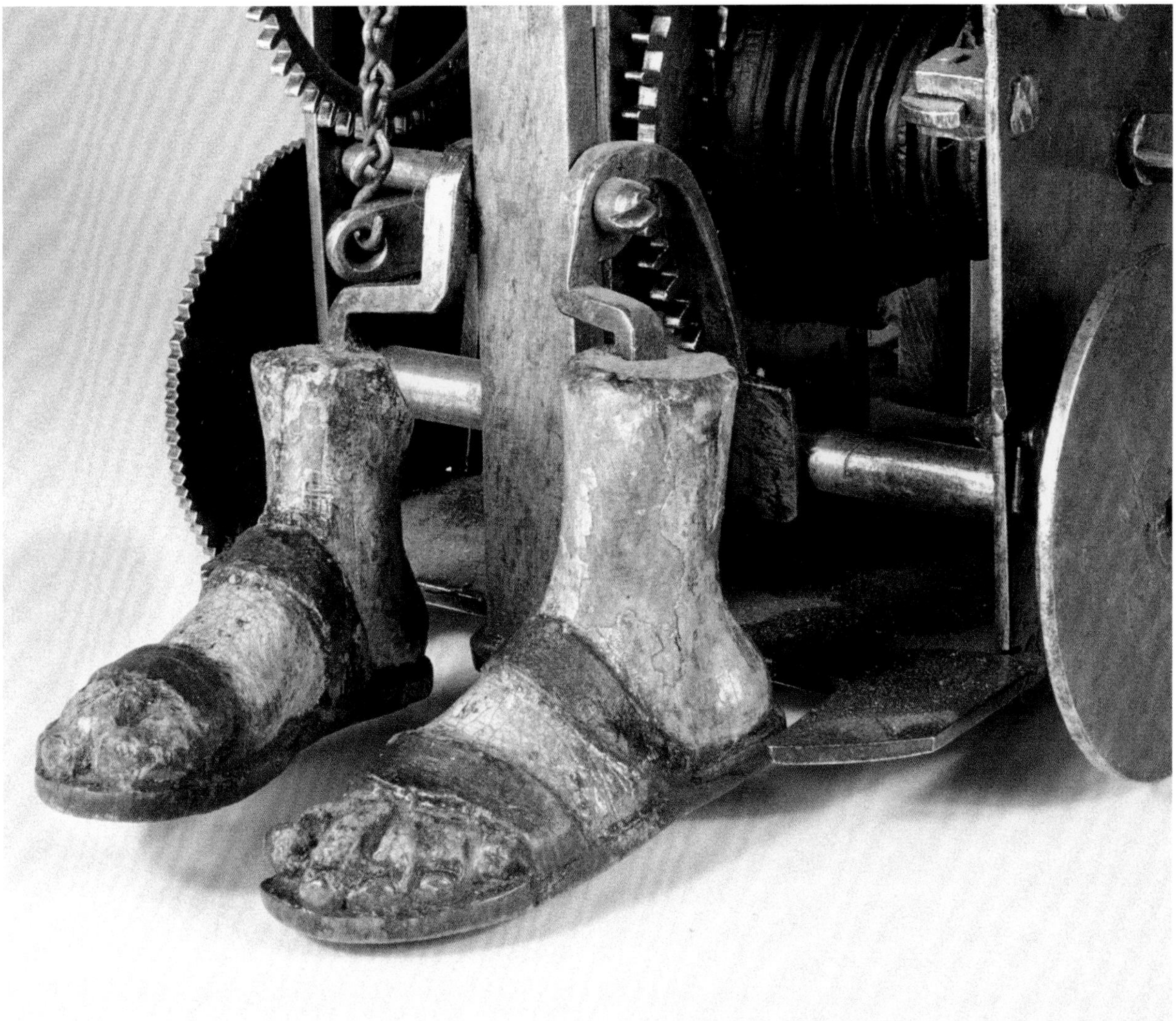

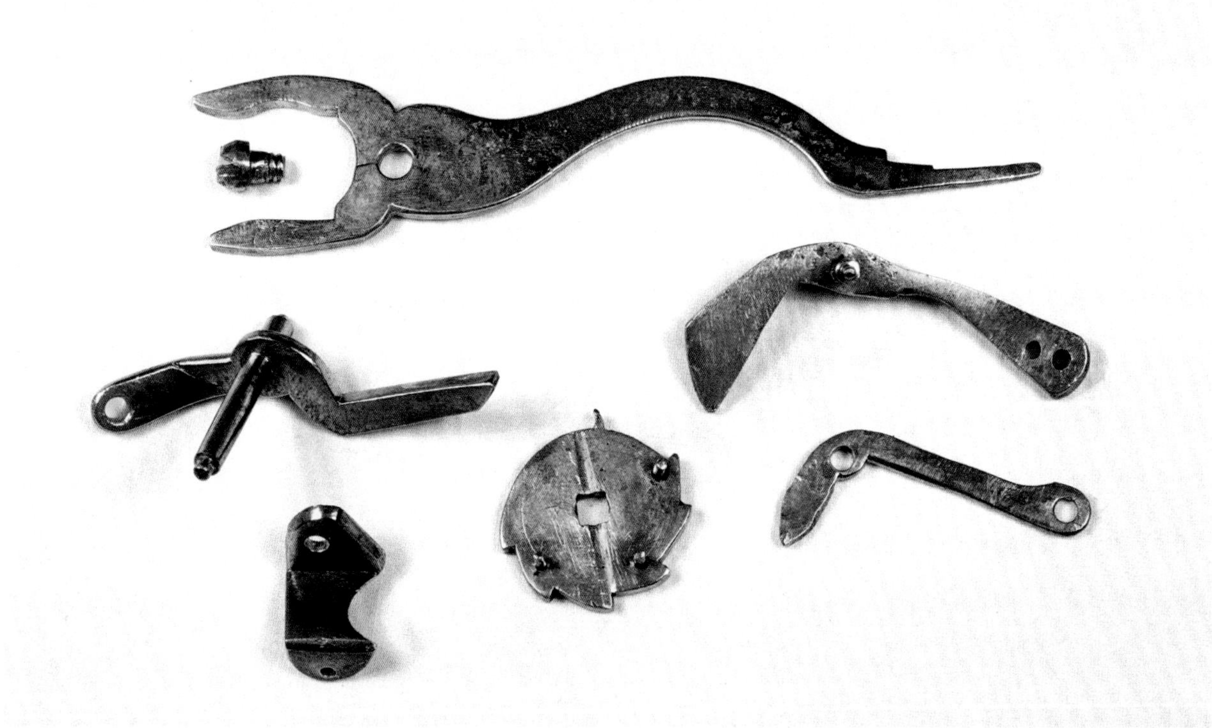

FIGURE 10.9A.
Levers and multipurpose cam, laid out.

CLOCKWISE FROM TOP: the serpentine lever (for turning the head) and the shouldered screw on which it pivots; the combined eye and right arm lever (a cranked lever with its two holes, one for linkage to the eyes for line-of-sight motion and one for linkage to the right arm for the beating of the chest); the smaller cranked mouth lever (for speaking and for the kiss); the multipurpose cam; the bracket that supports and pivots the eye and right arm lever; and finally the double-ended lever, which is associated with the simultaneous raising of the left arm (which bears the cross) and bowing of the head (for the kiss). Note the single pointed spike on the multipurpose cam. The lever to the mouth, riding along the profiled edge of this cam as it turns, reaches that spike and abruptly opens and closes the mouth to perform the kiss.

Built directly onto the fusee arbor are three cams (see FIGS. 10.6A, 10.6B, above). As the arbor carrying these cams rotates, the cams push levers that generate the turning of the monk's head, the bowing of the head, the motion of the left shoulder, and the monk's walking turn. A fourth, more complex cam, the *multipurpose cam*, is secured to the arbor's far end, outside the supporting train bar. FIGURE 10.9A shows the various levers and the multipurpose cam. FIGURE 10.9B shows two levers and the multipurpose cam in place. A disk with steps and pins, the multipurpose cam governs the actions of the mouth, the movements of the eyes, and the action of the right arm, activating levers and chain linkages to each part. These four cams, along with additional input from the upper steps of the gear train, produce the choreography of the monk's head and arms and his walking turn.

The cam is a crucial development in the history of technology. An ancient device attributed to Archimedes, the cam reached broad use in the fifteenth century in the striking trains of clocks. A cam is simply a barrel or disk

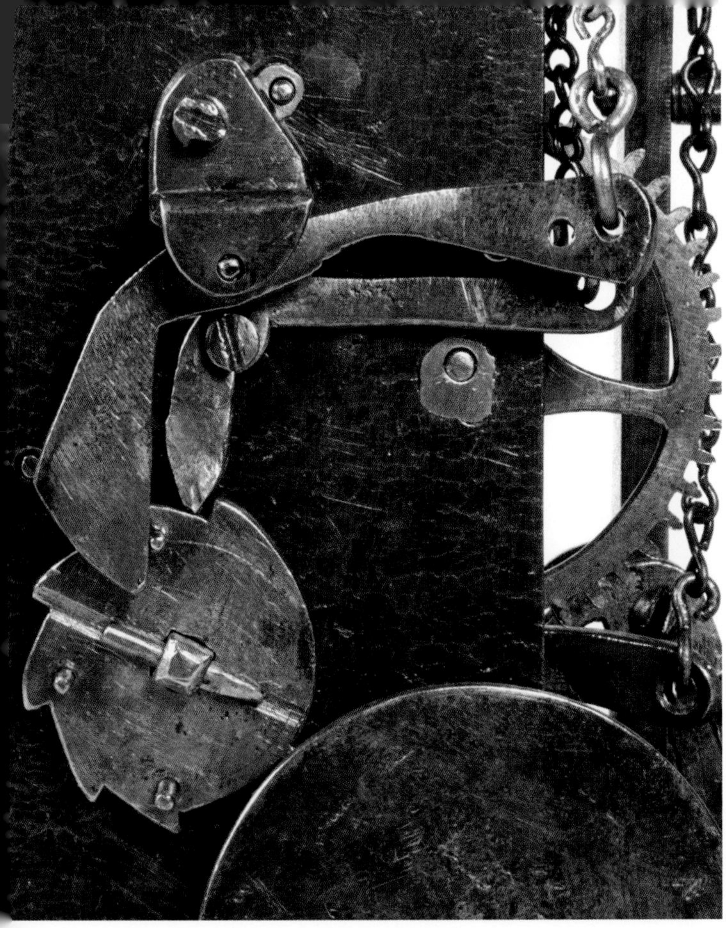

FIGURE 10.9B.

Multipurpose cam in place with levers to mouth, eyes, and right arm.

The multipurpose cam is pinned onto the squared end of the fusee arbor. In this view the cam will turn counterclockwise in rotation. The smaller cranked lever to the mouth runs along the steps of the cam profile to generate the "speaking" movement of the mouth. The "kiss" spike is hidden behind the perambulation wheel in this photograph. The larger cranked lever to both the right arm and the eyes is activated by pins on the cam face. Note two holes at the end of this lever: one (with chain in place) is for the linkage to the eyes, the other is for the linkage to the right arm. Putting eyes and arm on a single lever assures their synchronization. The shape and position of this lever, as each pin pushes against it in turn, is designed for a gradual tilt followed by an abrupt drop off the pin, and this accounts for the gradual backswing of the arm followed by its violent beating of the chest. Likewise, the eyes gradually move to the left and suddenly snap back to center with each blow of the arm to the chest. No pin is located on the cam near the spike for the kiss: eyes and right arm must be entirely still for this moment.

of metal rotated by the gear train. Its outer edge is either studded with short pins or cut to a calculated profile, and as it turns, one end of a lever riding against that uneven edge is set in motion. Called a following-arm, this lever translates the cam's calculated profile into reciprocating movements that can be highly precise and perfectly timed. Taken together, the monk's cams are the central memory of the machine, nothing less than a small analog minicomputer, rotated by the fusee arbor. A look at the multipurpose cam in FIGURE 10.9A shows the profile that generates the movements of the monk's mouth. Note the thin solitary spike at twelve o'clock on the disk, distinct from the other steps cut around the perimeter. This is the kiss!

The same cam bears the three pins that generate the all-important movement of the eyes. As the cam turns, the pins one by one come into contact with a lever so shaped that it rides gradually against each pin and then abruptly drops off it. The lever's linkage to the eyes thus draws them very gradually to the left, and when the lever drops off a pin a return spring in the head snaps the eyes back to center. Note that the pins are located well away from the cam's spike for the kiss: this is so the eyes stay focused on the cross at that penultimate moment. But, the kiss done, the monk raises his head and continues his advance across the table, and now his eyes again move slowly to the left, giving the uncanny appearance of a sustained sight line fixed on a single point as the monk walks past it. Brief as this is, if an onlooker happens to occupy this spot, he will likely feel as if the monk has singled him out for special attention, in effect keeping an eye on him. The surveillance lasts just so long, then the eyes dart forward, as if in sudden dismissal. For each revolution of the fusee, the eyes drift left and then spring forward three times. The third time, as the monk pivots to march in a new direction, his eyes stay momentarily locked on the person toward whom he had been headed, and he holds the gaze as he turns away.

Putting It All Together

FIGURES 10.10 through **10.15** describe the five separate actions for the head: its bowing, its turning, the movement of the eyes, and two different movements of the mouth. All these actions, plus those of the arms (**FIGS. 10.16-10.18A, B**), come together at the moment of the kissing of the cross. The monk must turn his head to the left, tilt it forward, and simultaneously raise his left arm at both elbow and shoulder to bring the cross to his mouth; at exactly the precise moment, the jaw opens and snaps shut to perform the kiss.[7]

All the while, the monk continues his advance across the table, step by step. **FIGURE 10.19** details how he performs his walking turns.

Finally, the all-important *fly stopwork lever*, essential to the staging of the monk's performance, can be seen protruding from the rear of the clockwork frame, where it would be hidden beneath the cloth of the cassock (see **FIGS. 10.3C, 10.11**). Pivoted near the bottom plate, it is operated by a small crank, visible in figure **10.19** at top left. **FIGURE 10.20** shows the crank too, in the shadowy interior within the cutaway of the wooden body shell, just to the right of the

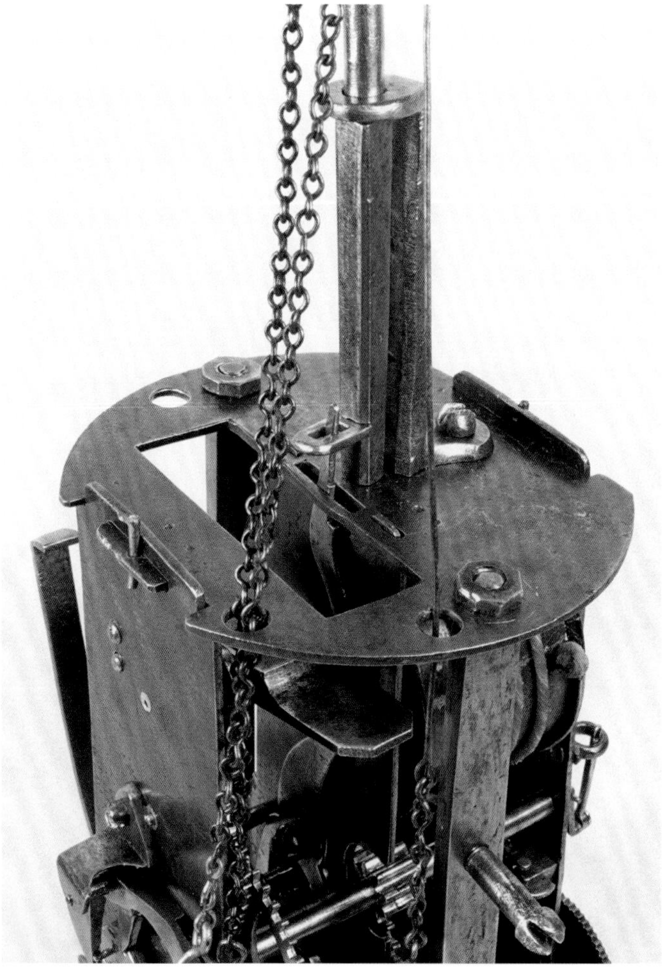

10.10.

FIGURE 10.10.
Top plate and clockwork from above.

A large cutout in the top plate accommodates the spinning fly. The small, narrow, rectangular slot alongside it admits the tip of the head-turning lever, called the "serpentine" lever, so named for its nimble detour around the cross traffic of the gear train arbors deep inside the clockwork. Other openings allow pass-through for linkages to arms and head. Vertical train bars attach to the plate at right and left, mortised and slotted, held in place with tapered pins. The large slotted screw protruding from the front pillar, at lower left in the photo, is for securing the monk's wooden body shell to the clock frame.

FIGURE 10.11.
Clockwork and linkages to the head.

Note the multipurpose cam and its levers in place just above and behind the perambulation wheel. At rear, the long lever protruding outward from the base of the clockwork is the fly stopwork lever, visible also in **FIGURES 10.10** and **10.3C**.

FIGURE 10.12.
The head and its linkages.

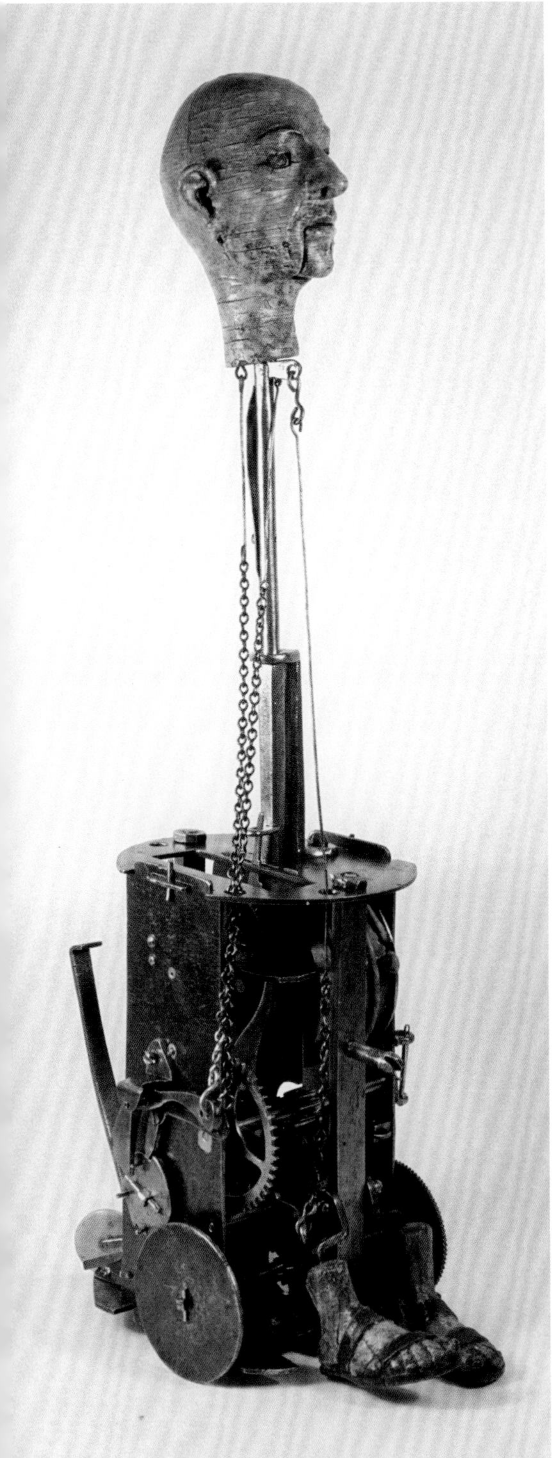

10.11.

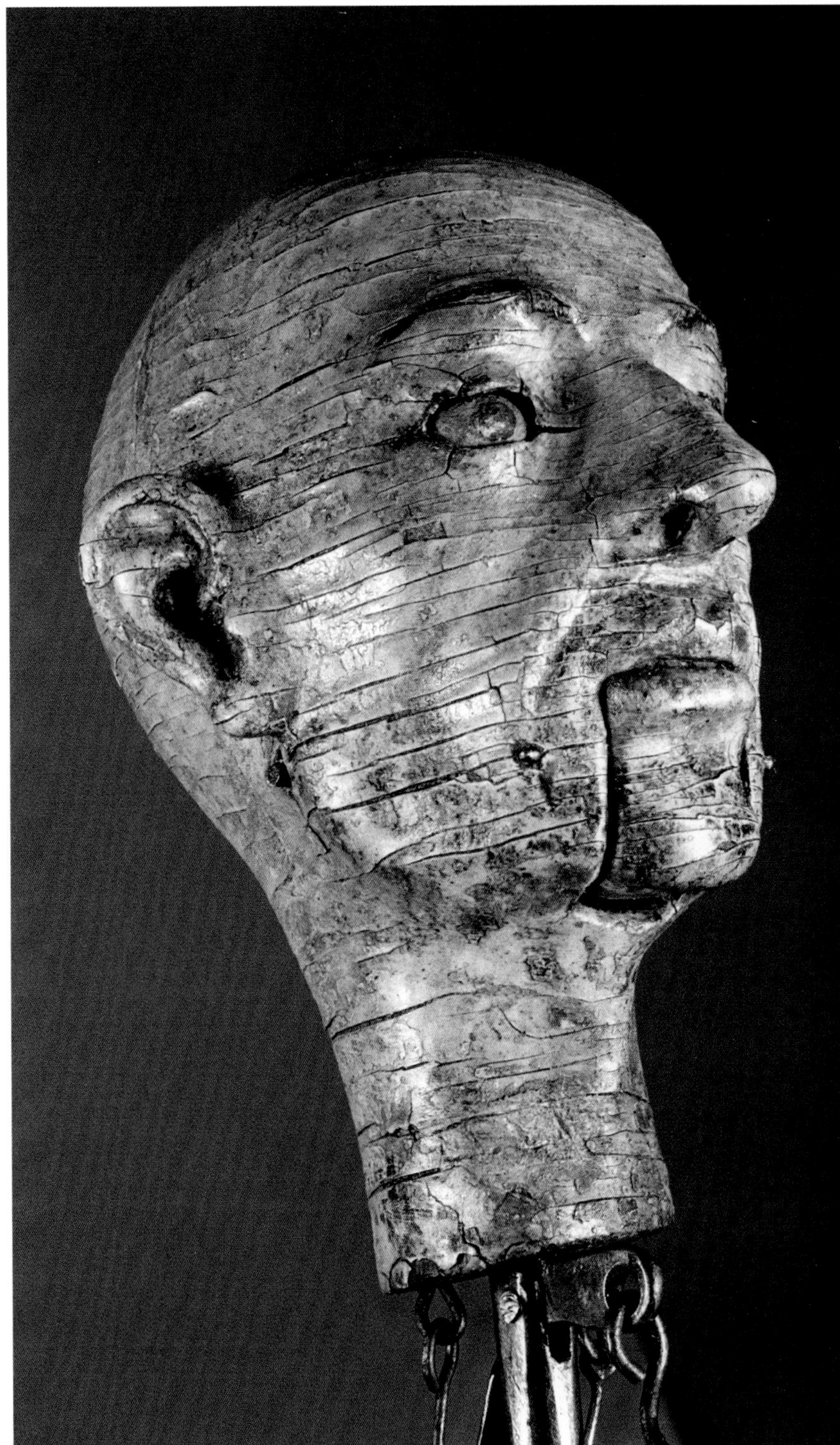

10.12.

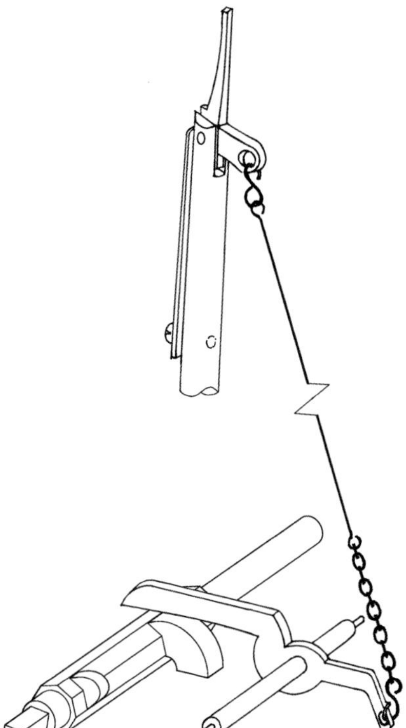

10.13A.

FIGURE 10.13A.

Head bowing mechanism, three-quarter front view.

The second cam on the fusee arbor (see **FIG. 10.6B**) lifts the back of the double-ended lever, causing the chain and wire linkage to pull down on the head lever and tilt the head forward. A strong return leaf spring (just visible in the drawing, secured to the back of the vertical head turning arbor) returns the head to its upright position. The head is mounted on the pointed end of the head lever; the linkage to it is seen at the front of the neck in **FIGURES 10.11** and **10.12**.

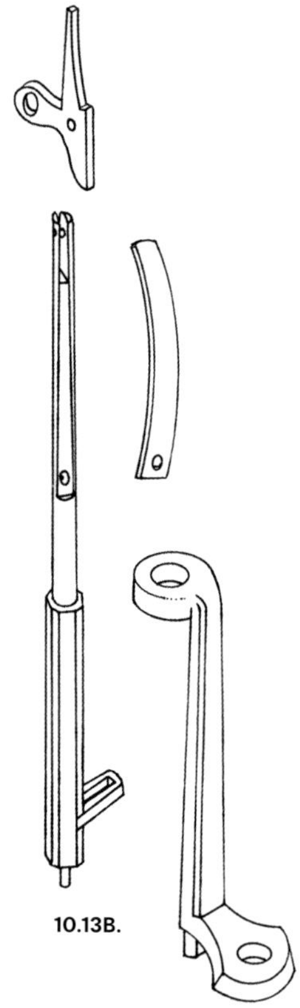

10.13B.

FIGURE 10.13B.

Head bowing and turning mechanism disassembled, rear view.

The component parts of the head bowing mechanism, the leaf spring, and the strong well-sculpted bracket, or "cock," for the head turning arbor, seen from behind. In assembly the bracket is secured to the clock frame's top plate. The bracket must withstand the forces of all the linkages to the head, together with their return springs.

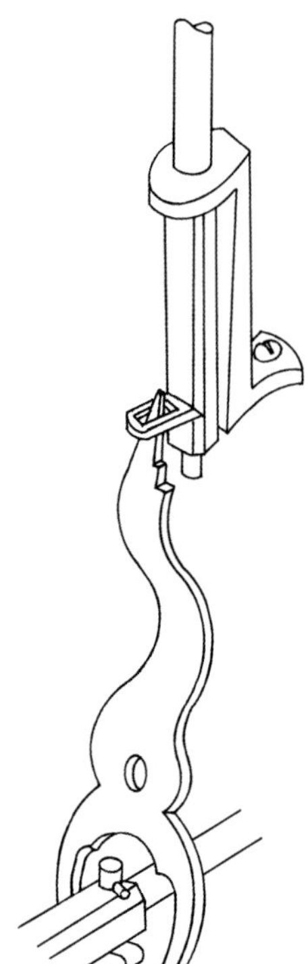

10.13C.

FIGURE 10.13C.

Head turning mechanism, three-quarter front view.

The upper tip of the serpentine lever engages a slotted tab on the vertical head turning arbor. The bottom of the serpentine lever forks into two curved tines. As the fusee arbor slowly revolves, the first cam on that arbor pushes one tine, then the other, oscillating the lever back and forth, to turn the head from side to side. The gap between the tines ensures that the head turns, pauses, turns back, pauses, and so on, in intermittent motion. (The upper tip of the serpentine lever is seen in **FIG. 10.11. FIG. 10.10** shows clear glimpses of the full serpentine lever deep within the clockwork.)

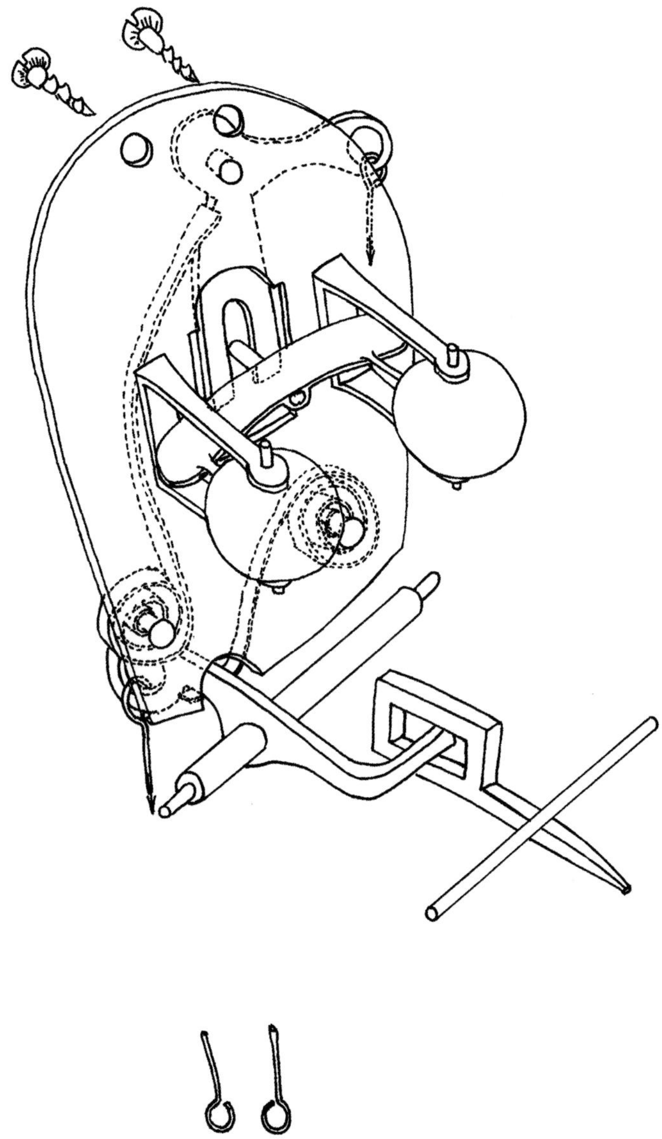

FIGURE 10.14.

Inside the head: eyes and mouth.

David Todd's original diagram, 1979. As the monk advances across the table, his eyes move slowly to his left, giving the impression of a sustained sightline fixed on a single point as he walks past it. For each revolution of the fusee, the eyes move from center to left and back again three times, controlled by three pins projecting from the face of the multipurpose cam. The pins in sequence push and then release a cranked lever connected to a long chain and wire linkage that ascends to the inside of the head, where it operates a reciprocating eye-movement assembly. Mounted on a plate inside the head, the spherical painted iron eyes are pivoted top and bottom in brackets, and wire rods from the back of each eye are connected to a horizontal sled that can move from side to side. As the chain linkage pulls the sled to the monk's right via a pivoted iron fork, the eyes move to his left. A coil spring attached to the back of the plate returns the eyes to center.

The monk opens and closes his lower jaw as he walks, in imitation of speech. As the multipurpose cam turns, a cranked lever rides along its profile, rising and falling in six asymmetrical steps. The motion is transferred via a chain and wire linkage that passes up into the head to the hinged jaw. The steps of the multipurpose cam, in concert with the calculated contour of the lever that rides it, allow the mouth to open naturally, while the cam step's gentle backslope closes the mouth slowly. Inside the head, a coiled spring puts tension on the short lever to the jaw to close the mouth between pulls. The kiss is delivered by the seventh step of the multipurpose cam, an abrupt spike. As the cam rotates, the lever encounters that spike and leaps on its pivot to deliver a sharp jerk on the linkage to the jaw: the mouth opens and shuts with a snap.

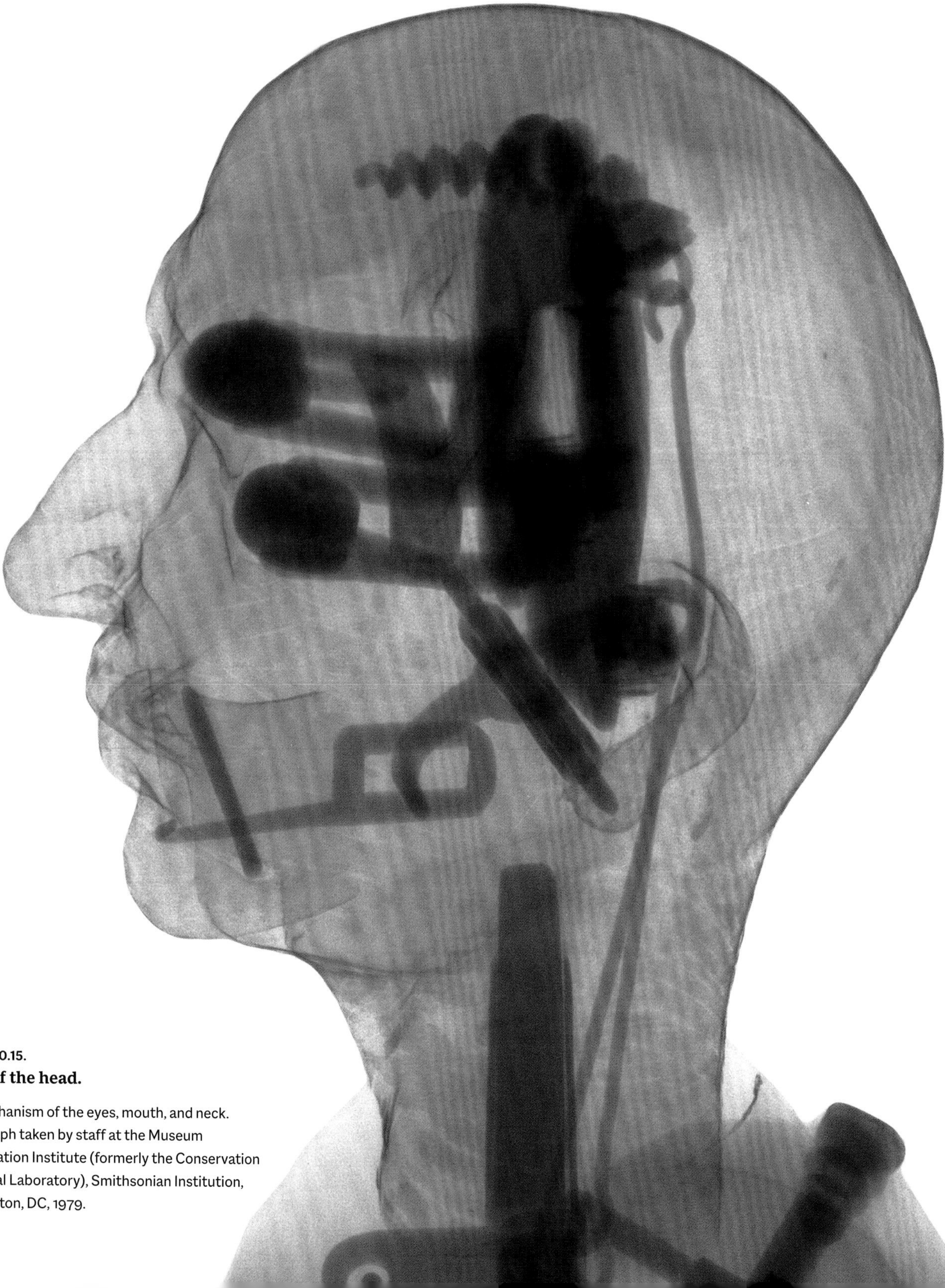

FIGURE 10.15.
X-ray of the head.

The mechanism of the eyes, mouth, and neck. Radiograph taken by staff at the Museum Conservation Institute (formerly the Conservation Analytical Laboratory), Smithsonian Institution, Washington, DC, 1979.

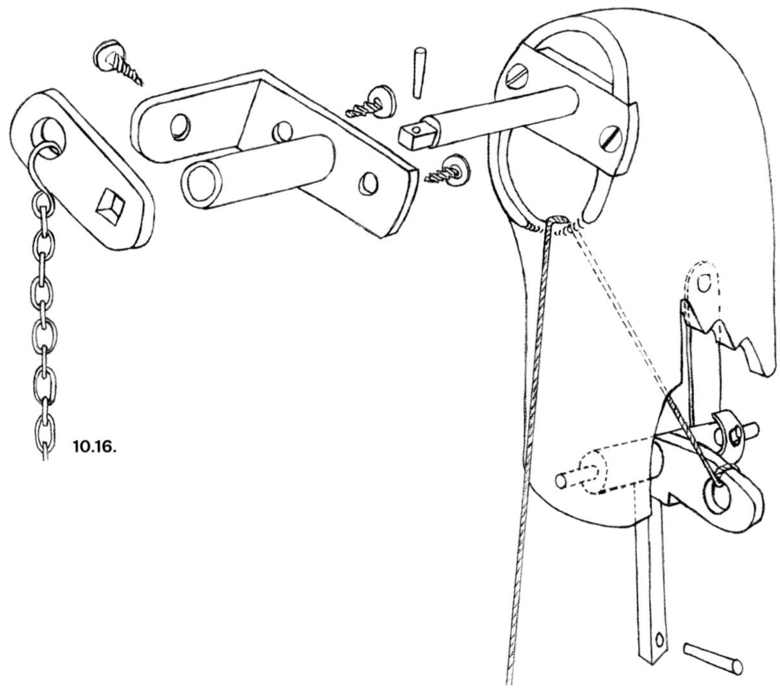

10.16.

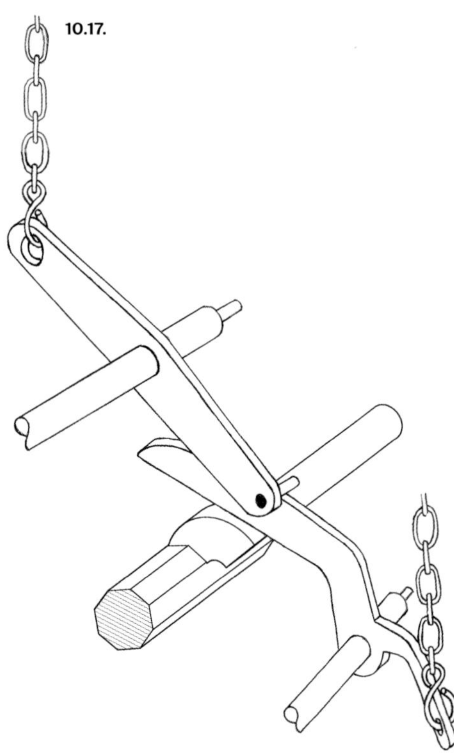

10.17.

FIGURE 10.16.

Left arm shoulder and elbow, three-quarter front view.

The elegant and complex mechanism of levers governing movement of the left arm. The **shoulder** joint, seen here partly exploded, has two brackets: one screwed to the inner edge of the wooden shoulder and bearing a long pivot, the other screwed to the wooden body and bearing a strong iron pipe that receives the pivot. The end of the pivot carries a small lever with a chain linkage that passes down through the monk's body to a pair of levers riding on the large second cam on the fusee arbor. The cam pushes the lever, which pulls the chain to raise the upper arm, making it pivot forward and upward from the shoulder. Separately, the mechanism just above the **elbow** activates the forearm. It shows a simple sprung pivoted lever for raising the forearm, via a gut line linkage that passes up through the shoulder and armpit and then down to the gear train below, connected to an eccentric crank on the second pinion's arbor (visible in **FIG. 10.3B** and color plate on p. vi). The monk holds the cross in his left hand and raises and lowers it using only his elbow and forearm: six times for every revolution of the fusee. But the kiss requires the combined movement of elbow and shoulder together, to raise the cross higher as the head bows forward.

FIGURE 10.17.

Double-ended lever, three-quarter front view.

A portion of the fusee arbor shows its large second, or central, cam, with the head bowing lever in direct contact with the surface of the cam (see **FIG. 10.13A**). Riding atop the head lever is a second lever that controls the movement of the monk's left shoulder (**FIG. 10.16**). The cam lifts the head bowing lever, which pulls down the chain linkage (at right) to bow the head for the kissing of the cross. Simultaneously, the head bowing lever itself lifts the left shoulder lever (hence the name "double-ended lever"), which pulls down the chain linkage (at left) to elevate the arm at the shoulder, and thus the monk bears the cross to his mouth for the kiss. This double-lever action assures the paired motions of bowed head and raised cross.

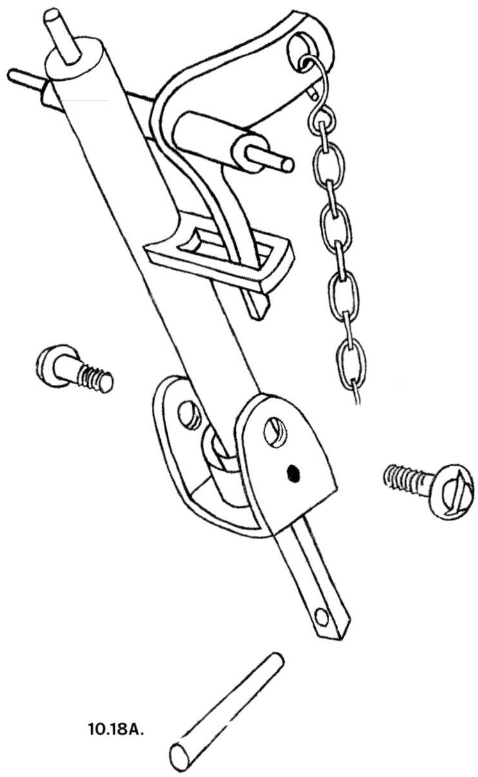

10.18A.

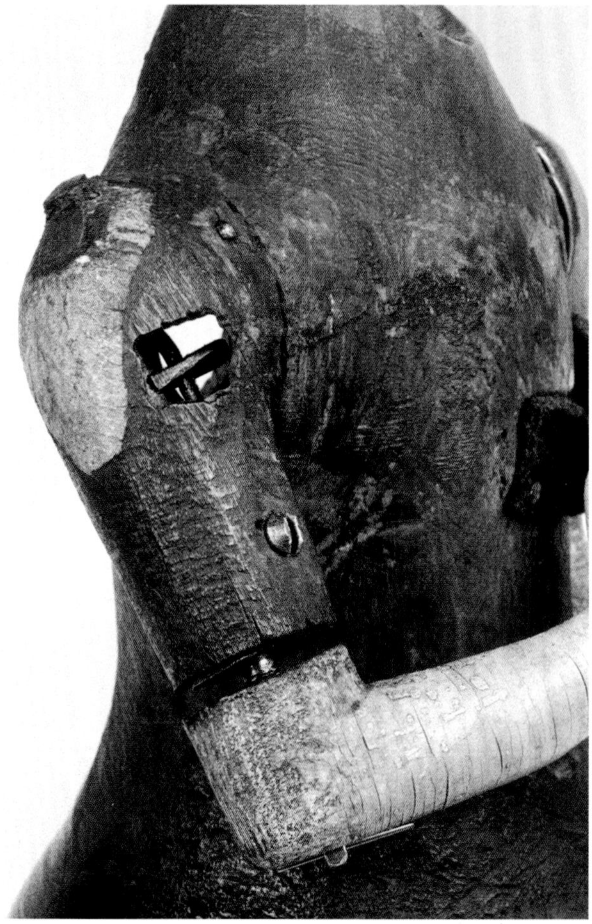

10.18B.

FIGURE 10.18A, B.

Right arm mechanism.

The "mea culpa" of the right arm beating the chest: a cranked lever is set within the wooden shoulder; its lever tip engages the slotted tab of a vertical arbor extending the length of the upper arm from shoulder to elbow. The vertical arbor pivots in a metal bracket above the elbow and is held in tension by a small spring coiled inside the bracket. The wooden forearm is pinned to the lower end of this arbor, just above the elbow. When the chain linkage is pulled downward (by the right arm lever at the multipurpose cam; FIG. 10.9B) the vertical arbor rotates and the forearm moves away from the body, against the tension of the coiled spring. As the multipurpose cam turns, its lever abruptly drops off the cam's pin, the sprung arm is released, and the beaked hand strikes the chest.

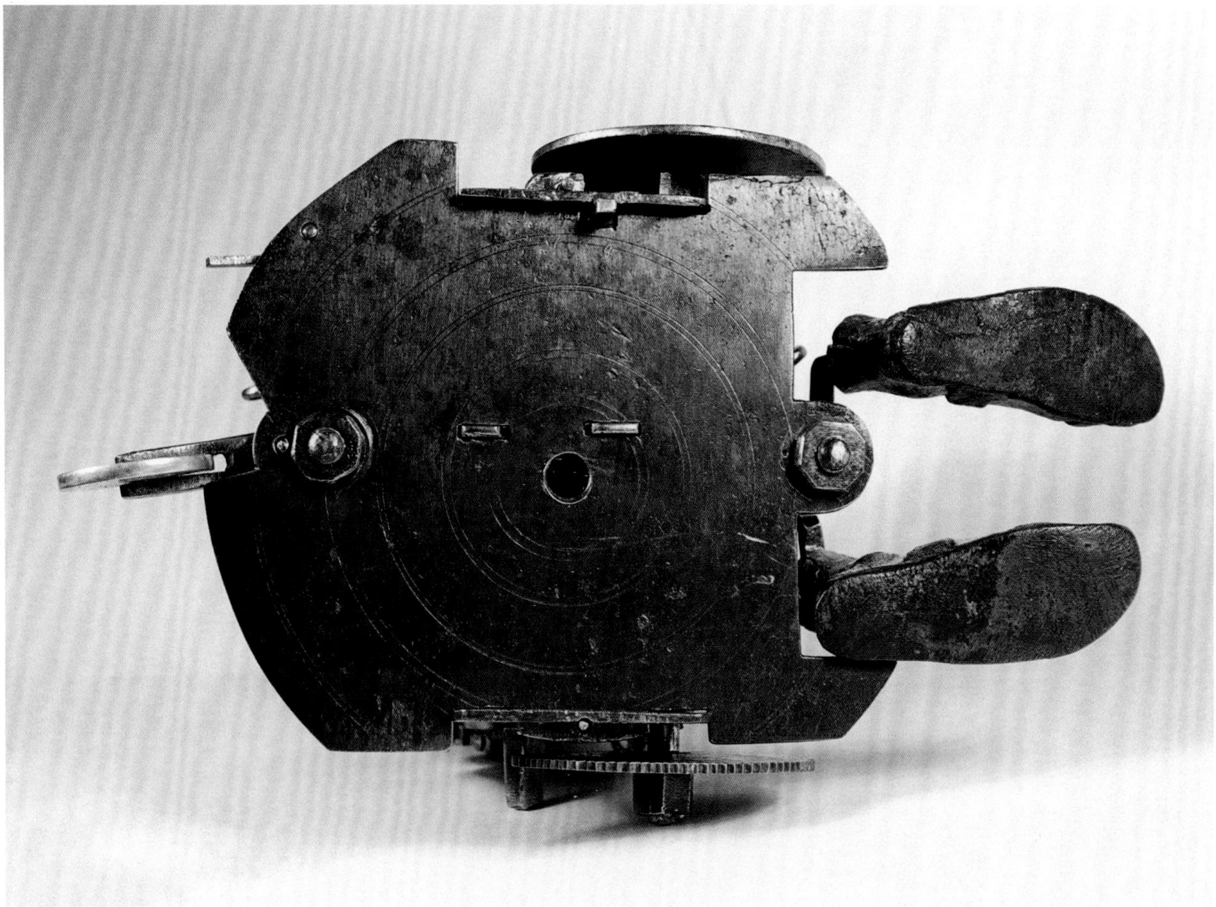

FIGURE 10.19.
Bottom plate.

Note the cutouts for the feet, the two perambulation wheels, the outer train bars, and the trailing wheel. The small rectangular bar seen end-on near the smooth perambulation wheel (at photo top) is the turning pivot: it periodically projects through the plate—pushed down by the third cam on the fusee arbor—to make contact with the tabletop. The push is slight, just enough to lift that perambulation wheel off the table surface, and the bar becomes the pivot for the walking turn. Separately, the thin rod projecting from the rear of the plate (at ten o' clock in this view) is the crank on the hidden stopwork lever. **FIG. 10.3C** shows a good view of the fly, the stopwork lever, and its crank. The color plate on pp. 74–75 gives a beautiful close-up view of the crank, just to the right of the trailing wheel. Discreetly pressed with a finger, it will tip the stopwork lever back to release the fly, freeing the clockwork to run. Lifting the crank will tilt the stopwork lever forward into the path of the fly, halting the machine. Other details: the two narrow slots near the center are for locating the central train bar. The octagonal nuts secure the lower ends of the front and rear pillars, which add strength to the clockwork frame. The feet are pivoted on the front pillar. The inscribed rings, present on both surfaces of the top and bottom plates, are a decorative feature widely seen in period clockmaking.

trailing wheel. The monk's mainspring may be fully wound and ready, but only with the release of the fly stopwork lever can the fly take off, the gear train begin to run, and the monk come to life.

The principle of the lever and the physical laws of velocity and ratio govern all the monk's movements. The individual parts by themselves are simple, but their orchestration delivers an irreducibly complex performance. The relationship of the fusee arbor cams, the multipurpose cam, and the crank that moves the left forearm is critical to the coherence of the monk's act. In assembling the parts, if the meshing of the fusee wheel with the second arbor pinion is just one tooth out of place, the concert of moving parts required to raise the cross to the lips is lost. Which of the four possible positions of the multipurpose cam on the squared end of the fusee arbor is the right one? Or the four possible positions of the left arm crank on the second pinion's arbor, or the many possible wheel/pinion synchronizations? It is relatively easy for the conservator to get it right, as the ratios between parts have already been established. Cam profiles have been filed, lever lengths decided, gear train ratios calculated. The fusee itself has been shaped and cut to address the unique behavior of the mainspring. Synchronization for the conservator is just a matter of trial and error, disassembling and reassembling until everything behaves as it should.[8] But the maker started with iron bars raw from the chafery. He had to calculate the ratios and create the relationships before he could begin the labor of filing the cams and gear teeth. In his mind's eye he had to visualize the monk's performance, seeing not only what the onlooker would see but also the hidden interactions within the machine, of part with part. We, the onlookers, can only try to peer back into time and picture the clockmaker working his way forward through the maze. Look again at the X-ray of the monk (**FIG. 10.1**). A thin iron spike runs inside the length of each wooden forearm. These add strength to the wood so that the arms will never break. The clockmaker was thinking about the future.

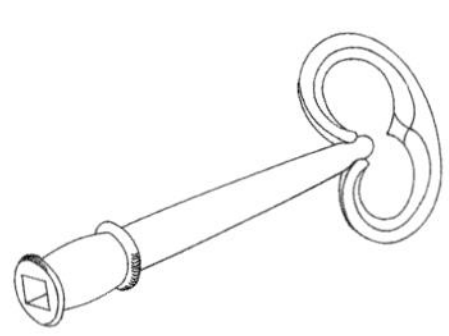

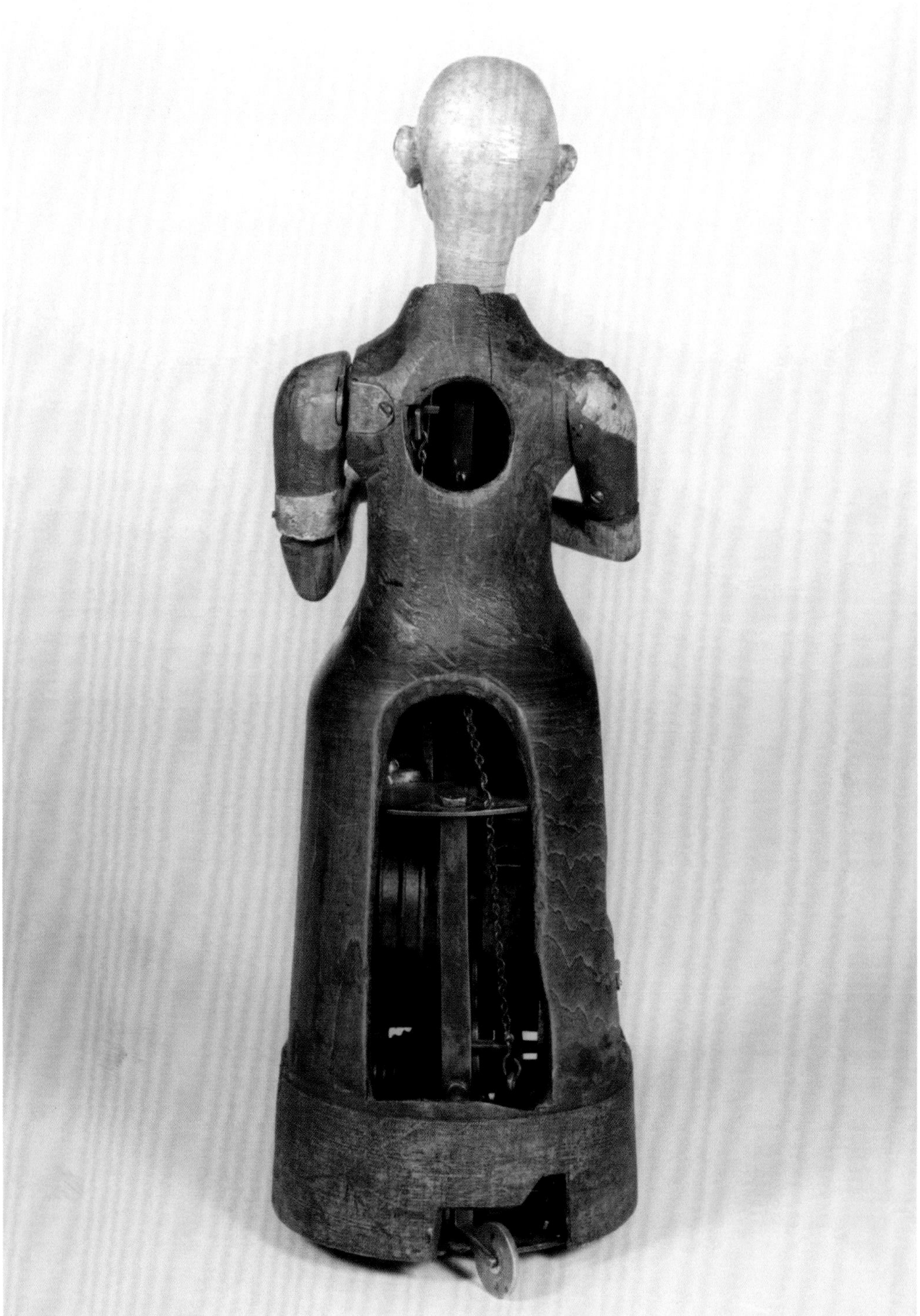

FIGURE 10.20.
Stopwork lever released, the monk heads out.

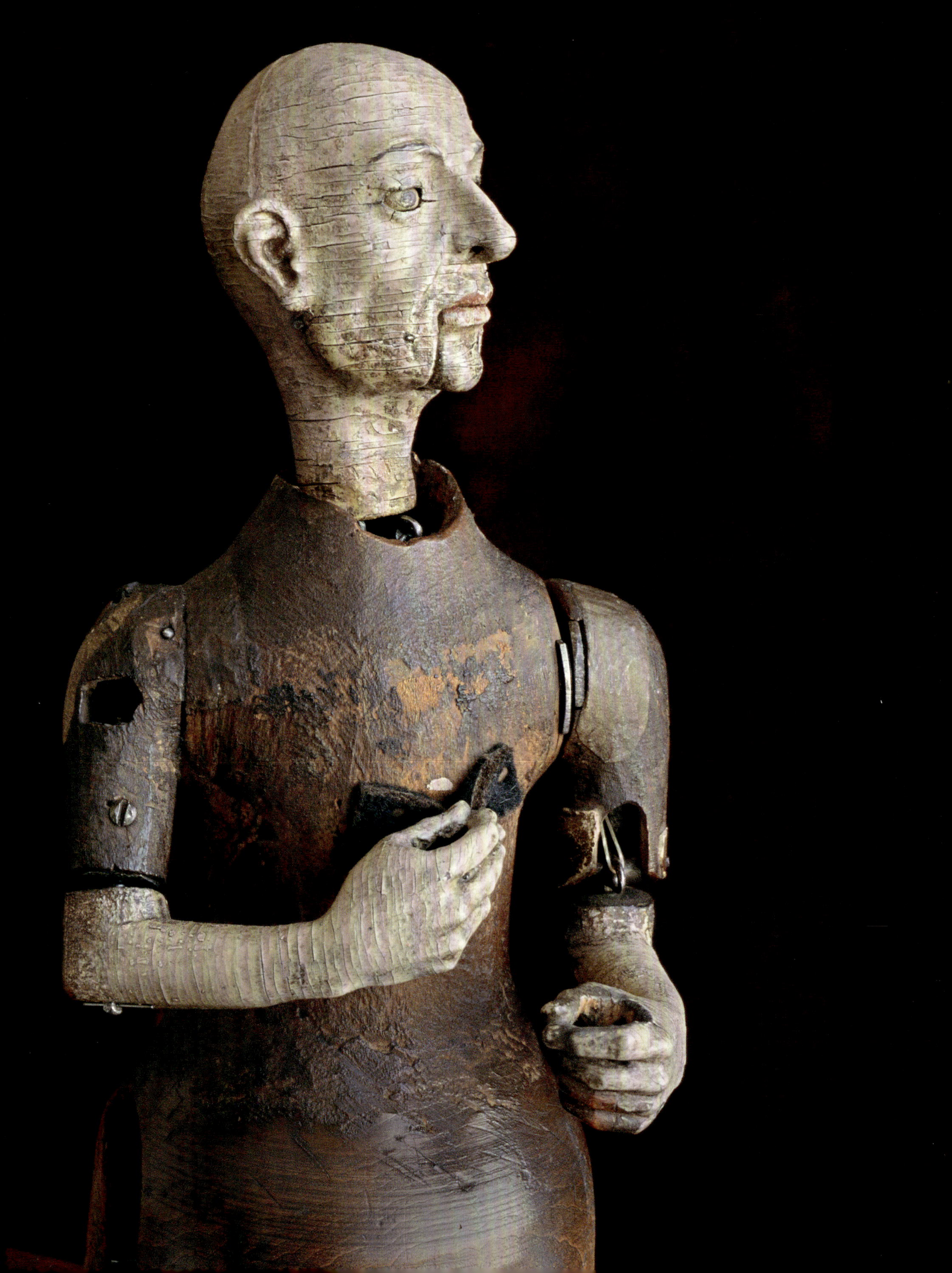

PART III
Eight Androids

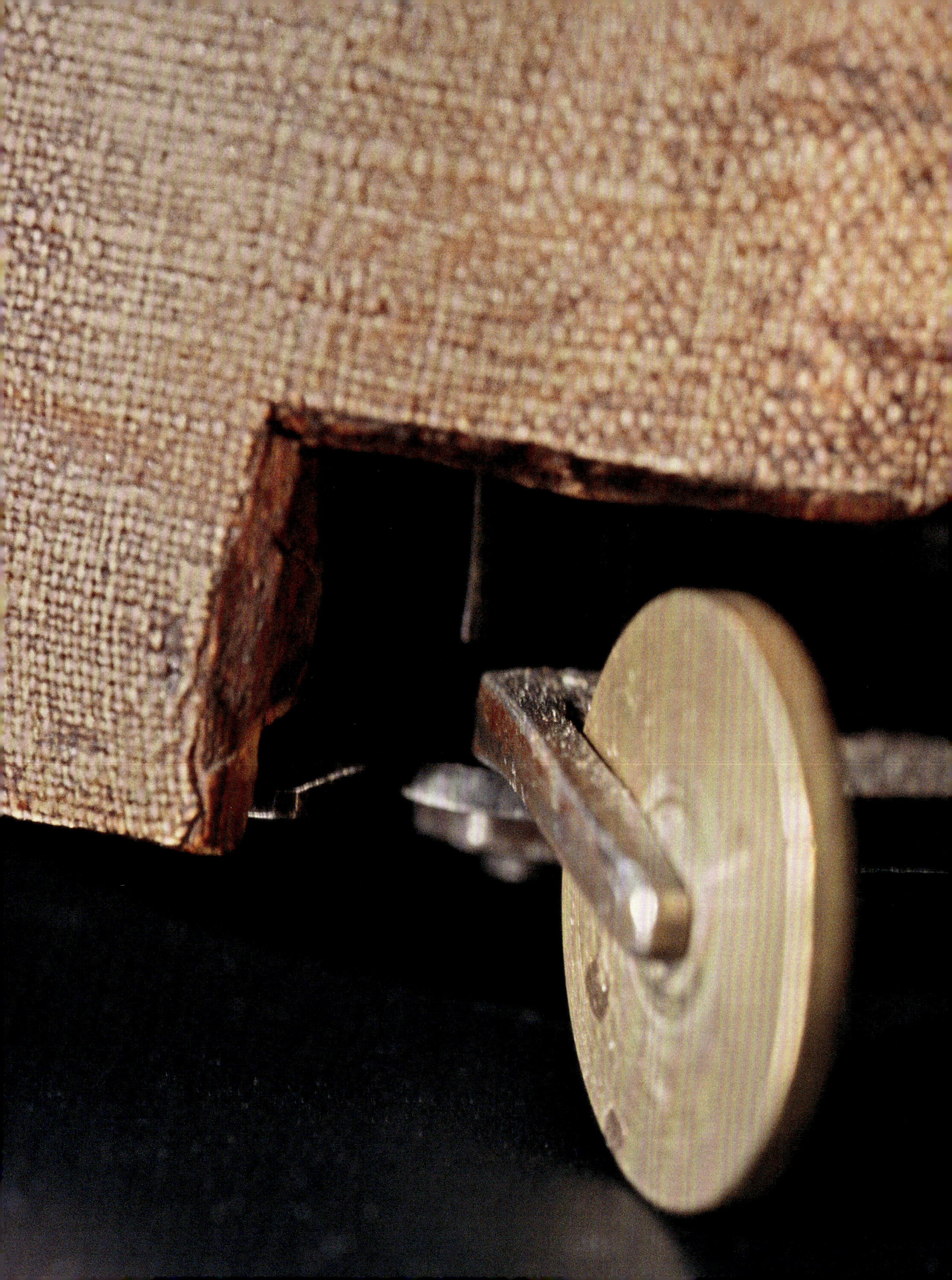

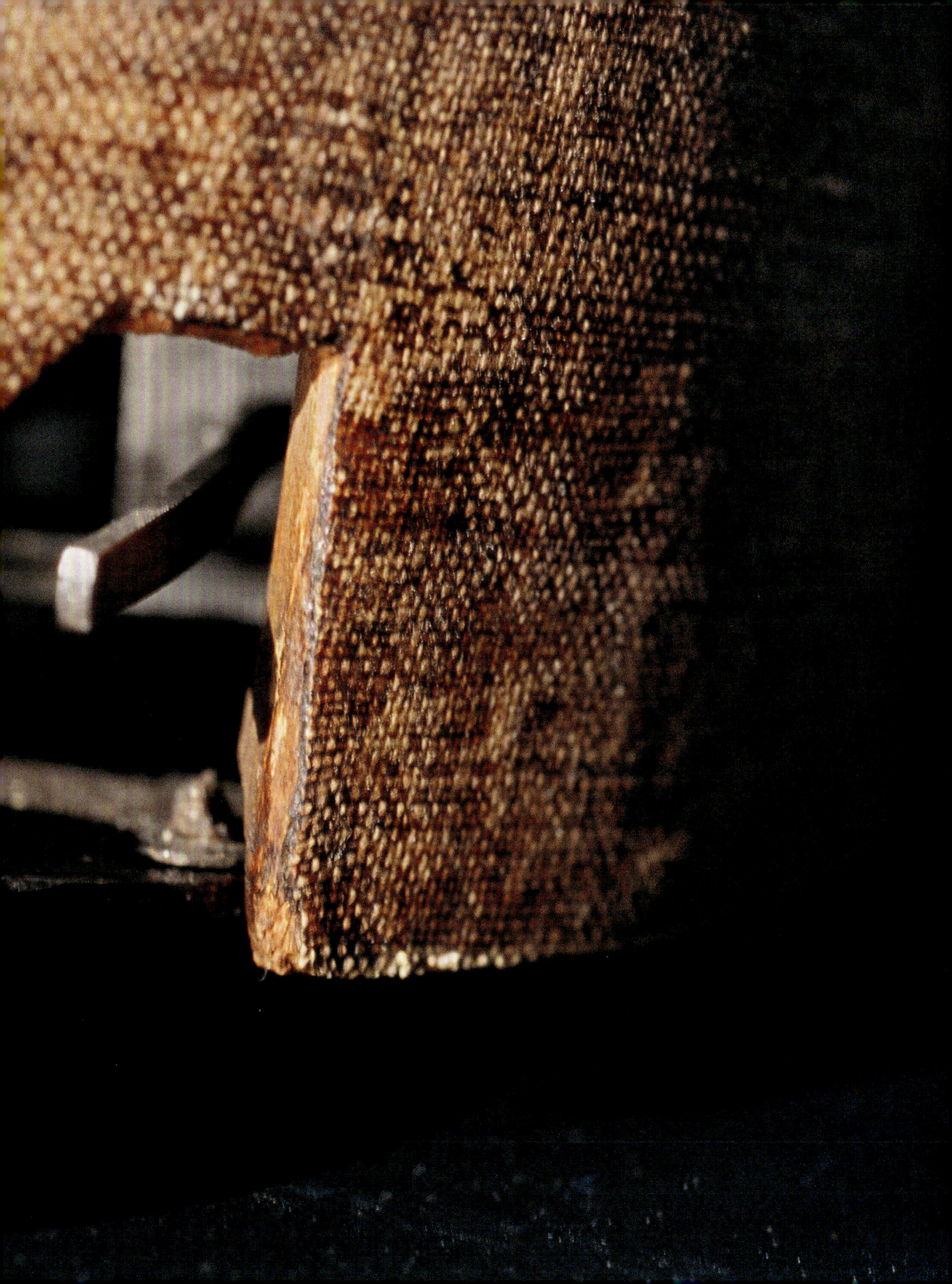

11 Monks, Saints, and Musical Ladies

The automaton monk made its public debut in the 1980 exhibition *The Clockwork Universe: German Clocks and Automata, 1550–1650*. Co-curated by Otto Mayr at the Smithsonian Institution and Klaus Maurice at the Bayerisches Nationalmuseum (Bavarian National Museum) in Munich, the show opened in Munich and traveled to Washington the following year.[1] To the finest pieces from their host institutions Maurice and Mayr added masterworks borrowed from sixty public and private collections in Europe and the United States, assembling a landmark show of clocks, mechanical celestial globes, music-playing machines, and automata from the golden age of German clockmaking. The exhibition catalog, with essays by fourteen European and American scholars, remains an essential text in the field of horological history.

Pride of place among automata on display went to figure clocks and composite pieces: stationary clocks surmounted with figures performing at the strike of the hour, and ornate machines in the form of carriages, ships, and towers, some with ensembles of soldiers, musicians, and animals performing on miniature stages. Made by celebrated clockmakers and goldsmiths, many of these were signed and thus documented. The monk, though with uncertain national pedigree, was included as the show's earliest example of a tabletop automaton. Compared with costlier entertainments, the monk was a humble anomaly: faded paint on a wooden body, lowly cloth habit—materials genuine to a representation of monkish poverty. Yet its inclusion brought closer attention to a dispersed group of similar clockwork figures that have also survived to the present. Mayr's recognition of the significance of the automaton he acquired for the Smithsonian and David Todd's analysis of its mechanism helped set a new standard for the study and conservation of these figures. Later restoration reports for automata similar to the monk cite the catalog entry in this exhibition.

At present, six self-propelled walking figures kindred to the monk are known. Made in Europe between 1550 and 1650, all are tabletop performers of similar scale, their heights ranging from 11 to 17 inches. All wore garments of real cloth over jointed bodies of carved wood. Heads, hands, and feet were painted in lifelike colors, in imitation of flesh. Hidden clockwork is iron or brass. Like the monk, none bear identifying stamps to establish provenance, though Juanelo Turriano has been proposed in the recent past for several of the figures. Dates of origin are based primarily on analysis of internal clockworks. A seventh figure closely resembling the others but considered eighteenth-century work is included in our study for comparison. The value of these automata both then and now lies not in precious material or the pageantry of craft but in their close imitation of human appearance and motion.

Six of these seven automata may be found in public collections in Europe and the United States, some on view well before the monk appeared in the 1970s. The cittern-playing lady in the Kunsthistorisches Museum in Vienna (see **FIG. 5.1**), bequeathed to that museum in 1934, is the best known. Three more figures are variations on a lute player: one in the Smithsonian itself, a small maiden in good original condition, acquired in 1984; another in the collection of the Bavarian National Museum in Munich, given to that museum in 1927; and a third in the Conservatoire national des arts et métiers in Paris, there since 1853. Two male figures more closely resemble the monk. One is a bearded figure, also called a monk, purchased at auction by the Deutsches Museum in Munich in 1985, just eight years after the Smithsonian monk arrived in Washington. The other is a haloed saint in the collection of the Iparművészeti Múzeum (Museum of Applied Arts) in Budapest, acquired in 1915. Finally, an unusual figure has recently emerged from a private collection in Milan: a "noble matron in the act of playing a drum," seen for the first time in a 2005 exhibition in Italy on the history of the Italian clock.

José A. García-Diego, the first recent biographer of Juanelo Turriano, must be credited with the early comparison of several of these figures, which he traveled to see in the 1970s and early 1980s in his search for works by the master. The first comprehensive study of a group of sixteenth-century walking automata was a master's thesis written in 1990 at the University of Hamburg by Adelheid von Herz, "Androiden des 16. Jahrhunderts" (Androids of the sixteenth century).[2] Von Herz carried out research on seven of the eight figures we compare here. To learn about the Smithsonian monk and its sister lute player, she corresponded with David, and when she later sent him a copy of her completed thesis he in turn benefited from her work on the others. We have compiled data on our group of eight automata for a table of comparisons (see appendix), based on direct observation, photographs, museum archives, restoration reports, catalog entries, and correspondence with museums and owners.

The physical condition and restoration records of the automata vary profoundly from one to the next. Exposed wood and cloth have suffered most, as might be expected, particularly in the figures that show frequent performance and handling. Iron and brass outlive wood and cloth in these machines—and an automaton's wooden body protects its metal clockwork, as any clock is protected by its casework. In the Smithsonian monk, the iron clockwork itself is responsible for wear on the wooden body: a small felt pad is found nailed to the monk's chest to protect him from the blows of his own hand—and to protect the hand too. Many of the automata have had wooden hands and feet replaced and heads repainted, and at least one has a new head entirely. The two Smithsonian figures are considered to be in relatively untouched original condition.[3]

As a group, the automata exhibit broad differences in sculptural style and character. Carved wood faces and limbs carry the most weight in our initial response to a figure, but a look at the clockwork inside the wooden bodies of the eight automata is startling. For all their exterior individuality, the hidden clockwork of each figure consists of a posted frame and drive system whose layout clearly comes from the same prototype. Vertical train bars, mainspring and fusee, the illusion of stepping feet glimpsed under the hem of a cloth robe that in turn hides a three-wheeled rolling frame—seven of the figures share these features. The size of the clockwork mechanism itself, and thus the height of the automaton, can vary by several inches, and the material used to fabricate it also varies: brass in the case of the lutenists in Munich, Washington, and Paris; iron for all the others. But the structural and functional similarity of the clockwork is unmistakable.

Most revealing is the comparison of the three male figures: the Smithsonian monk, the Deutsches Museum figure in Munich, and the saint in the Iparművészeti Múzeum in Budapest. The sixteenth-century technology of ironworking is everywhere apparent, giving stylistic and methodological clues to the age. Clockwheels, plates, pillars, and train bars reveal the kinds of tool marks characteristic of the period, with discs and plates hammered out by hand from cast-iron bars direct from the chafery, then filed and scraped to a finish. Decorative elements such as the framework's chamfered pillars and securing nuts are traditional sixteenth-century work. The three automata also share a similar bottle-shaped wooden body shell. Heads, hands, and feet, according to conservators, were carved from linden or poplar, both species widely used by period sculptors.

Elizabeth was able to view all but one of the automata in European collections in 2019 and early 2020 and meet with the curators and conservators who care for them. (Budapest was out of reach, the saint in storage since 2017 while its home museum is renovated.) David had already made detailed conservation documents for the clockworks of three of the full family of eight: the Smithsonian monk, the Smithsonian lute player, and the lute player in the Bavarian National Museum, which he helped conserve in 1980. In comparing them for clues about the monk, both of us have come to believe that these figures together comprise a class by themselves in the history of artificial life: the first solo free-walking self-propelled androids. In the following brief chapters we introduce each automaton and sketch what is known of its history. With their similarities and differences in view, chapters 17 and 18 address what they reveal as a group and assess their historical significance.

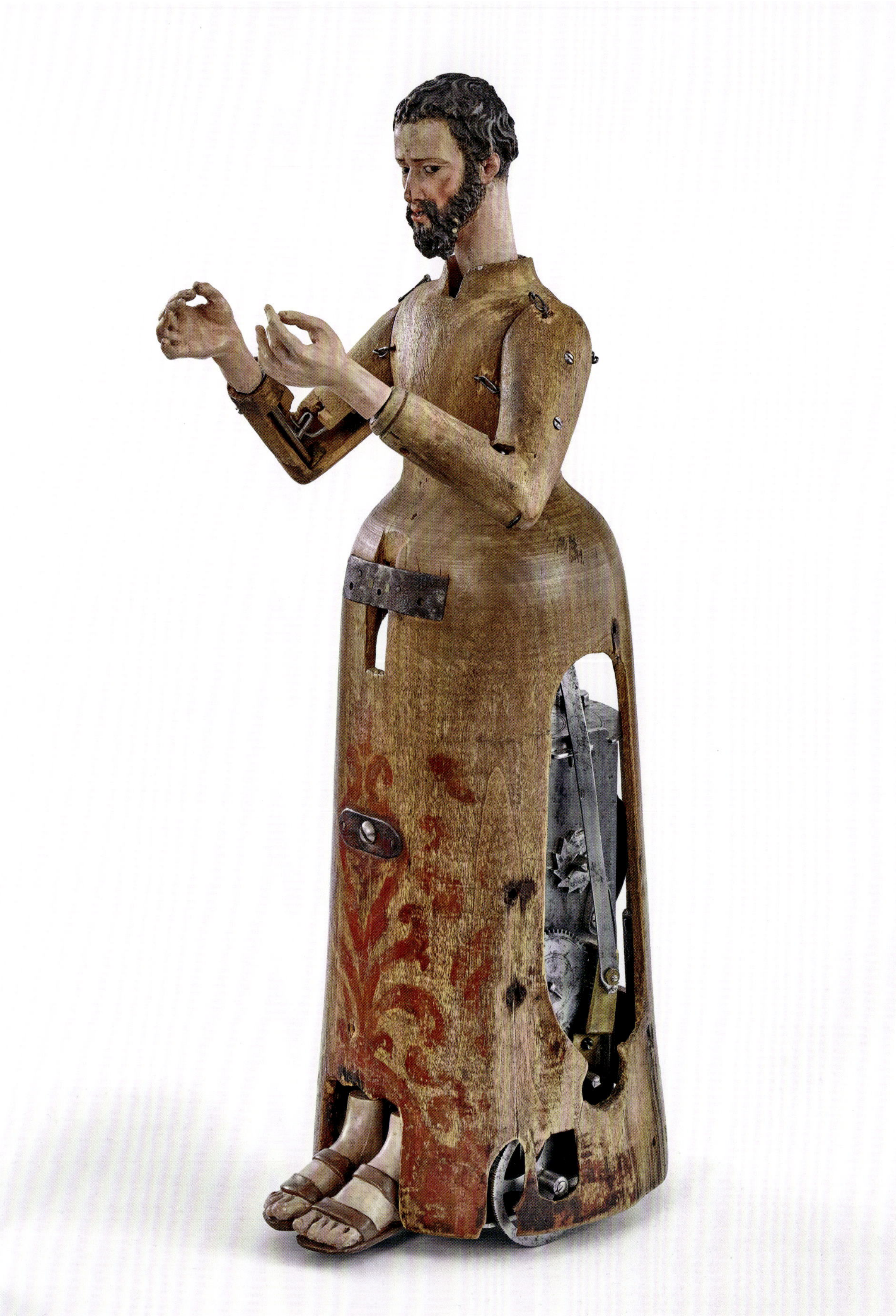

12 A Second Monk?

In October 2019 the conservator Thomas Rebényi, master clockmaker and head of the Deutsches Museum Restoration Workshop for Scientific Instruments and Clocks, wound up the museum's automaton and filmed the performance. Automatenfigur: Predigender Mönch, the figure is identified at the museum: preaching monk automaton (**FIG. 12.1**). It had recently returned from loan to an exhibition in Vienna, and during the journey's vibrations, the original glue in the seam between the front and back halves of the wooden head, brittle with age, gave way. Discovering the back cap loose in the crate, Rebényi was first horrified, then fascinated, for he suddenly had a forbidden glimpse inside the hollow head, where the now-exposed micro-machinery that animates the automaton's mouth and eyes was sealed. It is more than likely that he was the first person to see inside that head since the day in the sixteenth century that its walnut-sized clockwork was installed and the halves of the wooden skull were glued tightly together. An X-ray of the head by an earlier conservator had shown a ghost silhouette of the mechanism, but here was the real thing (**FIGS. 12.2-12.4.**). With the agreement of museum curators, Rebényi elected to leave the back cap off. He would later mount it on its own stand and display it on a slender rod at the right height, just behind the automaton in the vitrine, allowing museum

FIGURE 12.1.
Automatenfigur: Predigender Mönch (Automaton: Preaching monk), ca. 1560. South Germany or Spain. Wood (poplar), iron clockwork. Height: 39 cm (15 5⁄16 in.). Deutsches Museum, Munich. Its name notwithstanding, the identity and specific nature of the performance of this automaton remain a mystery. Different in character though it is from the Smithsonian monk, it harbors a clockwork and wooden body shell remarkably similar in design.

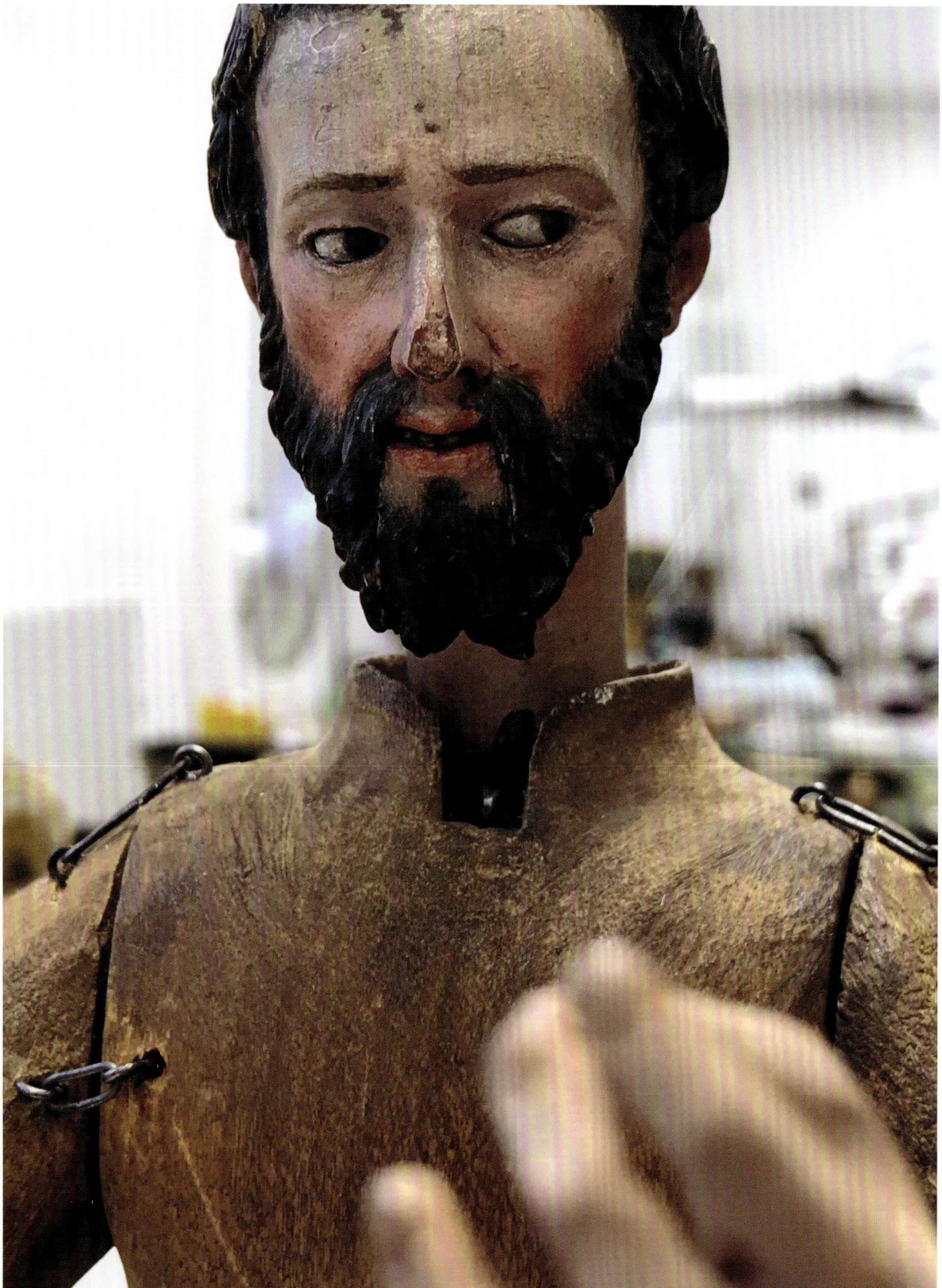

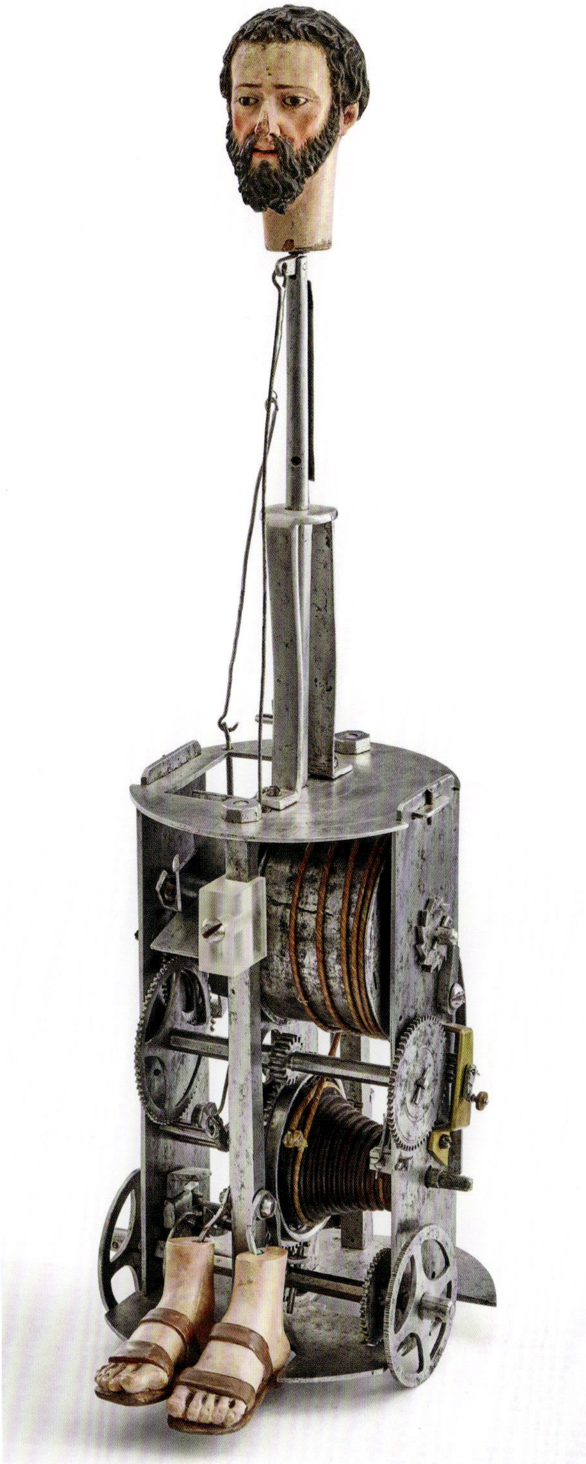

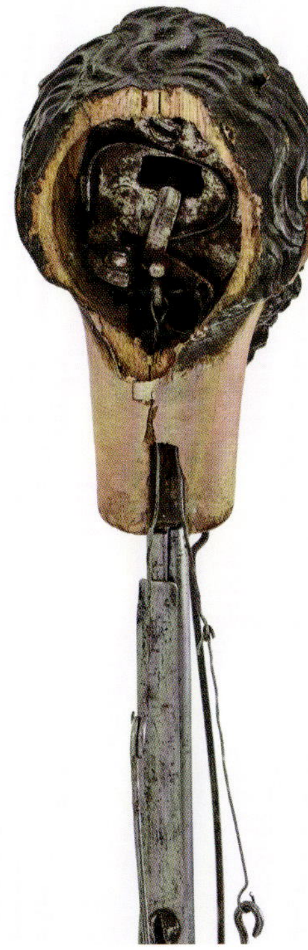

visitors too to see inside the head (**FIG. 12.5**). But first, with the figure still in the lab, he thought, "This is surely the time to wind the automaton's mainspring and set it running for the camera." Freshly reminded of its fragility, he proceeded with utmost caution, giving the mainspring a fraction of a turn with the key and restraining the fly. The automaton performed without a hitch, as if it had been waiting. One gentle quarter-wind at a time, Rebényi captured the performance piecemeal, and from all angles, distances, and depths of field. He did not fail to zoom in to the back of the head to shoot the levers, cranks, pivots, and return springs as they clicked and whirred to animate the automaton's face. The detailed footage is an unprecedented look inside the "live" head of an early android.

FIGURE 12.2. (opposite)
The automaton in the Deutsches Museum conservation lab, March 3, 2020. A wide-awake face and a mind of its own. The elegantly carved beard hides the seams of the movable mouth.

FIGURE 12.3.
Clockwork and linkages to the head. A comparable view of the Smithsonian automaton (**FIG. 10.11**) shows the close kinship of the two machines.

FIGURE 12.4.
A rare view inside the head: visible are the cranked lever and return spring that animate the motion of the eyes; beneath these, a glimpse of the linkage and lever to the jaw that opens and closes the mouth.

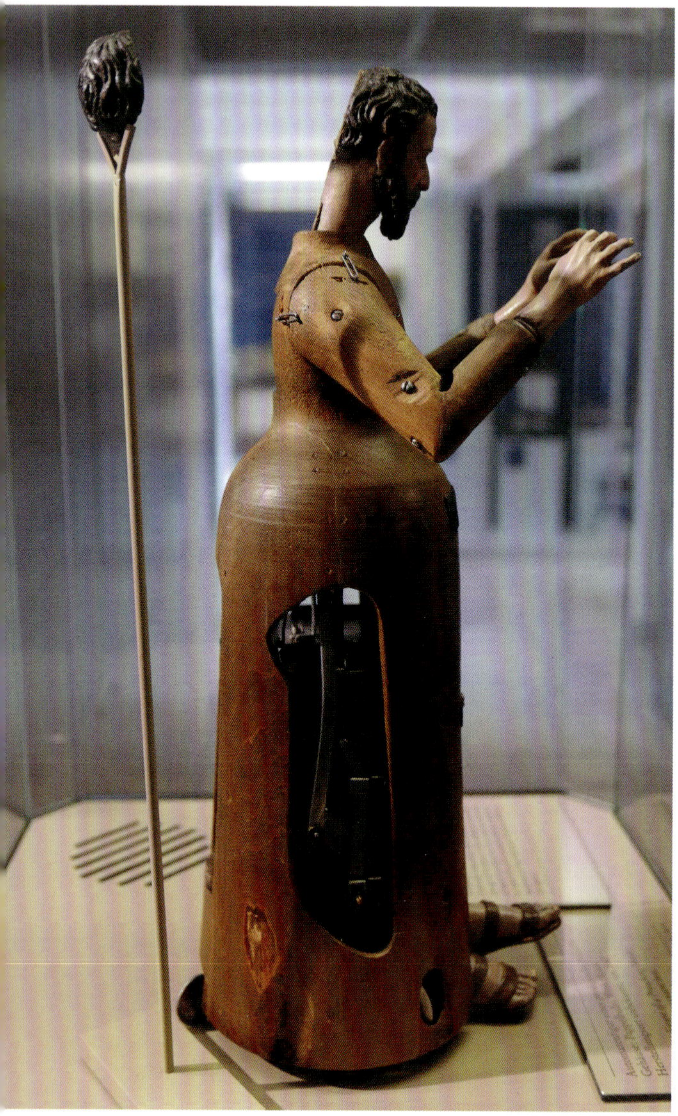

FIGURE 12.5.
The automaton on display at the Deutsches Museum, Munich. The back of the head wins its own stand inside the vitrine.

Fully wound, the automaton would travel a distance of 40 centimeters (15¾ in.), turn to its right, and follow the path of a variable polygon, accomplishing eleven walking spans and turns. The walking turn would be executed via the same arrangement seen in the Smithsonian monk: one perambulation wheel lifted by a cam-lowered pivot, letting the other wheel roll in a partial arc around it (the third trailing wheel at rear keeping the figure steady). Along each walking span, the head turns once to the left, once to the right, and nods once. The mouth opens and shuts four times. Eyes glance right to left and back eight times. Measurements were first made by Peter Friess, a former conservator at the Deutsches Museum, who undertook the restoration of the figure and published a detailed report of his work in 1988.[1] The parallel with the Smithsonian monk's performance was immediately apparent. The biggest differences appear in the motion of the arms. The Deutsches Museum figure is all wrist. With both arms held up before the body at fixed angles, the right wrist rotates back and forth forty times between each walking turn, and the left wrist, twenty-two times.[2] Rebényi's video footage shows the right hand oscillating more than once a second, the left hand every second and a half, an almost 2:1 rhythmic performance. What can these very fast movements represent? Hard and fast calculation went into the choreography of these pivoting hands. As noted in chapter 10, the timing of these motions is determined by the numerical relation of wheel teeth, pinion leaves, and cam profiles in the gear train. Once the metal is cut, the resulting actions are set in iron. Von Herz speculated that the automaton may have manipulated, or appeared to manipulate, the beaded loop of a rosary in these busy hands.[3] Friess had noted that inside its wooden body shell this figure originally contained a miniature hidden glockenspiel: a tiny set of bells struck by spring hammers, also driven by the mainspring—as if there wasn't already enough machinery in so small a chamber! Could this automaton have once grasped some kind of musical instrument, miming a performance coordinated with actual sounds from within?

Rebényi, pressed for an opinion, doubts that the figure represents a preacher or a monk.

Though much corroded, the iron clockwork of the figure still functioned when it entered the museum collection, according to Friess. It lacked only the lift rod for implementing the walking turn, and this has not been replaced, hence we cannot be certain of the angle of the walking turn. Friess described cleaning the mechanism and replacing the gut line joining the fusee to the mainspring barrel. But the figure's wooden body shell, limbs, and head had suffered much prior damage. One hand and both feet were missing, Friess noted, two fingers were gone from the remaining hand, and chipped paint on the face disclosed layers of original color now overpainted.[4] Following David's lead, Friess had the head X-rayed in order to view its hidden mechanics without breaking the head open. The two X-rays show the close similarity of the machinery inside each head. In both, the original wooden head would have been carved, then cut or split vertically behind the ears into front and back halves, the back half smaller, like a cap. Each half would then be hollowed out, the jaw cut and hinged, and the eye sockets opened. The iron mechanism, built to fit, would be installed within the front half: the eyes (prepainted) snug in the ready wooden sockets, and the jaw lever engaged. The whole tiny assembly would then have been fastened in place with two wood screws. With wire linkages hooked in place for later connection to the gear train below, the two wood halves would then be glued back together and the head given a chalk primer and painted.

While Friess's report closely details the process of replacing absent sections of the wooden body shell—necessary for the figure to be stable enough to wind and fully move again—he makes no mention of the replacement of the missing feet, hand, and fingers, which appear in bright finished form in the photographs illustrating the report. Such information on the exterior sculpture is what is most missed in efforts to identify the provenance and past restorations of this and so many of the other extant automata. Their repair requires the skills of a clockmaker but those of a sculptor too.

And how does the sculpture compare in these two monkish automata sharing such similar clockwork? The turned shape of the wooden body shell is of different proportion: the Smithsonian figure has a lower "waist" to allow clearance for moving arms; the Deutsches Museum figure, with its permanently raised arms, has a higher waist, likely to provide additional space for the internal glockenspiel. The Smithsonian shell includes discreet "bumpers" at the wooden hem to cover protruding perambulation wheels (see introduction, **FIG. I.2**); the Deutsches Museum shell encompasses a larger circumference at the hem, hiding wheels within. Wheel coverage was important: when the figure was dressed in cloth robes, the automaton's running wheels had to be protected from entanglement in fabric. Perhaps the Smithsonian monk's feet, sandaled in the Franciscan style, served as Friess's models (thus influencing its identification as a monk).

The carved head of the Deutsches Museum figure presents a visage very different in style and representation from that of the Smithsonian. Facial features are less exaggerated in size relative to the face: observe the higher forehead, smaller ears, and straight nose, with carved hair and beard expertly concealing the mechanics of the jaw. Delicate eyebrows are carved as slightly gathered at the center brow, in concentration. This more refined appearance comes from the use of layers of a chalk ground, or gesso, to further articulate the surface of the carved wood prior to painting. In Renaissance polychromed wood sculpture, gesso functioned not only as a ground for paint (in the same way that panel painters prepared and stabilized a wood surface to accept pigment) but also to build finer detail, such as ringlets of hair or delicate creases of flesh. Made as a fine paste of chalk and hardening agents, gesso permitted the carver some shortcuts for achieving detail in any piece where paint would cover wood and gesso alike. The Smithsonian monk's original faded and cracked paint shows no signs of an underlying gesso ground.[5] With his deeply fluted upper lip, protruding ears, and large direction-finding nose, he is almost but not quite a caricature by comparison. It is even possible that his hooded cowl was meant to be drawn, shrouding his face in shadow. He looks fiercer and more ascetic than his brother in Munich. But we are also looking deeper into time when we study this untouched face.

One unique feature of the Deutsches Museum automaton: an ornamental flourish in red paint decorates the front surface of the wooden body shell (see **FIG. 12.1**). Worn and faded, the paint appears original. No other figure among those we compared carries this embellishment, which would anyway be hidden beneath a garment of actual fabric. In fact, only this automaton lacked fabric dress of any kind, or remnants of such dress, when it entered the museum collection.[6]

Elizabeth visited the Deutsches Museum in Munich in early March 2020. Thomas Rebényi had taken the automaton out of its humidity-controlled vitrine in the museum's hall of early computers and brought it into the lab for the meeting (leaving the back of the head on its pole to command the vitrine). He secured the figure on a cylindrical mount, like a car raised on a jack, to free the perambulation wheels from contact with the table so the figure could "walk in place" when wound. His new video footage was running on a computer screen nearby. Frank Dittmann, head of the Department of Automation, and Carola Dahlke, curator of Computer Science and Cryptology, were present. Joining the group was Raphael Beuing, curator of Timepieces and Scientific Instruments at the nearby Bavarian National Museum, where its automaton, one of the lute players, is kept.[7] The two museums are within blocks of each other, and staff members from each museum wanted to see the other museum's figure. We had met that morning at the BNM to view the lute player (more on this visit in chap. 14), and we had just walked together to the Deutsches Museum. Elizabeth's first glimpse of both automata after years of studying them only in photographs brought the sensory jolt one gets from the gulf—the abyss—between a photograph and the thing itself. Actual size forces the immediate re-register of an object in its native scale. Walk around it in real time, and its three-dimensional nature is irreducible. True color, in the here and now of the eye instead of the camera: the physical integrity of a thing cannot be mediated, though a great photographer can teach us how to look. These are the scribbled first impressions of the Deutsches Museum automaton in Elizabeth's notes:

> So much smaller than I thought! Fiercely focused on something in its hands. The mouth and inside it the teeth, meant to be seen. Other people here today also riveted. Missing paint on the nose. Awe at time's eloquent damage. Amazement that it has survived. How did it look long ago, seen by candlelight?

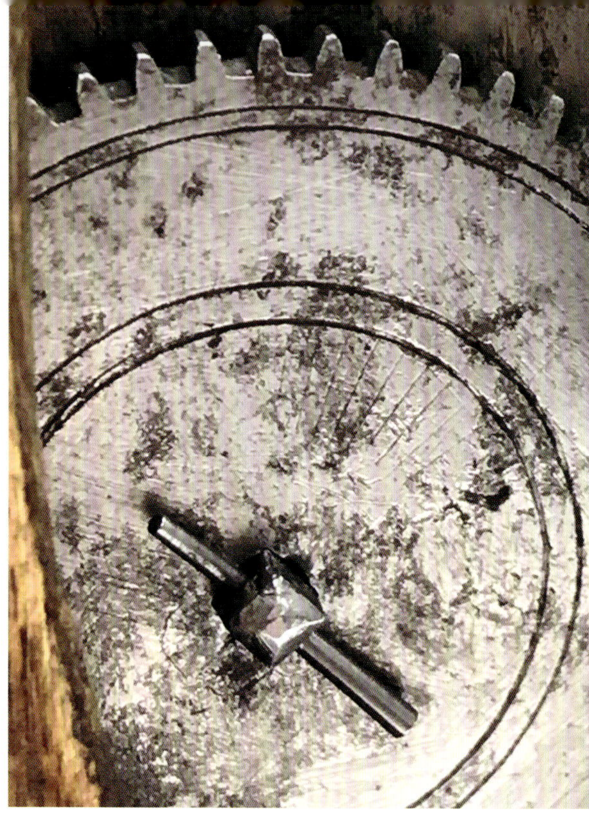

12.6A.

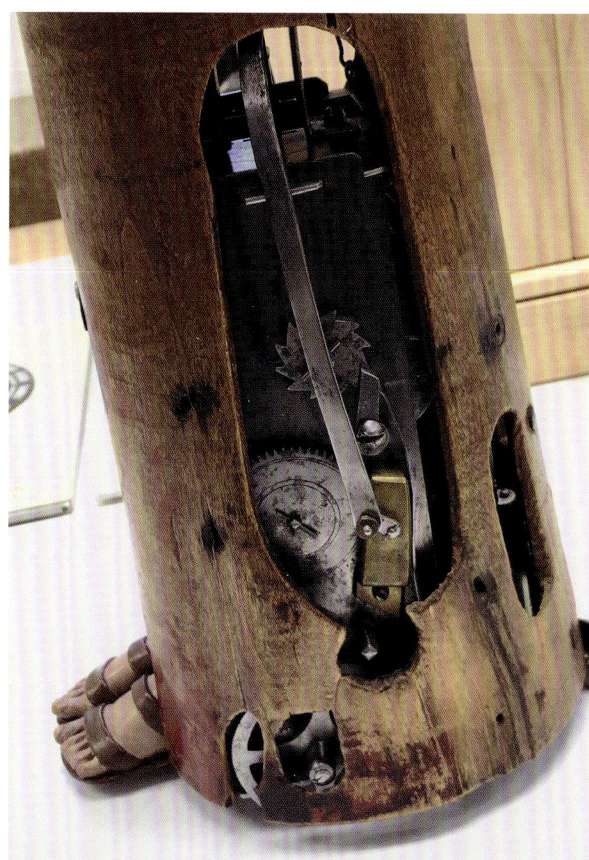

12.6B.

FIGURE 12.6A, B.
In the Deutsches Museum conservation lab, March 3, 2020. A close-up photograph (A) of the sixty-tooth clockwheel on the second arbor (B), showing the clockmaker's radiating layout marks for dividing the wheel perimeter into sixty equal parts. The teeth would then have been individually cut, one by one, with a handsaw and file.

With the key in his gloved hand, Rebényi wound the mainspring a quarter turn. The automaton's hands began their flutter. The eyes shifted, the head turned, the mouth opened…

Across two days in the lab, the group searched for clues to the mystery of what this automaton was meant to perform. Were the arms always fixed in this raised position? Yes: the iron joints inside the wood, transferring rotational motion from gear train to wrists, dictate the angle. The ingenuity of getting this motion to the wrists is itself a mechanical accomplishment: the linkages must navigate right-angle turns at both the shoulders and the elbows. Rebényi explained that the iron joint assemblies would not mate if the arm positions were changed. Hence the loose wood seams at the shoulders, and the wire loops holding the arms to the torso, are not signs of adjustable positions or later adaptations but were designed for ease of assembling the wooden parts of the figure around the iron clockwork.

Conversation moved to the iron strap mounted front and center in the monk's wooden body shell, surely a means for attaching a former prop of some kind, visible in **FIGURE 12.1.** (The lute players have versions of such brackets to support their lutes.) The strap spans a slot cut into the wood, beveled to allow the insertion, from above, of something that clearly called for substantial support. A drum? But the angle of the arms makes the wielding of drumsticks difficult to envision. Nor do the positions of the hands suggest this, though we recall that one hand—the left hand as was now apparent—had been missing and was remade by Friess or someone working under his supervision. What was the pose of the original hand?[8]

The inability to name what the automaton is doing lends it a kind of unique power. Rather than locked in a repetitive task we can parse, it keeps us guessing. Its purpose hidden, the purposeful gesticulation itself becomes the performance. This is a messenger from the past, the message yet to be deciphered.

Rebényi lifted the automaton and tilted it to allow us to see the bottom plate. Noting a squared opening in the plate, Dittmann pointed to its fore and aft edges, which were beveled at angles, suggesting the accommodation of a now-missing cylinder in the narrow space above it, directly under the fusee. He speculated that here was part of the missing glockenspiel. Rebényi pointed out two pivot holes in the train bars that would have supported an arbor and indicated several other now-empty holes in the bottom plate and rear pillar. Other observations: Sets of inscribed concentric rings are present on the top and bottom plates, which David Todd has found in all the other automata sharing this clockwork design, and many early German clocks as well. Decoratively nicked square nuts, surely original, secure the front and back pillars to the bottom plate, a custom that David associates with South German tradition. Mouth and eye movements originate not from shaped cams on the fusee arbor, as in the Smithsonian monk, but from pins on the perambulation arbor below it. Elizabeth was excited to discover unmistakable marks in one of the clockwheels, showing how its teeth had been laid out and cut one by one, by the original clockmaker (**FIGS. 12.6A** and **B**). Above all, everyone in the lab that day spent time wordlessly peering inside the automaton's head from the back. Newly exposed, the working parts of the eyes and mouth allowed a gaze into untouched history.

In his 1988 restoration report, Peter Friess adopted the same suggested date and attribution for his automaton as that of the Smithsonian monk. At present, the Deutsches Museum retains only the estimated date of 1560 and states simply, "place of manufacture unknown."

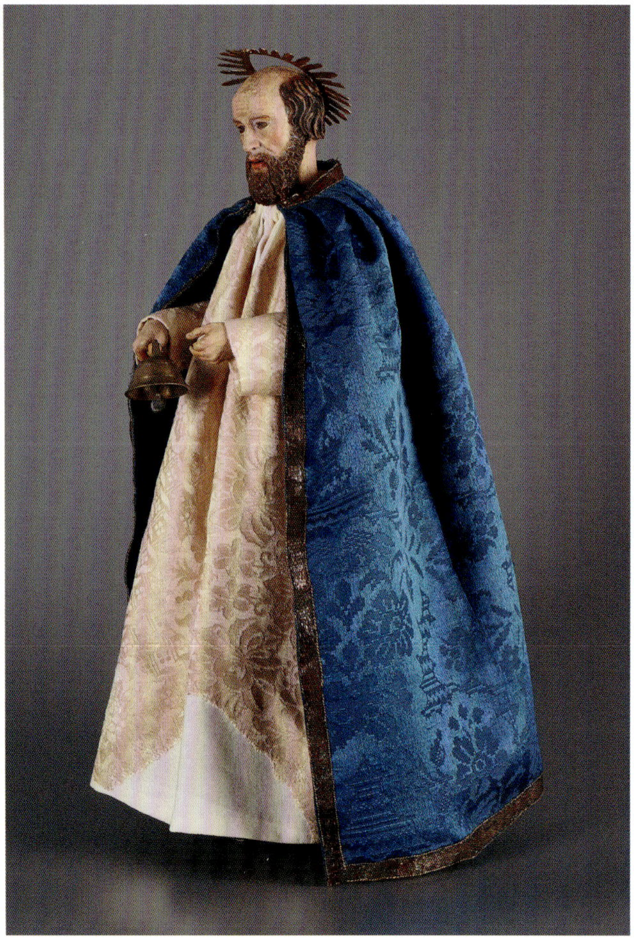

13.1.

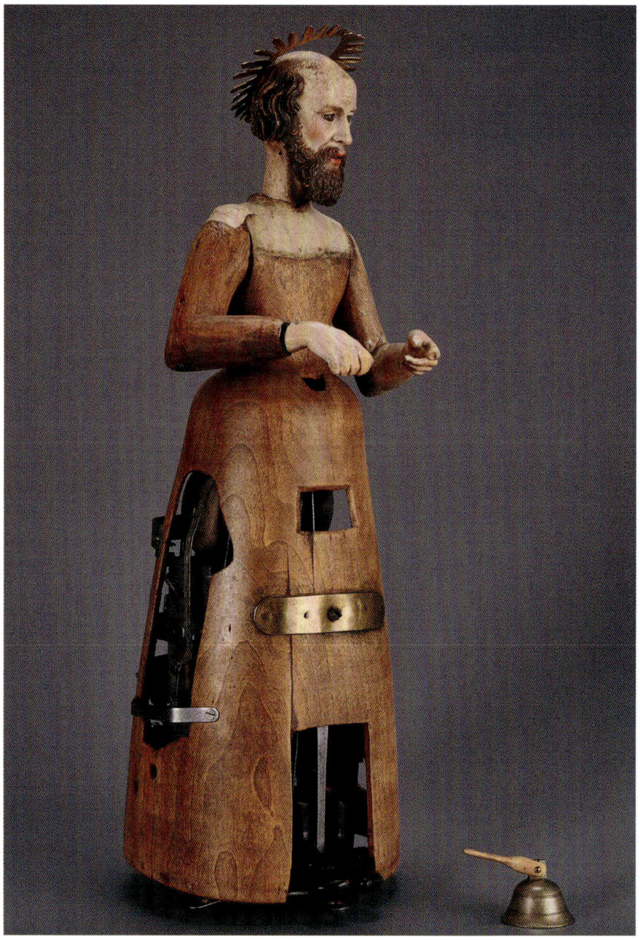

13.2.

FIGURES 13.1–13.3.
Zenélő automata, férfi szent (Musical automaton, male saint), sixteenth century. Spain. Wood, silk damask, iron clockwork, miniature glockenspiel. Height: 41 cm (16 ⅛ in.). Iparművészeti Múzeum (Museum of Applied Arts), Budapest. A musician-saint, with an onboard glockenspiel operated by the clockwork.

13 A Saint in Budapest

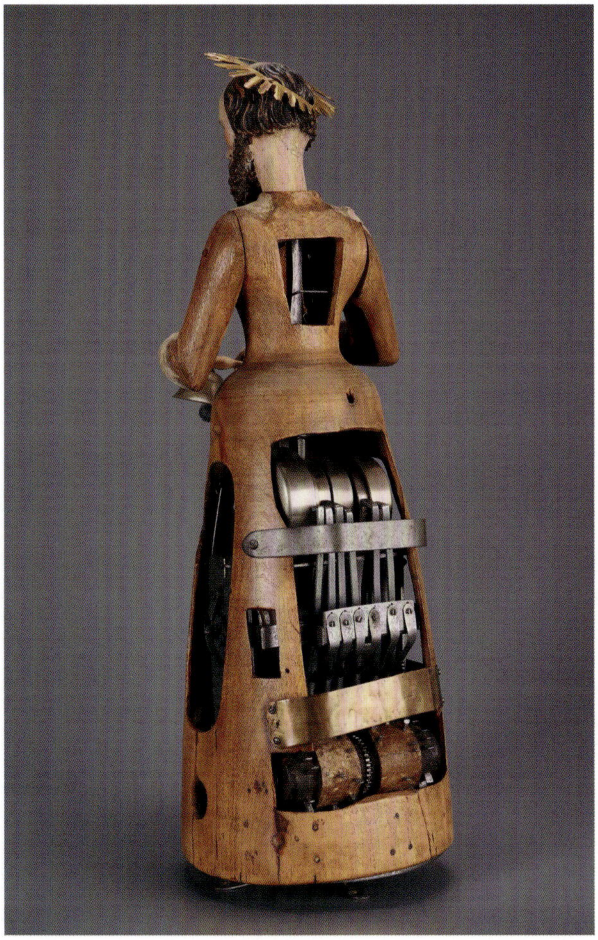
13.3.

Deemed a saint by the gilded halo that frames the head, this automaton in the Iparművészeti Múzeum in Budapest indeed possesses a bespoke glockenspiel, intact and functional, custom-designed to fit inside the narrow body shell with the animating clockwork (**FIGS. 13.1-13.3**). Photographs show its substantial real estate, far outsizing any similar instrument that might have once fit inside the Deutsches Museum automaton. **FIGURE 13.3** shows the horizontal array of six bells at the figure's backside, each with a corresponding strike hammer. The hammers, a set of vertical finger-like levers, are plucked by an arrangement of flat iron pegs driven into a rotating wooden barrel at the base of the figure. The barrel is turned by the same mainspring and fusee that animate the automaton's body. Museum conservators Katalin Soós and Jenö Rácz note that the wooden barrel shows evidence that it had been repegged in the past, hence the original melody is considered lost. The music historian Ilona Ferenczy has suggested the current notes play a secular melody, possibly a folk song.[1]

Significantly, this is a stationary, not a walking, figure. The internal clockframe reveals the characteristic cutouts in the iron baseplate for perambulation wheels, and the saint's individual wooden feet, as in the other automata, emerge from an opening cut in the turned wood body shell.

But photographs show no cutouts in the baseplate to make room for the stepping motions of the feet and no slot for a turning pivot (**FIG. 13.4**). Instead, a curious role is assigned to one foot. As with all the automata, a knowledgeable attendant must wind the mainspring's fusee and then initiate the performance by releasing a hidden stopwork lever. In the Smithsonian monk this lever hovers just above the baseplate at right rear, hidden by the monk's cloth tunic. In the saint, this secret lever is his right foot! After winding the spring, the attendant gently presses this foot, like tapping a key on a typewriter, and the action begins.

With mainspring fully wound and foot tapped, the saint mimes many of the same gestures as the monk and the monk's Munich brother but in a single cycle all his own. His head turns once right, once left; his mouth opens and shuts twenty-six times; his eyes move from side to side sixteen times. The motions of the eyes and mouth originate from the same wooden barrel that commands the glockenspiel: iron pegs in a dedicated section of the barrel activate their own levers to transmit action to the head via wires and gut strings. As eyes and mouth and head move, the glockenspiel rings its melody and repeats it twice. All the while, the saint wields a hand bell in his right hand and rotates the hand from the wrist to ring the bell a total of thirty-six times, in rhythm with the glockenspiel. When the song is finished, the saint bows his head. Net run time is about one minute forty-eight seconds. It seems safe to assume that the power of the mainspring—rather than driving the weight of a three-wheeled rolling chassis across a table—is here spent on the glockenspiel and its bell concert. But in every other respect, including the mysterious cutouts for wheels, the saint's clockwork design and wooden body is so functionally similar to that of the walking figures that its inclusion in the group is essential.[2]

The saint entered the museum collection in 1915, long before the automata acquired by the Deutsches Museum or the Smithsonian were documented or known. The museum presumed the saint an eighteenth-century figure, based on the style of its ornamented damask cape and cloak. But Soós and Rácz discovered a much older machine under those garments when they undertook the restoration of the saint in the 1980s. Their conservation report, published in 1990, referenced the work of both Todd and Friess as the impetus for "rediscovering" the automaton in their own collection.[3] Dismantling the entire automaton to its separate parts, they conserved each material—cloth, wood, and metal—according to best practice. Though a later addition, the garments were carefully conserved too, with new material clearly identified. The wooden head, they wrote, was in relatively good condition. Though the report made no mention of opening the head, the authors described the moving parts for mouth and eyes, noting that the eyeballs were made of pewter and painted.[4] The wood itself they identified as linden, with gesso primer on head, hands, and feet. But the hands were much damaged, especially the bell hand. The conservation report states, "A replenishment of the fingers was necessary because of the attachment of the bell[,]...[and] the grip of the bell was replaced."[5] The conservators clearly indicated which additions were new, as they did with the cloth garment. The left hand was allowed to remain as found, missing the forefinger. Splits in the body shell, formerly joined with brass bands, were repaired with glue. Their report made little mention of the feet; the photographs show simple generic shoes, painted black. The conservators made an inventory of drawings of each part of the clockwork, removed corrosion from all iron surfaces, replaced the mainspring, and fashioned a tiny brass fragment to repair a missing ray in the halo. The glockenspiel bells were cleaned and adjusted. The only previous repair they found in the clockwork, aside from wear and age, they wrote, was in the linkage to the figure's right arm with its bell hand. And as mentioned, the glockenspiel's wooden barrel had been repegged.

Contemplating the saint's performance (and recalling the signs of missing components in the Deutsches Museum automaton), David and Elizabeth continue to ponder the mystery of the cutouts in the clockwork's baseplate. Why would such cutouts for perambulation wheels have been made in a stationary figure? Was a formerly ambulatory automaton rebuilt for a new role? But this seems unlikely, since no cutout for a turning pivot is seen in photographs of the baseplate. Were parts originally intended for a walking figure adapted for a stationary one? Similarly, though the joints of the left shoulder and elbow are tightly glued together now, might the left arm once have moved, in

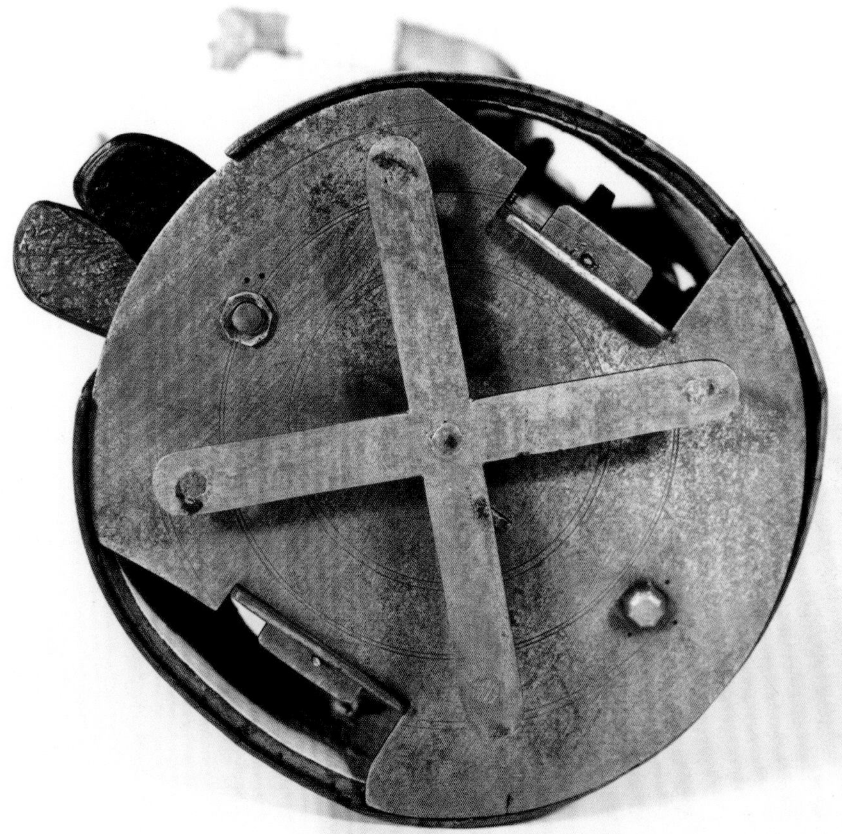

FIGURE 13.4.
The bottom plate of the saint's clockwork, showing cutouts for perambulation wheels, but no wheels or stepping mechanism for the feet are present. Compare with the Smithsonian monk's bottom plate, **FIG. 10.19**.

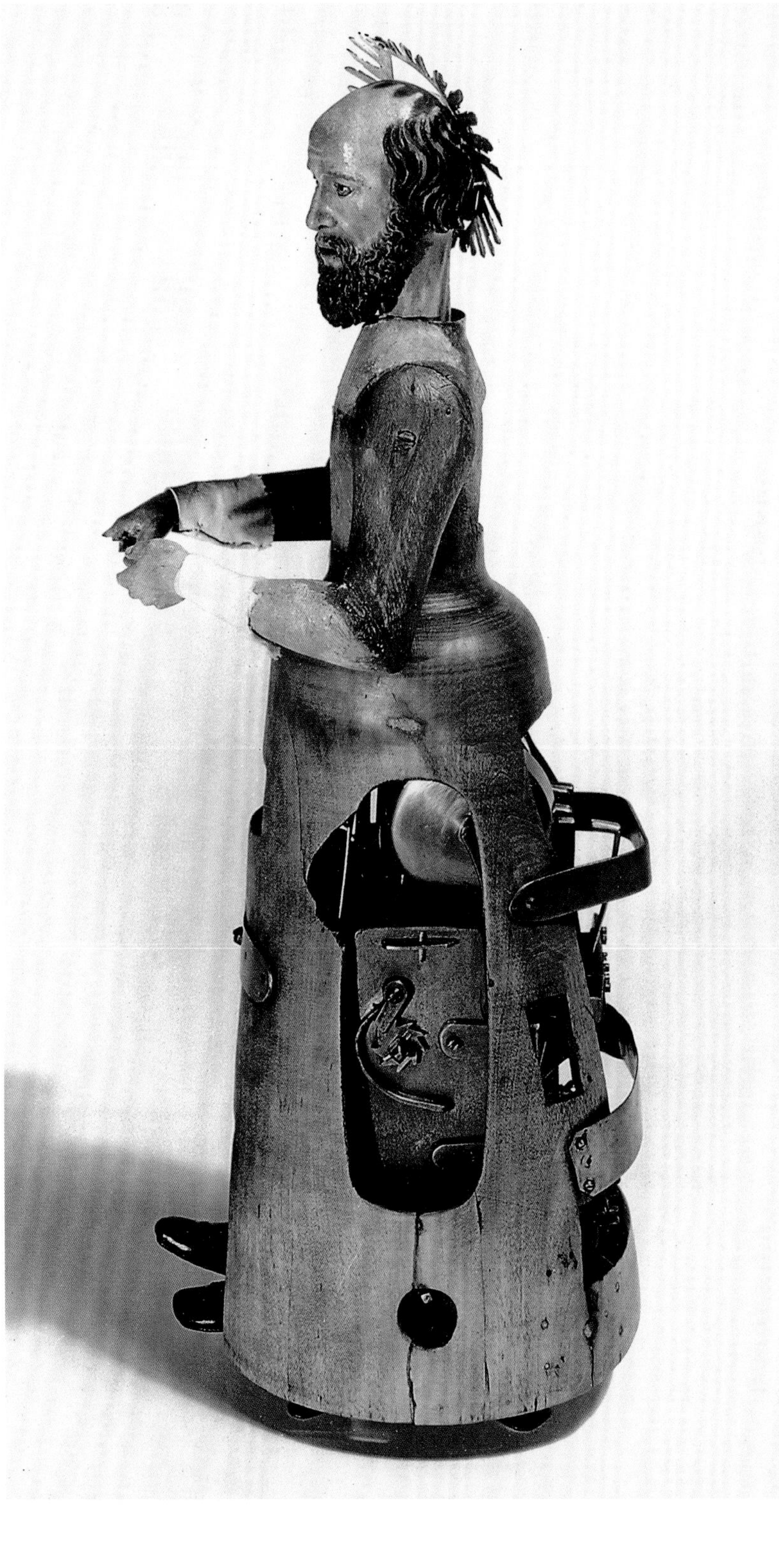

concert with the right arm? Adelheid von Herz gave the following elegant description of the arms and hands:

> **The upper arms of the figure are raised slightly to the sides; both forearms are angled forward and turned slightly toward the upper body. The preserved fragments of the hands are rendered in a naturalistic way. The finger stumps of the right hand indicate an encircling gripping position; the remaining fingers of the left hand, by contrast, are only slightly curved.[6]**

The still-visible original parts of the hands show that both were carved in active and specific positions (see **FIG. 13.2**). Further, there is an aperture at the saint's waist not unlike that in the Deutsches Museum figure, though without the rugged iron strap, where an additional prop of some kind might once have been secured. The edges of the opening, just as in the Deutsches Museum figure, are beveled in such a way that a bracket or post inserted from a steep angle would offer maximum support, as if to buttress something operated on by the hands. Such questions of course must remain unanswered. Any changes in the arms or in adapting a walking figure to a standing one, if they were made at all, would have taken place long before the saint reached safe haven in the museum.

The sculptural character of the saint's head is subtle, lifelike, and refined. This is the work of a sculptor well versed in the anatomy of the head and face. As von Herz writes, "The head shows the physiognomy of an older man. The figure wears a full beard that covers the neck, and long wavy hair; the upper part of the skull is bald. A wrinkled forehead, deep-set eyes, sunken cheeks, and pronounced nasolabial folds give the face a serious expression."[7]

A look at the saint's profile next to that of the Smithsonian monk (cf. **FIG. 13.5** and introduction, **FIG. I.2**) shows two very different character types, in style and depiction. As in the comparison with the Deutsches Museum figure, the monk's visage approaches theater, with its expressive features and pointed fervor. The three automata are virtually the same height: Smithsonian, 40.64 centimeters; Deutsches Museum, 39 centimeters; Iparművészeti Múzeum, with halo, 41 centimeters.[8] But how different they are in both look and performance! Perhaps the Munich brother once sounded a bell-note song, like the saint does now. The Smithsonian monk all but forbids such pleasure.

The Phonogrammarchiv of the Austrian Academy of Sciences once offered a recording of the saint's glockenspiel on the internet. The website link is still active but without the magic sound.[9] Elizabeth wrote to the Phonogrammarchiv in 2021 and was put in contact with Helmut Kowar, who made the recording himself in 1996, on site at the museum. With the kind permission of Christiane Fennesz-Juhasz of the Austrian Academy of Sciences, we were permitted to hear the recording again via private file, and we begged that it be restored on the website. In correspondence with Helmut Kowar, Elizabeth asked if the bell the automaton holds can also be heard in the recording. "At the very beginning of the recording you can hear a few strokes of the bell, very soft and faint," he wrote back. "Then, when the glockenspiel starts playing[,]... some bell strokes are perceptible if you listen closely."[10]

One tiny detail escaped David's memory for many years. When he dismantled the monk's clockwork in 1978 for cleaning, he had been surprised to find that the pinion on the arbor for the fly had an unusually low count of just four leaves, something he had not seen before (the teeth of a pinion wheel are called leaves). A four-leaved pinion would theoretically sustain tremendous sliding friction from the mating wheel teeth that drive it. The general rule in horology calls for no fewer than six leaves in a driven pinion. Further, he noticed that there was surprisingly little wear evident on either the wheel teeth or the pinion, and he remembers thinking that the mechanically impractical can sometimes be made to work—but also, more important, that the monk had seen little use. In making his steel model of the monk, he dutifully reproduced the impossible pinion. When the Soós and Rácz report appeared years later, he learned that the saint's clockwork, exactly like that of the monk, possesses a four-leaf fly pinion.[11] As he began compiling data for our table comparing the eight automata, the coincidence struck home. Four-leaf fly pinions are not supposed to run! And especially in a machine that had seen enthusiastic wear. The devil, even for a wooden saint, let alone a monk, is in the details.

FIGURE 13.5.
The saint's refined profile.

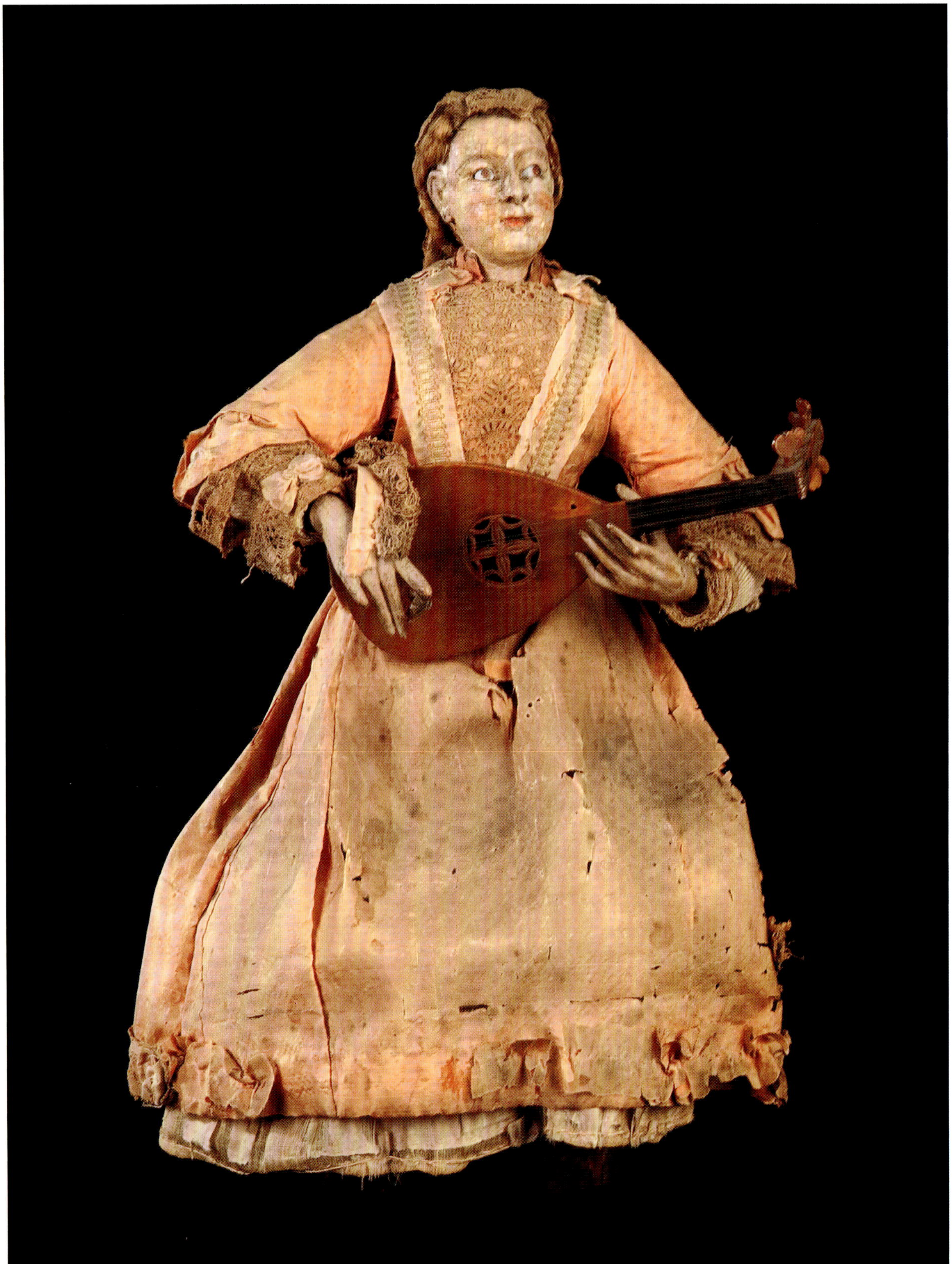

14 Three Sister Musicians

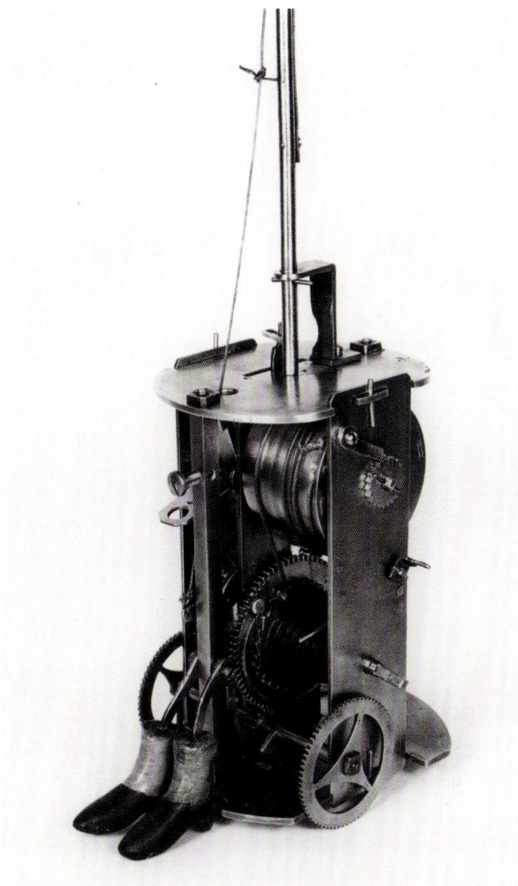

FIGURE 14.1A. (opposite)
Automaton: Lady lute player, ca. 1640–60. Germany. Wood, fabric, brass clockwork. Height: 31 cm (12 3/16 in.). A second walking automaton entered the Smithsonian collection, a few years after the arrival of the monk. National Museum of American History, Smithsonian Institution, Washington, DC.

FIGURE 14.1B. (above)
Automaton: Lady lute player, clockwork.

The Smithsonian Institution Lute Player

In 1979, two years following the acquisition of the monk, Otto Mayr brought a second automaton to the Smithsonian, on loan for consideration of purchase. The lady playing a lute arrived in working condition, original in almost all parts (**FIG. 14.1A**). The clockwork was fabricated not of iron but of brass, a significant difference arguing for a later date than the three male figures. David made a close study of her, as he had of the monk, and she was exhibited alongside the monk the following year in the Washington installation, *The Clockwork Universe*. Mayr launched a four-year campaign to raise funds for a purchase, with a successful acquisition in 1984.[1] Though she is smaller than the three male figures—the smallest of all the eight automata—the clockwork design and performance are nonetheless schematically similar if with less mechanical depth (**FIG. 14.1B**, and see **FIG. 18.3**, below). She steps along a roughly triangular course, the walking turn accomplished by means of the now-familiar lifted wheel and pivot. Between turns, she takes ten steps (each foot lifts and falls five times), rotates her right wrist to strum the lute (thirty times), and moves her left forearm from the elbow to carry her hand up and down the fretboard (five times). She turns her head once left, once right, and nods once, and her eyes move from front to right and back five times. The lute

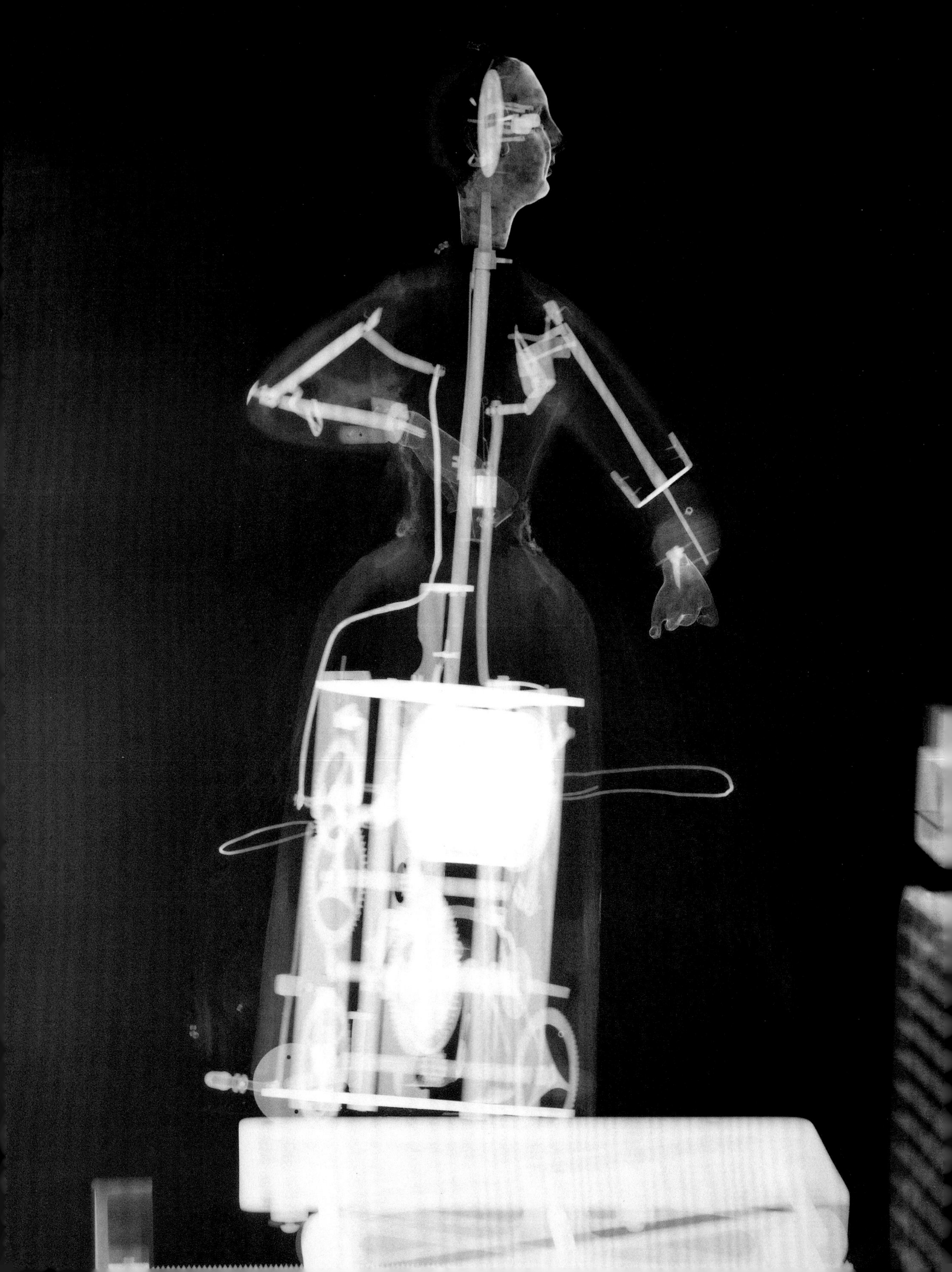

is played in mime of course; the performance is silent. There are no compound movements, as seen in the combined motion of the left arm, shoulder, and head in the monk, and no movement of the mouth. But the lute player has much poetry. She arrived in an eighteenth-century silk dress, and her lute too may have been replaced at some time in the past. The miniature lute itself is beautifully crafted, perhaps made by a period luthier. The wood fusee is more finely cut than that of the monk, with an eleven-turn spiral against the monk's seven. Radiography gives a glimpse inside the lutenist's head, revealing a layout similar to the monk's, though simpler, with no moving parts to the jaw. The head appears to have survived the centuries intact. The X-ray, something uncanny in itself, deserves pause (**FIG. 14.1C**). The high-resolution X-ray film shows the figure's exquisitely carved face and hands. The dress registers as a faint cloud around the body; the wire hoops that give the dress its shape leap out from the wooden body shell. Acquisition papers estimated a date of circa 1630 for the figure; at present the museum gives a mid-seventeenth-century origin, between 1640 and 1660.

The Bavarian National Museum Lute Player

The Bavarian National Museum in Munich, where *The Clockwork Universe* exhibition premiered in 1980, has its own lute-playing automaton, in the collection since 1927. When David traveled to Munich to assist in the exhibition installation, he was shown the lute player. Having just completed conservation documents for the two Smithsonian automata, he offered to produce an analysis of this one too, with detailed drawings and photographs. Peter Friess (who would later conserve the Deutsches Museum automaton) was then a young conservator at the Bavarian National Museum and joined David in the study of this figure. Larger than the Smithsonian lutenist (at 15⅝ inches, closer in height to the monks and saint), this figure too possesses a clockwork made mostly of brass (**FIGS. 14.2A, B**). But both clockwork and sculpture had suffered extensive damage in the distant past. Reconstruction, in the nineteenth century or earlier, had included a replacement head made not of wood but of wax, and the figure had been assigned an eighteenth-century date at the time it entered the museum collection.[2] David found evidence in the clockwork that the automaton had formerly possessed both moving mouth and eyes in an original wooden head. Though perambulation wheels and trailing wheel are still present, the figure's feet and stepping levers are missing. She would once have trotted across a tabletop in a lively tour, strumming the lute with her right hand, fretting it with her left, while turning and nodding her head, very much like the Smithsonian lute player (though one much-reproduced early photograph shows the lute incorrectly reversed in the automaton's hands).[3] Her wooden hands, original or not, were elegantly carved, with long graceful fingers, and this lute too is a first-rate miniature. The fusee arbor bears a cam that would once have depressed a pivot for the walking turn. The same arbor, David notes, carried remnants at its far end of another cam that would have tripped a spring-loaded hammer for a small bell on the train bar: a single strike just before each walking turn. The bell's muffled *ding!* from beneath the lute player's full-skirted dress would perhaps have reached the viewer's ear as if from afar, and as if to prove it, the automaton seemed to hear it too and turn on cue.

With clockwork thus undone by time, the figure remains inoperable. But the missing parts bear witness to the triumph of the lute player's survival. This automaton is the only one among the figures with clockworks made of brass to have had a moving mouth. Such mobility of the face required sophisticated engineering and represents more than mimed song or speech, as we show in chapter 23.

A full thirty years after David worked on this automaton in Munich, Elizabeth traveled to see it in 2020. As recounted in chapter 12, she had arranged two meetings in Munich, one to see the Deutsches Museum automaton and one to see the lute player at the Bavarian National Museum; and she had invited parties from one museum to also join her at the other. What made this especially valuable is that the Bavarian National Museum is a museum of art and culture, whereas the Deutsches Museum is dedicated to the history of science and technology. Each preserves a walking

FIGURE 14.1C.
A beautiful ghost. X-ray of the full lute player clothed. Radiograph taken by staff at the Museum Conservation Institute (formerly the Conservation Analytical Laboratory), Smithsonian Institution, Washington, DC, 1979.

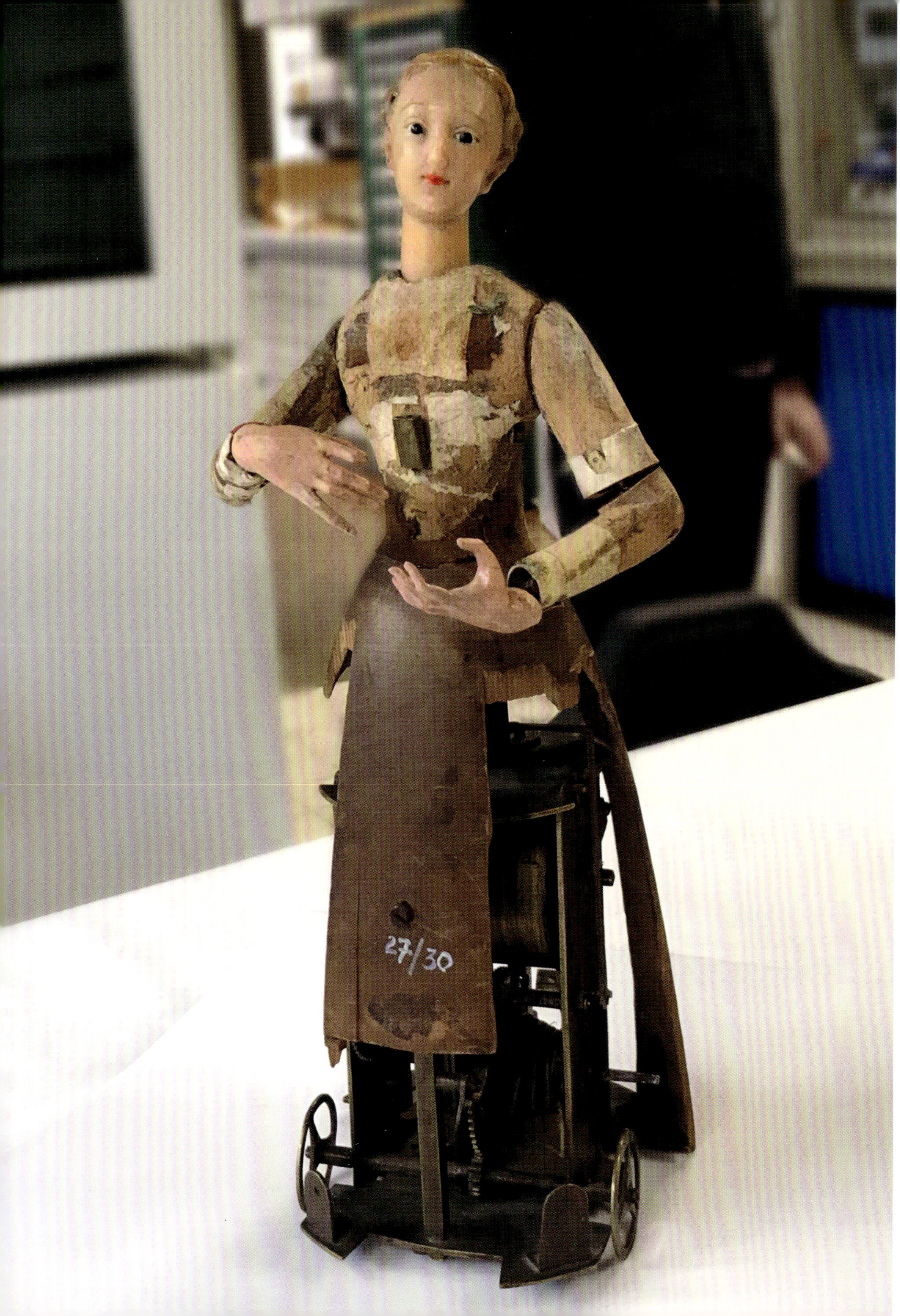
27/30

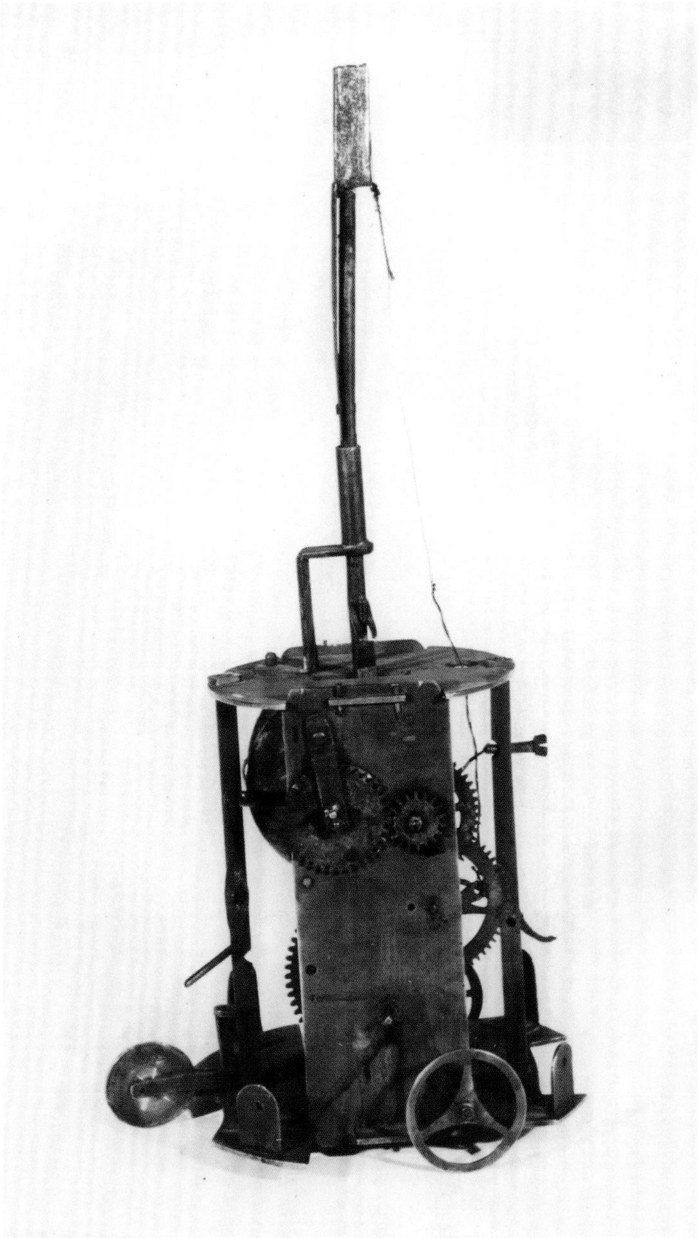

FIGURE 14.2A, B.
Lautenspielerin (Lute player), ca. 1600. Wood, brass clockwork, wax head. Height: 39.7 cm (15⅝ in.). Bayerisches Nationalmuseum, Munich. The original wooden head is lost, but surviving fragments in the core clockwork reveal cams that would have animated movable eyes and mouth.

automaton with a remarkably similar clockwork template but classifies it in a different historical context. The chance to compare them back-to-back in the company of curators and conservators from both institutions was a welcome acknowledgment of the hybrid nature of the automata.

The group met first at the BNM. Raphael Beuing and Joachim Kreutner, curator and conservator, respectively, had brought the lute player out of storage and set it on a turntable in the conservation lab. They had removed the lute and dress, to permit the close view of the wooden body and the clockwork inside it. The lower wooden body shell is missing much of its circumference—a misfortune for the automaton but a windfall for the visual access this gives to the clockwork from most angles. Mainspring, fusee, gear train, fly, cams, and linkages: 90 percent of the machine is intact. Though most metal components are brass, dimensions and handcraft are as robust as the iron works in the male figures. Kreutner had requested a microscopic sample analysis of the lower wooden body shell; it was identified as mountain ash.[4]

Unable to see the lute player perform, the group focused instead on the logic of the clockwork. Kreutner slowly revolved the turntable so each of us could study the machine in the round. We tracked the presumed actions of the figure's wooden hands back to source cams deep inside the machine, via levers, wires, and offset cranks. We counted the number of spiral cuts in the fusee arbor, which told us how many times the performed actions would have been repeated. We computed gear train ratios for the timing of separate actions. It was an exercise in empirical surmise, a collaborative effort to reverse engineer the imagined performance. From the sculpture to the mechanics and back again, the process showed us the clockmaker's genius. The lighthearted lute player was once animated by formidable calculation and craft.

That afternoon, schooled in the mathematics of one automaton, we walked as a group to the Deutsches Museum to see another automaton accomplish the performance. Could two figures be more different in character and play—a lady lutenist and an arm-waving preacher (if preacher he is)? But their clockworks are near copies.

The Mandolin Player in the Conservatoire national des arts et métiers

Another brass-works automaton, the Joueuse de mandoline, in the museum of the Conservatoire national des arts et métiers (CNAM) in Paris, has a lively history of shifting attributions (**FIGS. 14.3A-14.3C**). The 1882 edition of the Conservatoire catalog attributes her to the famous inventor Jacques de Vaucanson (1709-1782), with repair by Jean-Eugène Robert-Houdin in 1859.[5] A 1906 edition of the catalog retains this identification, adding that the figure entered the collection in 1853. In 1928 two photographs of the mandolin player appeared in *Le monde des automates*, the benchmark encyclopedia of automata by Alfred Chapuis and Édouard Gélis, but the authors move her to the nineteenth century, arguing that her mechanism is too simple for Vaucanson's hand.[6] The only sure information, they note, is the date she entered the collection, which they give as 1833. For its part, the Conservatoire kept its eighteenth-century estimate, though in a still later edition of its catalog the attribution shifted to one Pierre Gautier, in view of the automaton's similarity to another known lute player carrying this signature.[7] Gautier (also spelled Gauthier), according to *Britten's Old Clocks and Watches and Their Makers*, was an eighteenth-century clockmaker in Paris.[8] Indeed, Gautier's name apparently appears on a very similar automaton pictured in a volume by Alfred Chapuis and Edmond Droz, *Les automates, figures artificielles d'hommes et d'animaux: Histoire et technique*, published in 1949.[9] Finally, the 2008 CNAM online database dropped all identifications of an original maker; the figure is described as "original or reconstructed from original parts," dated 1775-99, with the Robert-Houdin restoration still noted, along with two additional subsequent restorations. The most recent database, which features three good photographs of the figure and its mechanism, patiently retains 1853 as the date of the automaton's entry into the collection.[10]

FIGURE 14.3A.
Automate: Joueuse de mandoline (Automaton: Mandolin player), 1775-99. Wood, fabric, brass clockwork (original, or reconstructed from original parts). Height: 35 cm (13¾ in.). Conservatoire national des arts et métiers, Paris. In the museum collection since 1853, this figure was the first of the eight walking automata to be identified and cataloged in a modern museum.

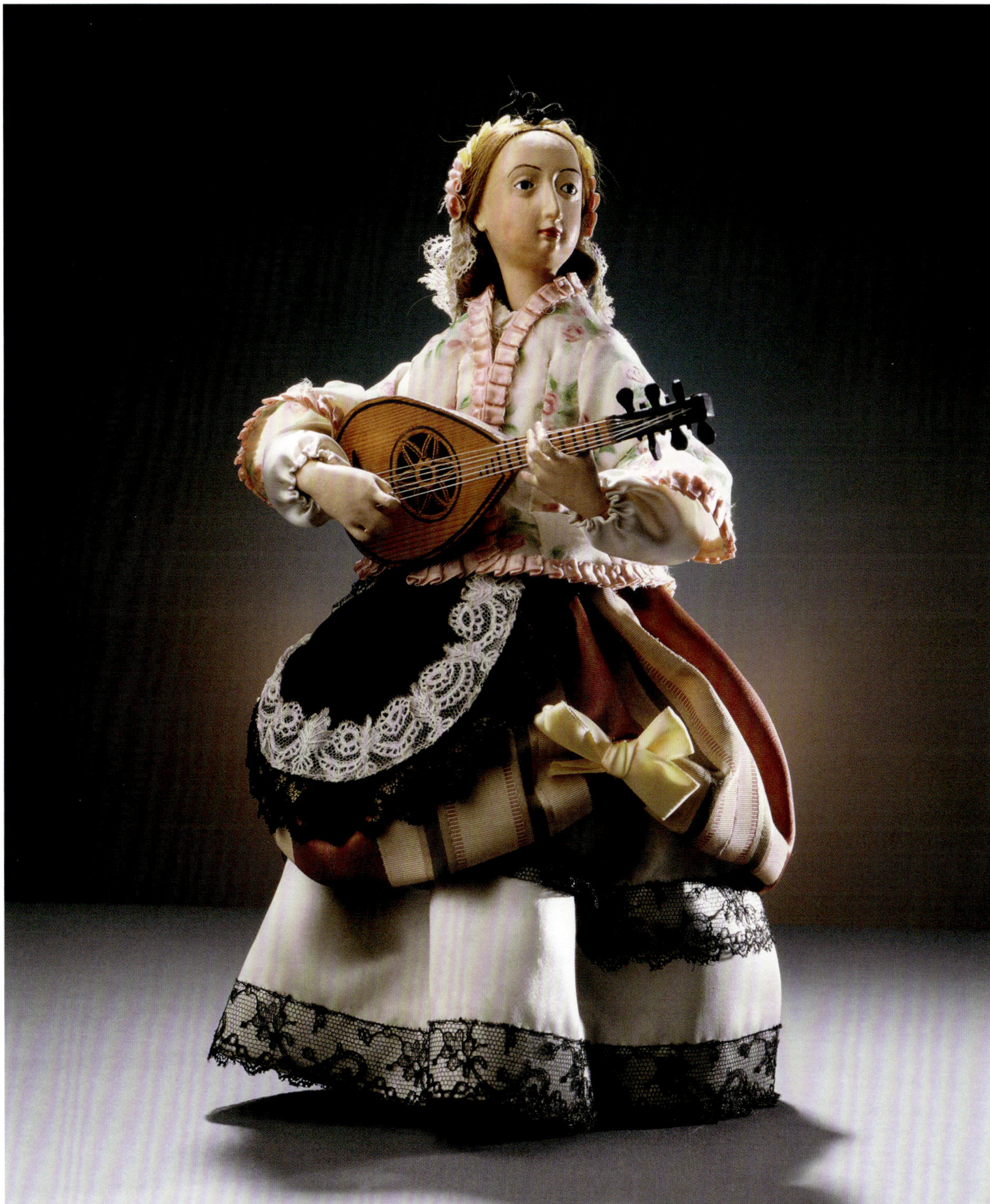

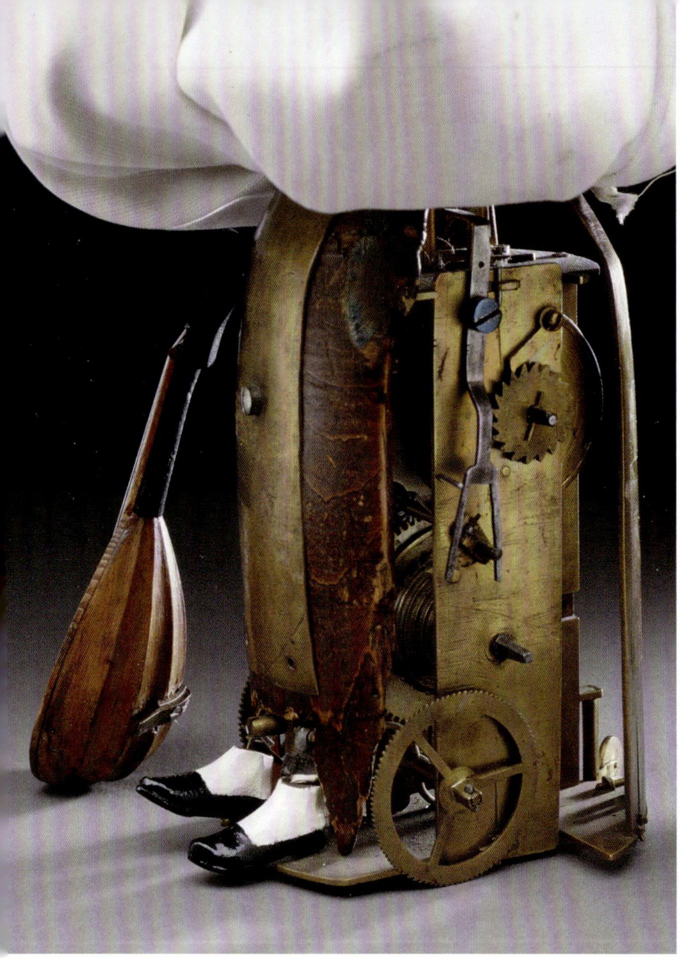

But the revisions are welcome, for this automaton carries the earliest published documentation of any of the automata we compare in this book. Though her origin remains in doubt, the record of restorations, going back to 1859, is invaluable for its candid disclosure of radical repair. Contemplating the lost wooden head of the BNM lute player, or the now-remade feet and hand of the Deutsches Museum figure, or the cutouts for perambulation wheels in the standing Budapest saint, or simply the parade of cloth garments in updated styles on earlier machines: all are signs of other hands in later shops. The one constant is the design of the clockwork. If the CNAM joueuse dates to the last quarter of

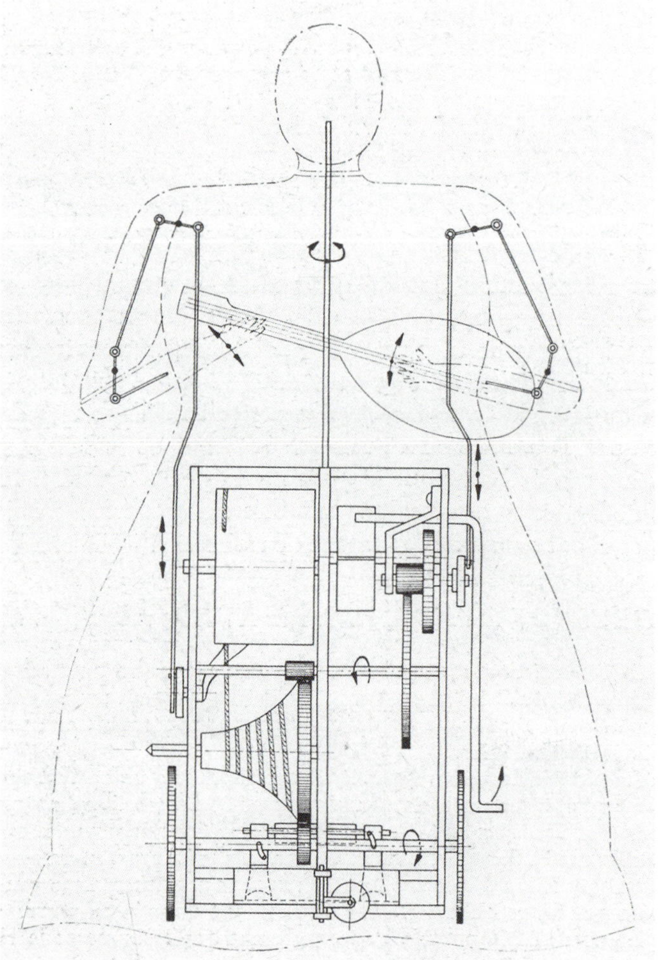

FIGURE 14.3B.
The brass clockwork of the mandolin player. The fusee too is made of brass rather than wood, an indication of the later date of this figure. Remnants of an original wood body shell remain in place over a much-operated machine.

FIGURE 14.3C.
Schematic diagram of the mandolin player's clockwork, rear view. *Conservatoire national des arts et métiers, catalogue du musée, section Z, automates et mécanismes à musique* (Paris, 1960), 54. An excellent if simplified illustration of the core gear train shared by most of the walking automata. Visible are the mainspring barrel with gut line to fusee, the fusee wheel to second arbor pinion, second arbor wheel to third arbor pinion, third arbor wheel to (hidden) fly pinion, and stopwork lever to interrupt the fly. But no cams are present within this automaton's clockwork: the arms are moved by offset cranks on the outer ends of upper arbors in the gear train.

the eighteenth century, the automaton is a tribute to the longevity of this design. She may be the most recently made of the eight walking automata in our study, but thanks to the Conservatoire she was the first to be publicly identified.

In January 2020 Elizabeth traveled to Paris to view the automaton. Cyrille Foasso, head of Scientific and Mechanical Instruments at CNAM, and Rémi Catillon, head of Restoration, kindly agreed to a meeting, and we were joined by museum conservators Perrine Starck and Marie Charet. The figure is displayed in the museum's dramatic, darkened Théâtre des Automates, home also to the famous dulcimer player, "La joueuse de tympanon," made by David Roentgen and Peter Kinzing for Marie-Antoinette in 1785.[11] Rémi Catillon unlocked the heavy glass vitrine door and carefully lifted out the small mandolin player and carried her to a temporary table nearby. The costume could not be removed, but the skirts could be lifted for a glimpse of the clockwork (*...la mariée mise à nu*). Armed with flashlights, we studied the body, the face and hands, the costume and mandolin, and the interior wheelwork, to try to divine original from repair.

She is small, just an inch taller than the Smithsonian lute player. She wears an eighteenth-century peasant costume and a bonnet over a wig of real hair, her garments in far better condition than those of the Smithsonian lady. But viewed in the flashlight's beam, her wooden face shows earlier paint layers partly chipped away, now overlaid with newer paint. A wooden torso to the waist appears intact, but only fragments remain of a lower wooden body shell. Instead, vertical brass strips had been added to the frame to keep the garments from catching in the gear train. Significant repairs to the clockwork are visible. As with the Munich lutenist, the missing body shell allowed an unbidden view of the clockwork. Unique among the kindred automata, the fusee in this machine is made not of wood but of brass, one argument for a later date of manufacture. Though no longer operated, her performance would match prototype: hands ply the mandolin, head and eyes move, feet step forward under hoopskirt hem, the walking turn won by lifted wheel and pivot. The motions of her two arms are similar to those of the Smithsonian Institution and BNM lute players: the right hand strums, the left hand frets, and the mandolin itself is supported by a hidden bracket attached to the figure's waist. The right arm is animated just at the wrist, the left at the elbow. Internal mechanics follow suit: the right wrist is pivoted via transfer linkages inside the fixed wooden shoulder and elbow. The left forearm moves side to side from the elbow, via a linkage through the shoulder, drawing the hand back and forth along the fretboard of the mandolin. The left thumb hooks behind the mandolin's neck to form a loose yoke so the hand would slide along the neck as if by intent. Further, the swing arc of the forearm requires some give at the wrist, and the wrist joint must hinge freely to accommodate it, the hand thus gracefully following the forearm lead. The resulting performance will be an elegant illusion.

"Original or reconstructed from original parts." So similar to the earlier machines is the clockwork inside this automaton that we wondered, is it a very good copy, a century on? Or could it be an earlier machine, redeployed inside a mostly newer body sometime before the museum acquired it?

One clue can come from close inspection of the clockwheels themselves, to determine whether the wheel teeth were hand cut, as were those in the earlier automata, or cut with a wheel-cutting engine, a faster production process that came into use in seventeenth-century clockmaking shops. Hand-cut wheels often still carry the clockmaker's layout marks on the wheel face (see **FIG. 12.6**), and teeth show slight irregularities, since each tooth was individually cut, freehand. Engine-cut wheel teeth are more uniform. The difference can be an indication of age, though not a binding one, since clocks and automata continued to be made by traditional methods well into the eighteenth century, particularly in smaller shops. For a closer look at the wheelwork in the mandolin player, Catillon took a series of microphotographs for David to study. The images show no layout marks on the wheel faces, for the original surfaces had been filed and wire-brushed either at the time they were made or in later cleaning. Yet David detects enough irregularity in the teeth to believe they were individually cut by hand.[12]

The caveat of course is that seeing something only in a photograph, devoid of scale (such clockwheel teeth can be less than a millimeter wide each), and above all not side by side with a comparable wheel of proven date, forces even a clockmaker's opinion into surmise. We defer to the long-standing expertise of the Conservatoire in assigning the date of the automaton.

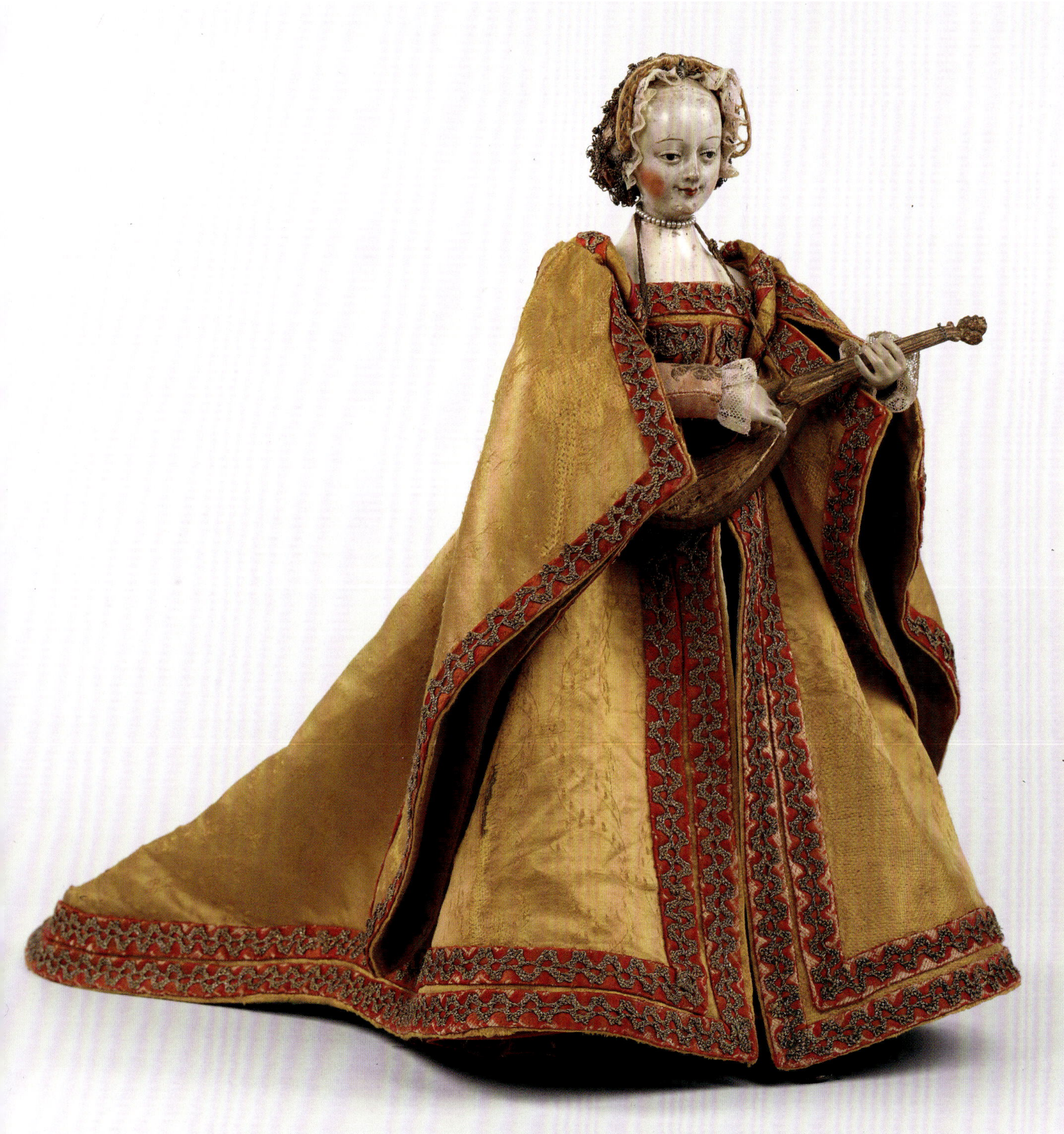

15 The Vienna Cittern Player

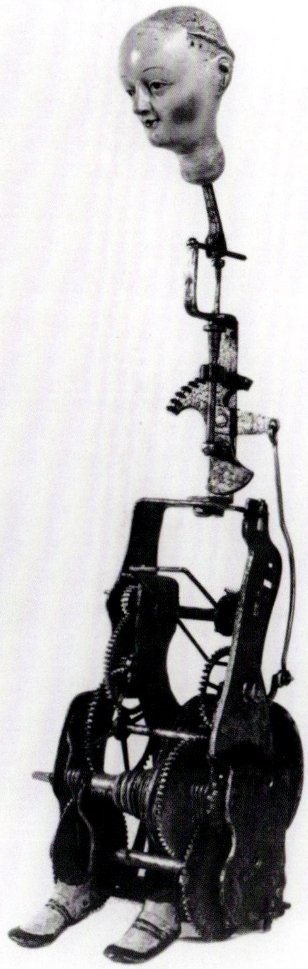

The venerable cittern-playing lady in the Kunsthistorisches Museum in Vienna **(FIGS. 15.1, 15.2)** received the first in-depth study of any of the walking automata addressed in this book, and partly for this reason she has come to be seen in recent scholarship as a prototype for the others. As introduced in chapter 5, her appearance in a private collection was reported in 1926 by the horological historian Ernst von Bassermann-Jordan. Citing the sixteenth-century testimony of the Spanish annalist Ambrosio de Morales, Bassermann-Jordan suggested this automaton might be kin to the dancing lady Morales described, made by Philip's court clockmaker Juanelo Turriano. "La dama que tañe y dança" (the lady that plays and dances), Morales had called Juanelo's little drum-playing figure, using the word *automaton* to list her among the mechanical marvels made by Spain's famous engineer.[1] Enough historians had already quoted this passage in referring to Juanelo's lost masterpieces that it seems

FIGURE 15.1. (opposite)
Automat, Cisterspielerin (Automaton, Cittern player), second half of the sixteenth century. Spain (?). Painted wood, linen, silk brocade, iron clockwork. Height: 44 cm (17⁵⁄₁₆ in.). *Kunstkammer*, Kunsthistorisches Museum, Vienna. See also **FIGURE 5.1**. In the collection since 1934, once attributed to Juanelo Turriano. Scholars writing about this regal figure in the first half of the twentieth century laid the foundation for the study of all the early walking automata.

FIGURE 15.2. (above)
The cittern player's iron clockwork. Arguably the earliest of the eight walking automata. The structure of the clockwork is unique to this figure: the mainspring barrel is placed not above the fusee but horizontally behind it; the arbors of the gear train step up in diagonal progression within a canted U-shaped frame.

as if the words were waiting for the Vienna figure when she appeared in the 1920s—and Bassermann-Jordan had the command to propose a match.

She plays a cittern, not a drum, but like Juanelo's *dama* she performs and walks in a circle on a tabletop. At first appearance she is similar to the lutenists described above: a rolling three-wheel chassis driven by mainspring and fusee, delicate feet stepping forth at the hem of an ornate robe, hand strumming the instrument, head gracefully turning. But her walking path is adjustable: the trailing wheel, like the rudder of a boat, can be set in advance by hand for a circular or a straight march. The other automata are programmed by hidden cams for specific paces and turns. Her iron mechanism places her considerably earlier in date than the brass-works lutenists and mandolin-playing lady. Even among the figures with iron mechanisms, the U-shaped serpentine frame of her clockwork is structurally different from the posted-frame configuration of the others (see **FIG. 15.2**). The diagonally stepped arrangement of the arbors of the gear train is singular to this clockwork, with the mainspring barrel sitting not above the fusee but horizontally behind it.[2] The feet are raised and lowered by pegs attached directly to the perambulation wheels. A rack-and-pinion assembly generates the turning of the head rather than the easier-to-make pin and slotted lever we see in the other machines. Her eyes and mouth are fixed: the head itself contains no moving parts. If this is a more primitive machine, it may well be the first of its kind, a prime innovation. The sculpture and costume meanwhile are resplendent. The great train of her brocade cloak is longer than she is tall.

Bassermann-Jordan's account of the cittern player is the subject of a full chapter in his 1926 opus, *Alte Uhren und ihre Meister*. Titled "An Automaton from the Era of Emperor Charles V," the chapter opens with the lament that automata from antiquity, from Arabic invention, from the wonders of the fifteenth-century Burgundian court, have survived only in the written record:

> The earliest preserved automaton [depicting a human] known to me dates from the mid-sixteenth century. It is a female figure, now dressed in the kind of ornate costume common around the end of the sixteenth century, . . . the height of a medium-sized children's doll of modern times. The torso houses the iron works, entirely constructed of components commonly used in mid-sixteenth-century clockmaking. The figure is propelled by two wheels, moving its feet forward and back, its right hand strumming a lute, which it firmly holds with its left. It turns its head alternately right and left. The date of origin of the doll—not after 1550—can best be deduced by the broad shape of its shoes. It was originally clothed in a low-cut white linen shirt. The automaton was treated carefully and kept with respect: when its original dress was worn out and no longer in fashion, a new one was made about forty years later, in the style of international court costumes of the nobility coming out of Spain at the time. . . . The figure wears a grand coat with a train, and with it a bonnet, which must once have been adorned with a feather. The predominant colors of the silk fabrics are red and yellow. The face, finished in enamel-like colors, has been modified in later times; the lute is still the original. Thus this little treasure has come down to us, likely the only remaining one of its kind, to this day still running about patiently—with much noise—and executing its movements, when a caring and knowledgeable hand winds its spring. The doll is old Austrian private property, having been inherited from the Netherlands.[3]

Bassermann-Jordan proceeds with a detailed deliberation on the origins of the figure, citing both Spanish and German spheres, and concludes that although there may be a Nuremberg precedent for the clockwork, the outward character of the lady seems distinctly Spanish.

His comments are referenced in turn in Chapuis and Gélis, *Le monde des automates*, published just two years later, in 1928, in which a chapter on the history of walking automata opens with this figure.[4] Chapuis and Gélis were the first to record that the automaton could advance in a straight line or follow the path of a circle, a design unique to this machine.

Whether or not the Vienna cittern player may claim precedence, Bassermann-Jordan's lingering description of her, his historical perspective, and his horologist's eye are foundational for the study of the walking automata.[5] We return to his analysis in chapter 20.

The cittern player entered the Collection of Sculpture and Decorative Arts (Sammlung für Plastik und Kunstgewerbe) in the Kunsthistoriches Museum in 1934. The 1935 edition of the museum catalog identifies her as the work of "Giovanni della Torre (Juanelo Torreano)...supposedly made for Charles V."[6] It seems safe to assume that Bassermann-Jordan's educated guesswork influenced the attribution. The next year, 1936, the art historian Fritz Saxl, head of the Warburg Institute (which had moved to London from Hamburg in 1933), gave a lecture at the British Academy titled "Costumes and Festivals of Milanese Society under Spanish Rule" and cited the costume of the cittern player.

> **We know very little about a dress, especially about a woman's dress, if we do not know how it looked when its wearer walked through a room, mounted a staircase, sat down, etc.... How did the wives of the Marchese del Vasto, Muzio Sforza, and their friends look as they moved about in their costumes?... [A]ll the art of Titian cannot—and is not intended to—give us an idea of the real form and grace of movement of his models. Fortunately, there is in the Vienna museum a mechanical doll, which serves to supplement our imagination. We know that the Marchese del Vasto introduced to Charles V a Lombard engineer, Giovanni Torriano from Cremona, who made similar mechanical dolls.[7]**

Saxl described the doll's motion in words only a witness seeing it in person could write.

> **A combination of stateliness and grace characterizes its movement. The feet move slowly, the figure seems to float. The skirt disguises the shape of the body completely. In contrast, the hands and head are in free and elegant motion. The head describes exquisite curves to the music produced by the sensitive delicate hands.[8]**

Apparently he had asked for a film to be made of the cittern player in performance, and he showed the film in his London lecture, thanking Leo Planiscig, then director of the Collection of Sculpture and Decorative Arts at the Kunsthistorisches Museum, for making it. In 2021, the film was recovered in the Warburg archives and sent to Paulus Rainer, current curator of the *Kunstkammer* at the Kunsthistorisches Museum, who kindly allowed Elizabeth and David to view an MP4 copy.[9] If reading Saxl's text brings a little wince at its unexamined liberty in seeking the body under the dress, is this not what we too are doing (with a bit less offense) when we see the automaton move? Watching the film is an extraordinary chance to experience this legacy figure in motion, a high point of all our work on this book. Indeed, her head describes exquisite curves to the music we imagine she is playing. We returned to the photographs of her clockwork to look again at the rack-and-pinion assembly that animates her head (see **FIG. 15.2**), realizing that this mechanism not only turns the head from side to side, but gently tilts and nods it in the process, a compounded motion that lends a dreamy lilt to her performance. She is listening as she plays! Meanwhile, her right hand is moving like a bird's wing on the strings of the cittern, too fast to see its form (frames per second battling teeth per clockwheel). As she slowly moves forward, her feet step forth from her hem, maybe two steps per second. There is enough residual momentum within the clockwork to give her left arm and the cittern some slight motion. The overall effect is fluid and hypnotic. There is nothing mechanical about it. We know very little about an automaton, any automaton, if we do not know how it looks when it moves.

The museum's 1966 edition of its catalog was more circumspect about the origins of the cittern player, adding the names Jakob Bulmann (a Nuremberg locksmith and clockmaker) and Christoph Margraf (court clockmaker to Emperor Rudolf II in Prague) to the name of Juanelo as makers of mobile automata.[10] Correspondence in 2000 with Rudolf Distelberger, then director of the *Kunstkammer* at the Kunsthistorisches Museum, informed Elizabeth that the cittern player functioned effectively until 1964. At 44 centimeters in height, she is the tallest of the eight automata we compare. Upper body, hands, head, and feet are polychromed wood. For her linen and silk brocade costume, the catalog suggests a French character but adds that the fluctuation of fashions could also point to Spain, the Netherlands, or Italy. Though no longer operable, the figure is beautifully cared for, and she holds pride of place among automata in the museum's collection.

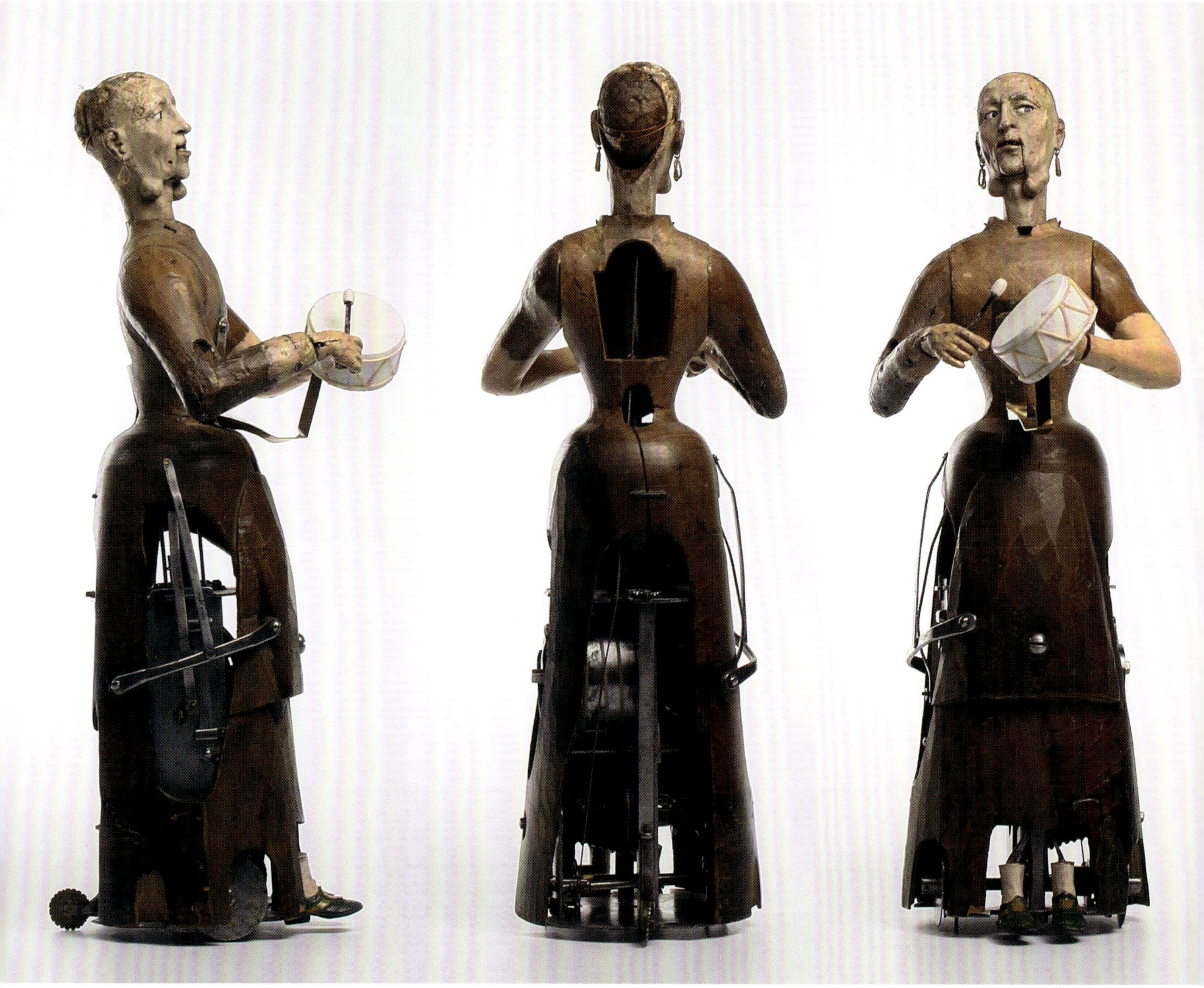

16 A Late Arrival from Milan

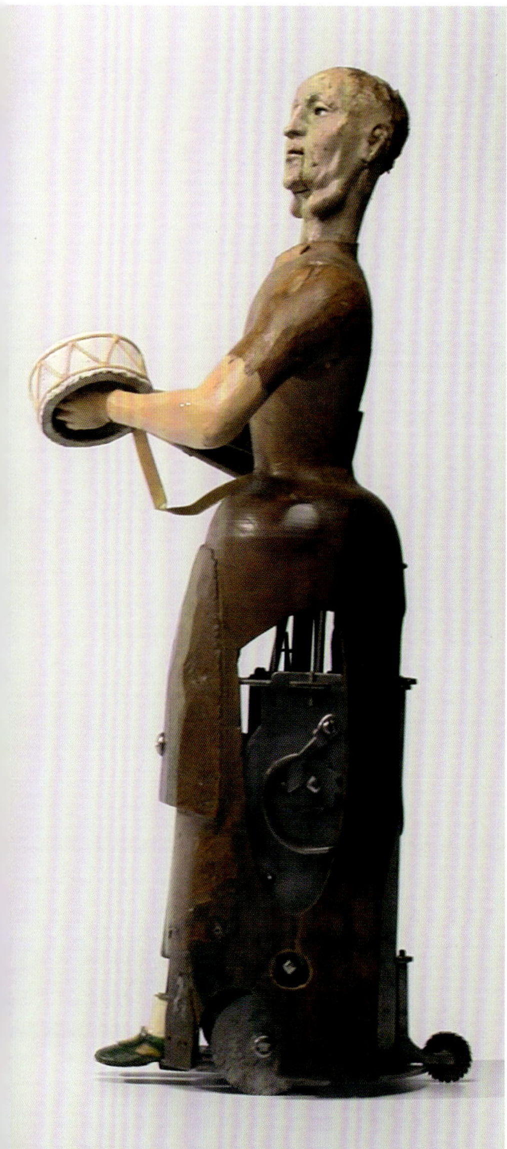

FIGURE 16.1.
Automa: Nobil matrona in atto di suonare un timpano (Automaton: Noble matron in the act of playing a drum), ca. 1560. Wood, iron clockwork. Height: 38 cm (15 in.). Private collection, Milan. Shown for the first time in 2005; restored by the Milan clockmaker Giorgio Gregato. A grande dame to be reckoned with. Remnants of a wig, fancy bonnet, and dress not shown.

Well into our work we were astonished to learn of the appearance of another automaton whose clockwork design and appearance locate it squarely in the family of early walking figures (**FIG. 16.1**). A lady musician, she represents an elderly court matron once playing a drum, we learn. On loan from a private collection, she had appeared in an ambitious 2005 exhibition in Trent celebrating the history of Italian horology, *La misura del tempo: L'antico splendore dell'orologeria italiana dal XV al XVIII secolo* (The measure of time: The antique splendor of Italian clockmaking from the fifteenth to the eighteenth century), curated by the historian Giuseppe Brusa. The exhibition catalog included essays by two dozen authors, among them Silvio Bedini, and it was Bedini who alerted us to the automaton.[1] In the catalog she is identified, "Automa: Nobil matrona in atto di suonare un timpano" (Automaton: Noble matron in the act of playing a drum). "Maker unknown, probably from Milan. Mid-17th century. Iron." We were thunderstruck. An iron-clockwork automaton like the monk and his brothers—and the photographs revealed what was clearly a movable mouth.

The carved wooden head is extraordinary: deep-set eyes, Roman nose, delicate cheekbones and ears, and a marionette-style movable chin with seams disguised by expressive wrinkles and pendulous jowls. This is no maiden lute player but a grande dame to be reckoned with. She is pictured without dress but adorned with beaded hoop earrings. Remnants of hair whisker her head. The wooden cap that forms the back of the head is not glued on, as in the other automata, but is attached with thread, laced through tiny holes behind the ears. The familiar lathe-turned body shell in wood is of a piece with the upper torso, the torso gracefully carved. Openings in the shell permit a glimpse of the iron clockwork within: mainspring, wooden fusee, pillars, cams, and elegantly detailed mainspring barrel clickwork, fully intact and startlingly similar to that of the Smithsonian monk. Even her height is close: 15 inches, to the monk's 16. The aperture in the wooden shell of her upper back is tapered and arched, like the handsome architecture of a clock casing. In the monk the same aperture (for access to the mechanism inside) is only a roughed-out hatch (see **FIG. 10.20**). Dainty feet step out from the body shell, carved and painted as buckled slippers. The catalog describes her performance: "The automaton walks, following a triangular path, the feet move, the head turns, the eyes roll and the mouth opens, the wrist of the right hand rotates as if to strike a tambourine [*tamburello*]."

The automaton was shown again in 2016, this time in an exhibition in Cremona, *Janello Torriani: Genio del Rinascimento*, celebrating Cremona's famous native son. Curated by Cristiano Zanetti and Cinzia Galli, sponsored in part by the city of Cremona and the Fundación Juanelo Turriano in Madrid, this show too was accompanied by an in-depth exhibition catalog, important for its new research by Zanetti and a host of distinguished authors on the life and legacy of Juanelo Turriano.[2] Here the date of the automaton is moved back to the second half of the sixteenth century, with a tentative attribution to Juanelo, now supported by Ambrosio de Morales's account of Juanelo's dancing lady. The catalog entry was written by Giorgio Gregato, Milanese horologist and conservator of antique clocks. Gregato declares that this automaton is the only specimen from the late Renaissance that corresponds in movement and appearance to Morales's report.[3]

Gregato, who also contributed the description of this figure in the 2005 catalog, further proposes that the automaton is mentioned in the inventory of the famous seventeenth-century collection of the Milanese nobleman Manfredo Settala.[4] Published in 1666, the inventory lists a musical ensemble of moving figures one of which is a "nobil Matrona in atto di suonare un Timpano."[5]

As we soon learned, Giorgio Gregato himself restored this automaton in 1989. Master clockmaker and expert in Renaissance clocks and scientific instruments, Gregato apprenticed with Luigi Pippa in the 1960s, and in 1972 he established his own restoration laboratory in Milan that he runs to this day.[6] He has conserved historic clocks for the Biblioteca Ambrosiana, the Museo nazionale della scienza e della tecnologia Leonardo da Vinci, the Castello Sforzesco, and the Museo del Risorgimento, among museums in Milan; Savoy clocks in the Royal Palace of Turin; and many other revered timepieces throughout northern Italy. A former expert for the auction house Finarte, he has also played coordinating roles in significant exhibitions of clocks, including the 2005 *La misura del tempo* in Trent, where the newly conserved automaton made its debut.

He welcomed Elizabeth to his Milan workshop on May 13, 2019, for a visit to see the lady. As would happen time and again, the encounter with the actual automaton brought a haptic jolt of misrecognition. Why is it that expectation, or memory, makes things larger than they are? In an instant, like Alice in Wonderland, the lady shrank to real size, in real space. Amazement at her small dimension is not unlike our wonder at the tininess of an infant's fingers but in this case an uneasy wonder, at a thing made long ago by human hands. The figure is twice aged: a portrait of a fierce old woman, the portrait itself is now fiercely old. (We hope she retains this double triumph, and no one dares repaint her face.) Fairly glaring at us, she held a makeshift drum in one hand, a short drum stick in the other (**FIGS. 16.2, 16.3**). Gregato set her on the immaculate floor of his showroom, wound the fusee, released the stopwork, and stood back. Instantly she took off, madly beating the drum, her head bobbing in

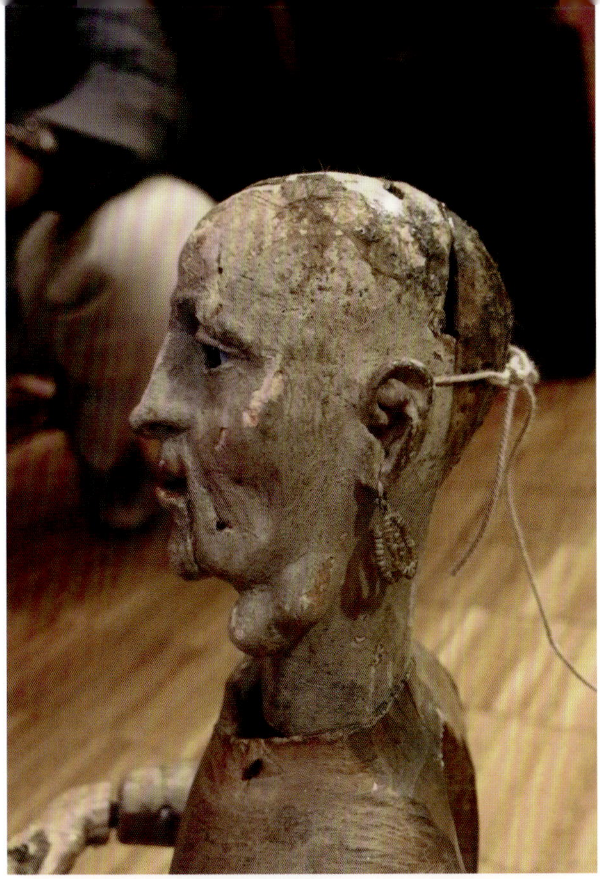

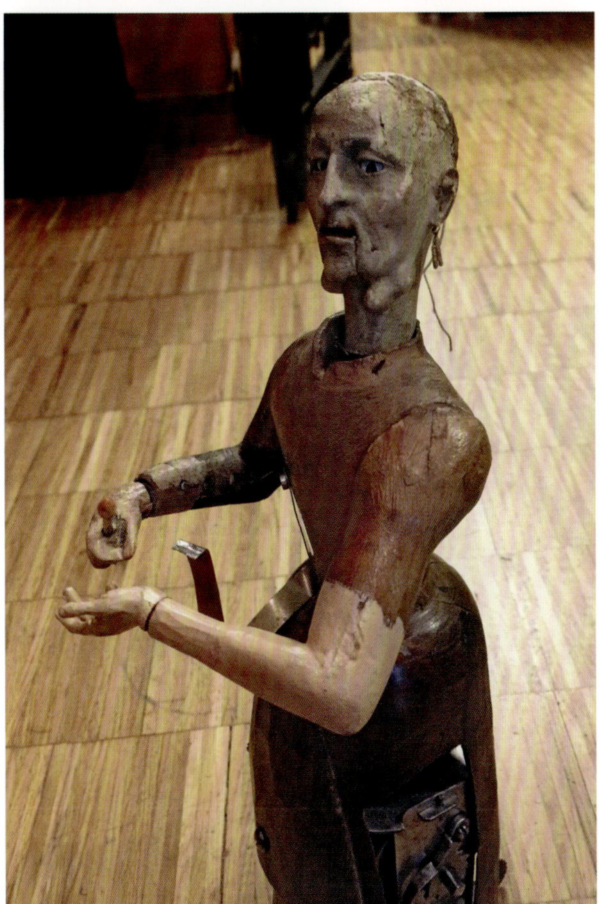

every direction, eyes rolling, mouth chomping like a piston. If the sight of her was shocking, the performance was more shocking still. The slow motion of the monk was a distant dream compared to this wide-awake creature. Seen from above, she charged back and forth on the floor in running steps and tight turns, Gregato pointing out that she tracks the path of a five-pointed star. Indeed, not a far cry from Ambrosio de Morales's description of Juanelo's automaton: "a lady more than one *tercia* high who, placed on a table, dances all over it to the sound of a drum which she meanwhile beats herself, and goes round in circles, returning to where she started."[7]

Gregato brought out photographs taken prior to his restoration that showed the clockwork intact but cemented under rust and corrosion. He had dismantled it down to arbors and pins, cleaned every surface, and carried out needed repairs. The wooden body was missing the left arm and hand, both feet, and large sections of the body shell. He engaged a colleague, a wood carver, to remake these; the original wood was identified as poplar. A provisional drum (made from a bottle cap) awaits further research. The automaton had arrived dressed in once-elaborate silks and petticoats now in tatters, with a bonnet of lace over real hair, possibly early eighteenth-century replacements. In her right hand she had held, not a drum stick, but a fan. Early video footage shows her feverishly fanning herself—perhaps more typical than drum pounding for a lady of that age (Elizabeth's opinion). Gregato identifies the sculpted swells at her neck as goiters, a story unto itself about art's imitation of life in this era. When she opens her mouth, a few teeth are now spied missing—but she has a tongue!

A video of the automaton in performance is now available on YouTube, with Gregato himself winding her up and illuminating her clockwork.[8]

FIGURES 16.2 and **16.3.**
The matron performing in Giorgio Gregato's shop, Milan, May 13, 2019.

17 The Category Challenge

Half of the family of eight automata considered here has appeared in the public record only within the past fifty years. Conservation carried out on the newcomers has shed new light on all of them, affording a better understanding of what is sixteenth-century craft and what is later repair, alteration, or copy. Clearly, a distinct class of early walking automata is just now coming into view. Their appearance puts new pressure on the search for archival references. Primary text sources—like the objects they describe—undergo histories of neglect and rediscovery. Uncertainty about makers and origins have deprived these automata of the reputation they deserve in the histories of art and technology. We argue that they are the first true androids: solo self-propelled at-large artificial beings in human form.

The very task of classifying them poses a category challenge. Current-day institutions, disciplines, and markets have named and sorted them in ways that have kept them apart. Housed in distant kinds of museums, or in separate departments of like museums, they suffer a mixed history of reclassification and taxonomy. Sometimes the host museums themselves change names and focus, as when the Smithsonian Institution in 1980 changed the name of its Museum of History and Technology to National Museum of American History. Where, then, did its un-American monk belong? Originally installed in the great Hall of Timekeeping as an example of early European clockwork technology, the monk was removed when this space was dismantled in 1997 to make way for a new exhibition, *On Time*, which examined timekeeping from an American perspective.[1] In 2008 the monk reappeared briefly as a side feature in an exhibition of robotic automobiles. (As we write, the monk is on view again at the museum in a fine show titled, *Discovery and Revelation: Religion, Science, and Making Sense of Things*, curated by Peter Manseau and Andrew Ali Aghapour.) The saint in the Iparművészeti Múzeum in Budapest is classified under the category "Decorative Arts." The Deustches Museum in Munich cared for its figure within the Department of Energy, Technology, Power Engineering, and Automation Technology, where it was exhibited in the context of the history of computers.[2] But a new home for this figure is being prepared as we write: the exhibition *Robotics* will open at the museum in 2022. The Conservatoire national des arts et métiers in Paris has addressed the problem in elegant style with its newly designated Théâtre des Automates, placing its mandolin player in the company of objects of mechanical theater and music. The history of religious icons could make its own claim to the ecclesiastical automata. Art history owes them allegiance as sculpture. The study of anatomy and morphology is still informed by models from the clockwork paradigm. The cittern player, acquired in 1934, was placed in the Kunsthistorisches Museum's Collection of Sculpture and Decorative Arts (Sammlung für Plastik und Kunstgewerbe), the collection renamed in 1990 the *Kunstkammer*, where treasures of every genre are gathered in the historical context of the great sixteenth- and seventeenth-century Habsburg art and wonder cabinets. Ancestors of the modern museum, *Kunstkammern* were repositories for as wide and mixed a compass of objects as possible; each princely collector categorized things in imperial if recondite order. An argument could still be made for classifying automata by their intended effect, as empirical *wonders*.

But it is no wonder now that so little opportunity has emerged for considering the walking figures as a group, and for comprehending them as a singular class. In this book we have helplessly defaulted to naming them by the museums or cities where fate brought them: a near abandonment of classification. We bring them together now.

Alfred Chapuis and Édouard Gélis give a clear definition of the word *android* in their 1928 encyclopedia, *Le monde des automates*:

> **An "android" is, in the broadest etymological sense, an "automaton of a human figure." However, the idea that this automaton moves by its own powers is the exact definition; we will adopt it therefore ourselves, adding that the particular androids that we will consider here are gifted with four clearly defined functions: they speak, they walk, they write or draw, or they play a musical instrument. A bell tower jaquemart [bell striker] we will not consider an android then, any more than, for example, the small figures we've seen moving in clocks or watches with automatons.[3]**

Chapuis and Gélis locate the makers of the first true androids in the sixteenth century, and they offer chapters on each of the four declared functions. Chapter 23, "Les automates et androïdes marchants" (Walking automata and androids), opens with the Vienna cittern and Paris mandolin players, the earliest walkers then known to them. Anticipating the protest that these two should be considered music-playing androids as well, the authors point out that the musicians only mime the play and that their instruments "make not the slightest sound."[4] One might argue that the act of walking too is only mime; the figures only appear to walk, their feet step forth in mock stride as they roll along on hidden wheels. Here, of course, is the ever-shifting distinction between the "real" and the "seeming" that characterizes all stages in the evolution of the android. From a theater of illusion to a scientific model of human movement, the increasingly sophisticated command of mechanics in the next two centuries led to the great works of Jacques de Vaucanson, the Jaquet-Droz family, Henri Maillardet, Abraham and David Roentgen and Peter Kinzing, and others, who made androids that indeed literally *do* what they appear to do: their fingers strike and sound notes on flutes and harpsichords, torsos heave in audible sighs, hands dip pens in bottles of ink and draw letters and pictures on paper, eyes follow the progress.[5] Such figures had their own advanced valence of pretense, however, best exemplified in the Jaquet-Droz writer, who pauses from time to time to look up and think (**FIG. 17.1**). Vaucanson's life-size duck was among the most famous: it fluttered its feathers, pecked and ate grains of corn, digested them, and

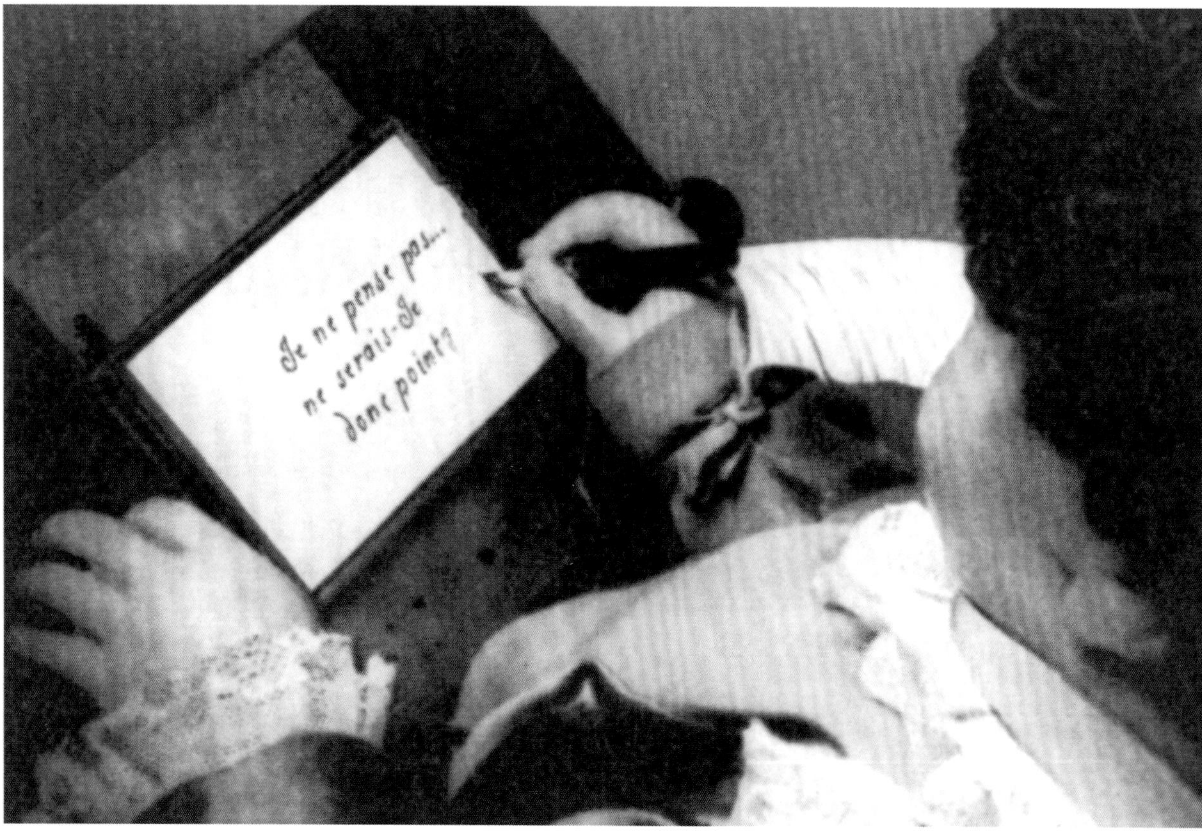

pooped. The duck—now known only through references and photographs (**FIG. 17.2**)—required some covert advance preparation for its performance, as Jessica Riskin relates in her essay "The Defecating Duck, or, the Ambiguous Origins of Artificial Life."[6]

If their wooden feet never quite touched the ground, the sixteenth-century walking automata were nonetheless ambulatory machines: peripatetic, processional, on the move. They were self-propelled, their means of motion, self-contained. "Self-contained" is key, because the size of the clockwork power source for all automata, then and later, demanded space. Cabinet-consoles acted as bases for automata, not unlike the sculptor's traditional solution to the problem of supporting an upright two-legged statue against gravity. For an automaton—a moving statue—the clockwork in the base was also the furnace in the cellar. Cellar space only expanded in the centuries to come as massive stacks of

FIGURE 17.1.
Pierre and Henri-Louis Jaquet-Droz, *L'Écrivain* (The writer), 1774, Musée d'art et d'histoire Neuchâtel, Switzerland. Photograph from Alfred Chapuis and Edmond Droz, *Les automates, figures artificielles d'hommes et d'animaux: Histoire et technique* (Neuchâtel: Éditions du Griffon, 1949), 404. In the theater of lifelike androids, how "real" is the performance? The boy dips the quill in the inkwell and pens a Cartesian riddle: "Je ne pense pas... ne serais-Je donc point?" (I do not think, am I therefore not?).

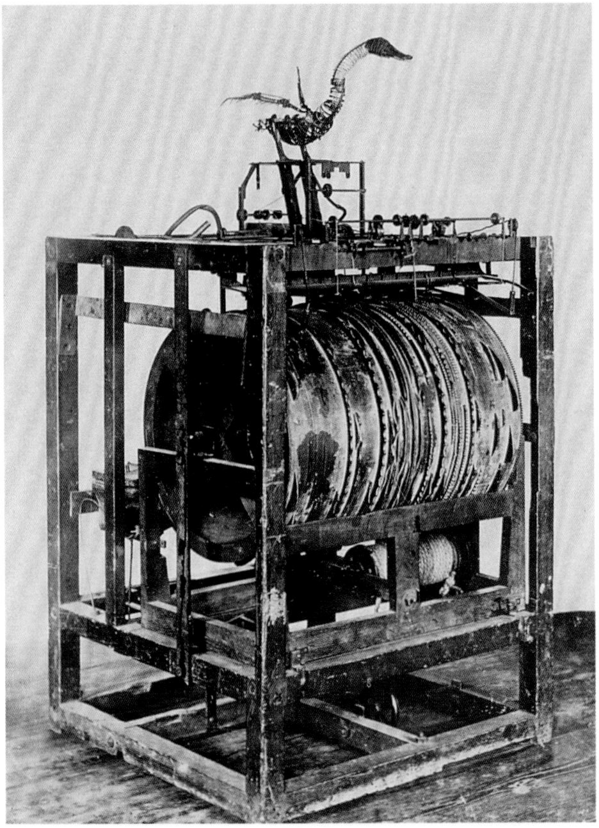

FIGURE 17.2.
A photograph said to show the ruins of Jacques de Vaucanson's legendary duck, ca. 1890s. The duck, first displayed in Paris in 1738, astounded witnesses for its visible as well as its invisible performance. It fluttered its gilded copper feathers, dabbled in a cup of water, quacked, stretched its neck to peck kernels of corn from the hand of its maker, swallowed the corn, and before long, produced a convincing excretion. Vaucanson declared his duck digested its food by chemical dissolution in its stomach, as in the real creature.

cams and linkages delivered an ever more lifelike repertoire of motion to the automaton perched above it (again, see the duck, **FIG. 17.2**). Of the best-known automata produced in the eighteenth century at the height of the clockwork paradigm, only the Jaquet-Droz writer and his brother the draftsman famously contain their mechanisms entirely within their bodies. As for the imitation of simple walking, the simulation of bipedal locomotion has been achieved only in the computer age, with on-board hardware to calculate weight transfer and balance. The later examples of walking automata in Chapuis and Gélis are primarily drawings and patent diagrams.

Thus in defining the android, Chapuis and Gélis ruled out bell tower automata, as well as small automata on figure clocks, for these gain their animation from external clockworks. The same distinction must hold for many of the celebrated composite automata made in the late Renaissance, the same period as the monk and the kindred walkers. As we mentioned in chapter 11, the showstoppers in the 1980 exhibition *The Clockwork Universe: German Clocks and Automata, 1550–1650* were a selection of marvelously ornate machines in the form of carriages, ships, towers, and processions, many bearing multiple animated figures. Larger pieces might harbor hidden mechanical organs comprising bellows and pipes, powered by more than one heavy mainspring, with separate gear trains. These were stationary machines, with figures moving on a stage or a platform—detailed settings extending the narrative while also providing space for the clockwork. A prime example in the show was the Trompeterautomat made in 1582 by the Augsburg master Hans Schlottheim (1544/47–1624/25): an ebony tower with ten small trumpeters and a drummer at the ramparts and an internal pipe organ, on loan from the Kunsthistorisches Museum.[7] But mobile constructions were produced as well, the pedestals themselves set on wheels to move the whole ensemble across a table: Diana astride a centaur, flanked by hounds; Minerva with satyrs, enthroned on a horse-drawn *triumphwagen*. Heads turn, eyes move, horses leap, the centaur shoots an arrow.[8] Mobile ensembles like these were intended for the banquet table, fashioned in dazzling silver and bronze, fire-gilded and set with gems. The materials themselves demonstrated wealth, but they also retained medieval associations with magic for many onlookers.[9] Some were designed for drinking games, popular

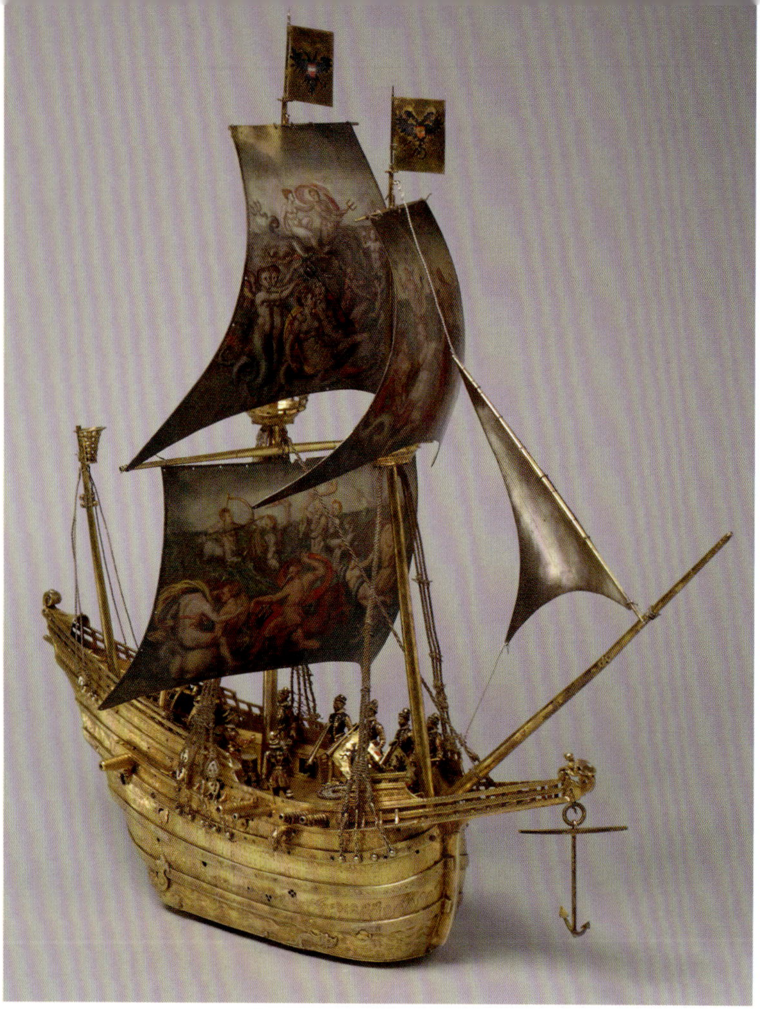

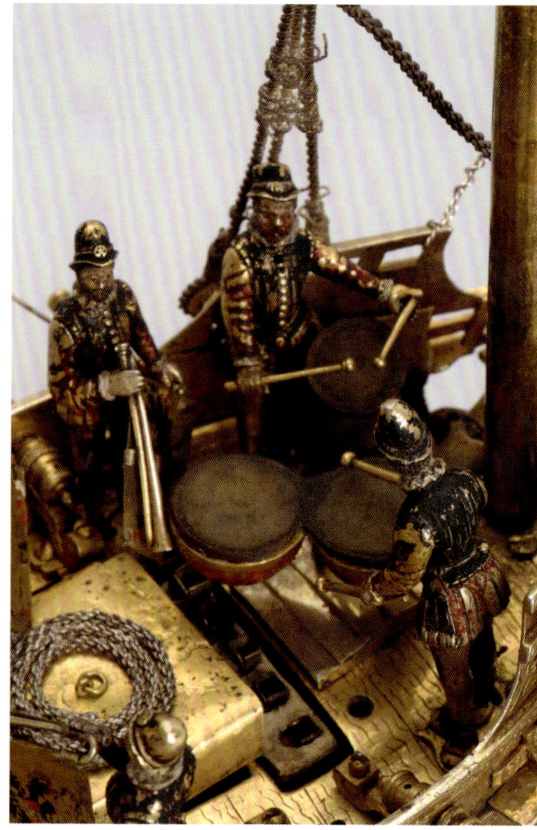

FIGURE 17.3A, B.
Hans Schlottheim, Automat in Form eines Schiffes (Automaton in the form of a ship), 1585, Augsburg. Gilded silver, copper alloy, enamel, oil paint; iron clockwork. 67 × 66 × 67 cm (26⅜ × 26 × 26⅜ in.). *Kunstkammer*, Kunsthistorisches Museum, Vienna. A mobile machine for an emperor's banquet table. Far from the material realism of the solo walkers, the gleaming ship is defined by its costly metalcraft and technical bravura, along with enough space in the hull for hidden clockworks to animate a spectacle of characters, music, and cannon fire.

in sixteenth- and seventeenth-century princely courts. The centaur's arrow would single out a guest who then had to contrive an erudite toast and empty his wineglass.

A sixteenth-century mobile automaton that dispenses with a base is the spectacular gilded silver nef, or table ship, also by Schlottheim, made for Rudolf II in 1585 (**FIG. 17.3A, B**). This marvel is kept in the Kunsthistorisches Museum *Kunstkammer*, displayed not far from the lady cittern player. Inscribed and dated, with the Habsburg double-headed eagle enameled on its banners, the 26-inch-long miniature galleon rolls along a table while multiple tiny musicians perform on deck, the figure of the emperor at the ship-of-state helm. Two internal pipe organs and a hide drum deliver the music. Discreet wheels advance the ship forward on the tabletop. Clockworks and organs are hidden within the ship's generous hull. The performance culminates in a firecracker burst from four of the ship's ten cannons, complete with flames and smoke. Guests must have leaped from their chairs.[10] With its clockwork components made of iron, not unsimilar in fabrication and date to those of the monk, the ship is an example of a mobile automaton with, among much else, simply a lot more room for a drive mechanism. And what a mechanism: three separate mainsprings, fusees, and gear trains, cued to operate in sequence.

The ship (and others by Schlottheim and peers) combined state-of-the-art technology and bravura metalsmithing to create a potent symbol of Habsburg seafaring imperialism.[11] Modest indeed are the solo walkers by comparison! But their own mechanical competence is just as sophisticated. What all the mobile machines share is a designed intent to interact in a direct way with spectators—on a table. At least the lute players too might have performed on festive occasions. Indeed, Adelheid von Herz included one additional walking figure in her thesis that we have not yet mentioned in this study—another cittern player—very much meant for the banquet table. This is a small early seventeenth-century figure, Joueuse de théorbe (in the collection of the Musée d'Horlogerie du Locle, Château des Monts, Le Locle, Switzerland) (**FIG. 17.4**).[12] Like the kindred cittern and lute players, she rolls along a tabletop, strums her instrument, and moves her head, tiny feet stepping in turn from beneath a bell-like skirt. But the full figure is entirely of cast bronze, fire-gilded to a high gleam. The dress is chased in a virtuoso imitation of brocade. Far from the polychrome realism of the wooden figures, this exquisitely fashioned bronze musician is more sculpture than android. Von Herz refers to her as an "unechte Androide" (an imitation of an android), suggesting that the clockwork was a later addition. Klaus Maurice, writing in Annette Beyer's 1983 volume, *Faszinierende Welt der Automaten* (Fascinating world of automata), describes her intended role: "The lute player was used to entertain guests at banquet tables. It walked down the table, and where it stopped, the chosen guest would have to deliver on a preordained pledge. Some of these machines were drinking cups filled with liqueur or intoxicating wine."[13] Here is an automaton that belongs with Diana and Minerva and the masterpieces of South German metalsmithing, this one meant to perform an entertaining roulette on a festive occasion. But here too is none other than a solo walker whose mechanism is hidden on board.

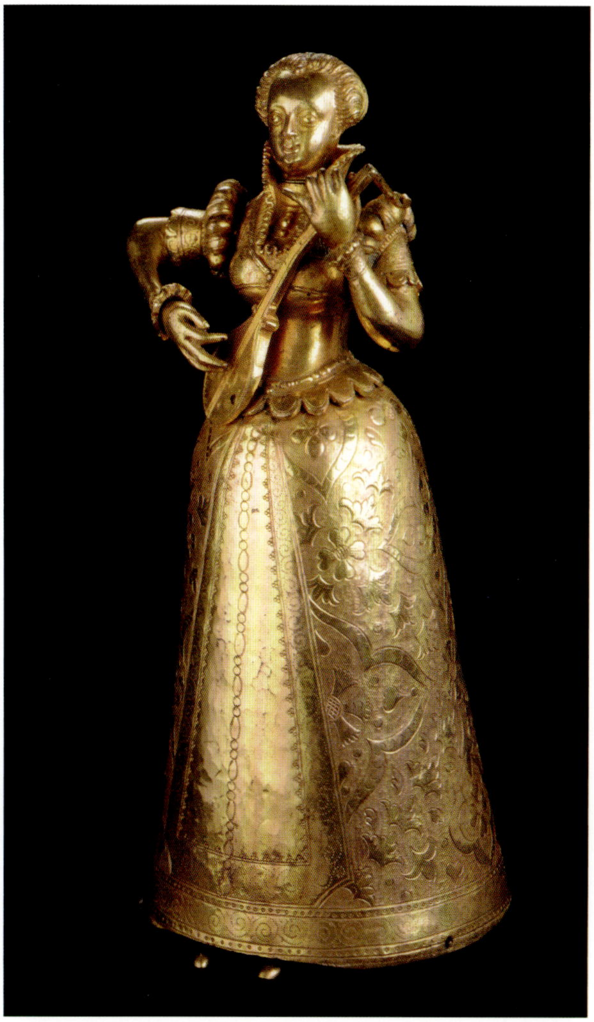

FIGURE 17.4.
Joueuse de théorbe, first half of the seventeenth century. South Germany (?). Gilded bronze and copper. Height: 25 cm (9 13/16 in.). Musée d'Horlogerie du Locle, Château des Monts, Le Locle, Switzerland. A solo walker with a lute or cittern (here called a theorbo), possibly once intended to contain a hidden drinking cup filled with wine. At the table where she stops, that guest had to discover the wine and drink without spilling. The body and dress are entirely of bronze; the maiden is more sculpture than android.

She advances on two perambulation wheels, with a third "steering" wheel, as von Herz calls it, that can be adjusted to set directional motion—very similar to the Vienna cittern player's adjustable trailing wheel. Beyer includes a photograph of the mechanism, a squat horizontal carriage, not much more than a mainspring to turn the two perambulation wheels, with a vertical rod to the figure's head. The clockwork also contains a bell. The figure is small, just under 10 inches in height.

The bronze Joueuse de théorbe recently traveled to the United States for inclusion in the stunning 2019 exhibition *Making Marvels: Science and Splendor at the Courts of Europe*, curated by Wolfram Koeppe and Ana Matisse Donefer-Hickie at the Metropolitan Museum of Art in New York. In the catalog for the exhibition, one of the contributors, Paulus Rainer, curator of the *Kunstkammer* in the Kunsthistorisches Museum, compares the figure to stationary objects called "maiden cups" that contained hidden reservoirs of wine that the appointed guest had to discover and drink.[14] Other examples in the show featured tabletop machines that combined the mobility of an automaton with a cleverly concealed vessel and drinking spout.[15] These were games of wits (and witnesses), with the stakes raised by the high-tech wizardry of machines whose apparent agency demanded a response in kind.[16] Could this curious late-Renaissance form of interactive clockwork have played a role in the origin of a much-adapted mechanical design that led to the first androids? The monk, on rare loan from the Smithsonian, was also featured in this wide-ranging exhibition at the Metropolitan Museum, taking part in the show's grand finale of automata.[17]

The walking automata were not unique, then, among machines made in the golden age of clockwork with enough spring power to move by themselves in locomotion. What singles them out and defines them as androids, following the ever useful definition of Chapuis and Gélis, is their autonomy as solo figures, machinery hidden on board, without recourse to exterior plinths or bases. Arrayed ensembles of human and animal figures mounted on pedestals or ships or carriages, while they may perform a more distributed repertoire of motions, adding sounds and music, more closely resemble miniature theaters. The individual android figures, in contrast, are solo actors that step across the boundary between theater and life.

Further, those in our study are fashioned of materials that are as close as art could muster to imitating the matter of the body. No fire gilding, gold, or gems mark them first as costly objects of royal prestige. They bear flesh of carved and polychromed wood in the tradition of trompe l'oeil sculpture, a tradition we say more about in chapter 25. They wear robes of real cloth that slip and crease over moving limbs. The female figures, coiffed with real hair, strum tiny lutes or citterns made of wood, pegs, and strings. Such materials may be humble, but they win their virtuosity by their mimetic realism: of a piece with the lifelike *primum mobile* of the clockwork inside them. The Greek mathematician and engineer Hero of Alexandria said an automaton had to be small enough so that no person could hide inside it to operate it.[18] Only the small size of the figures—their weight not more than a contained mainspring could power—contradicts their realism.

The Gondelautomat

One additional automaton of a boat must be mentioned, bringing us closer to the walkers. This is the Gondelautomat currently kept in the Ambras Castle in Innsbruck, site of the original *Kunstkammer* of Archduke Ferdinand II of Austria, nephew of Charles V.[19] A miniature gondola made of gilded bronze carries a gentleman and a lady, seated under a canopy, while a standing gondolier plies his oar at stern (**FIG. 17.5A, B**). The figures' heads and limbs are made of wood and polychromed; they are dressed in real cloth in late sixteenth-century style. The boat, running on hidden wheels, travels a distance of about 40 inches, then changes direction with the gondolier's stroke, to navigate a squared course on a tabletop. The gentleman turns his head and raises an arm while the lady strums what could be a lute or a cittern. The lady herself looks surprisingly familiar. High forehead under formal headdress, face brightened in delicate enamel colors, graceful hands in lacey cuffs: the Vienna cittern player in a Venetian setting? She strums the instrument with her right hand, as do all the walking lute players. The dimensions of the machine—14 9/16 inches in height, 28 5/16 inches in length—suggest that if the lady disembarked, she would stand close to the height of those lute players. Picturing the layout of the clockworks inside the walking androids, we wonder, could this automaton not employ the same design? A

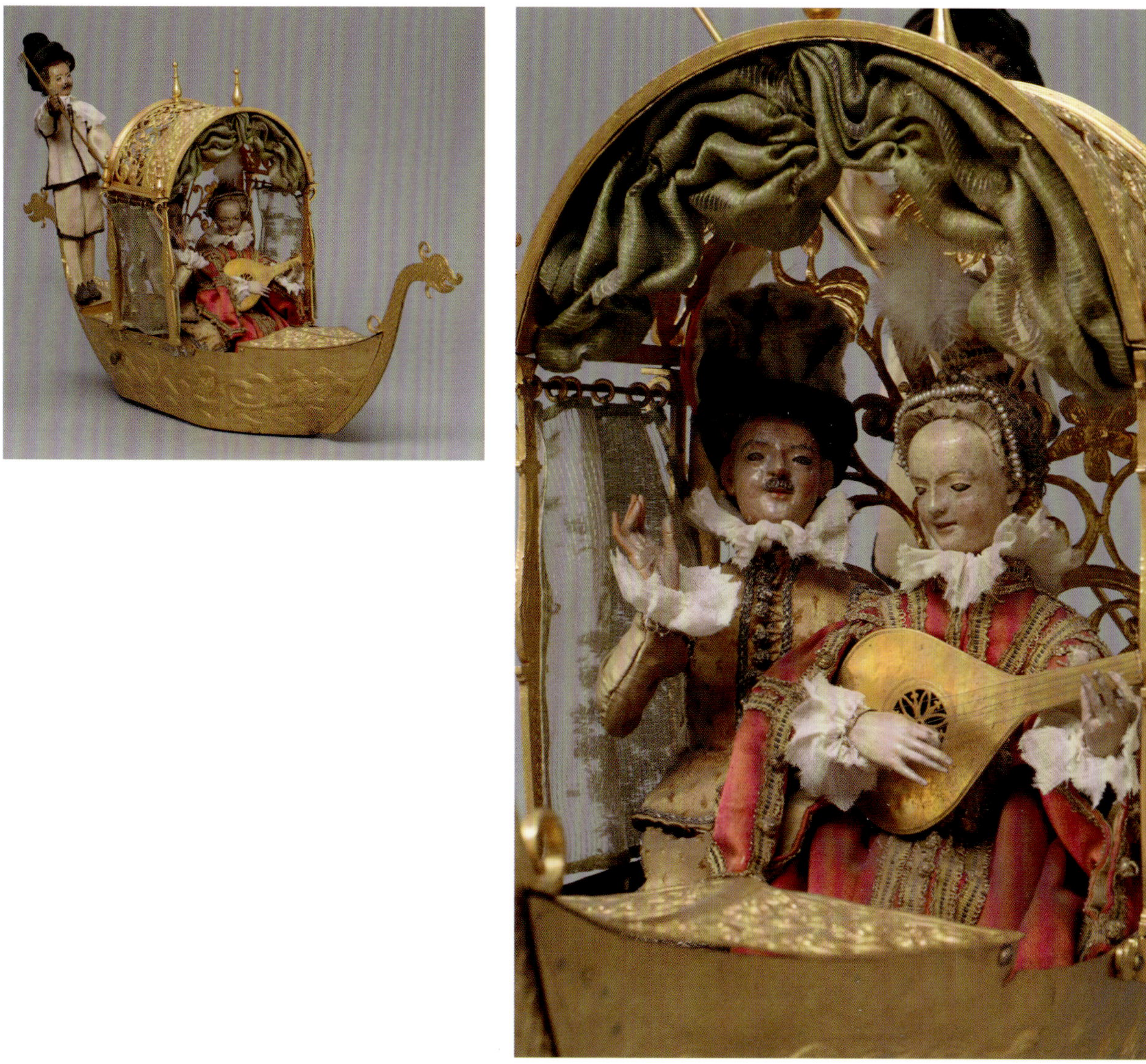

FIGURE 17.5A, B.
Gondelautomat, ca. 1600. Augsburg. Gilded brass, painted wood, linen and silk garments. 37 × 72 cm (14 9/16 × 28 5/16 in.). Sammlungen Schloss Ambras, Kunsthistorisches Museum, Vienna. Three moving figures in a gondola that plies the tabletop on hidden wheels The lady strums a lute or cittern. Could the internal clockwork be similar to that of the solo walkers?

three-wheeled rolling frame, now hidden by the gondola, which itself provides room for a few extra linkages and perhaps a larger mainspring. This can only be surmise, for we have not succeeded in learning more about the hidden works. But the gondola may conceivably be nothing more than a large "skirt" attached to one primary figure—the one wearing the full dress—with added links to the moving arms of the other two figures. The museum gives the piece a date of around 1600 and notes that the initials "J. F." appear inside the boat, thought to be those of the celebrated Augsburg goldsmith Joachim Friess (ca. 1579–1620). Future study is in order, to understand the potential relationship between this automaton and the walking figures, particularly as this machine is signed and connected to a well-known master.

Robed Figures

A final defining compromise the walking androids share: they are all *robed* figures. Those robes conceal the clockwork mechanism that produces their mobility.[20] Only persons properly wearing floor-length robes were represented among these automata! Is this why only monks and ladies constitute our corps? Period styles for male dress across the classes called for leggings or hose, waist-length shirts and doublets, and hip-length tunics. Such form-fitting dress would all but reveal the conceit of the imitation of walking. Tunics, cloaks, capacious capes, and voluminous skirts were essential for the walkers. Costume itself announces the final category challenge, between secular and religious spheres.

The cittern and lute players might well have found roles among the more elaborate machines deployed in festive settings: secular themes for affairs of state.[21] They may have also found their way to more intimate social settings or, less costly to produce, to modest estates vying for machines of their own. The three devout brothers, on the other hand, cued to the church, demand a different stage. Any attempt to envision how or where they were deployed conjures a separate mise-en-scène, perhaps no less theatrical but with different rules of decorum and propriety. The admonishing monk in particular: In what social setting would a clockwork performance of prayer have been appropriate or even acceptable, against a backdrop of Reformation politics?

But the church too had an ongoing history of theatrical mechanics in the service of conveying the drama of the Christian story. Processions, feast days, and sermons across medieval Europe featured reenactments of biblical episodes involving actors and apparatus. The very word *clock* derives from the Latin word for "bell," *cloca*, associated with the bell-striking mechanisms that accompanied early water clocks in monasteries, signaling canonical calls to prayer.[22] By the fourteenth century, weight-driven mechanical tower clocks were essential structures in cathedrals and wealthy monasteries, made possible by the invention of the *escapement*, a mechanism to intercept the motion of a falling weight and release its momentum in regular ticking intervals.[23] The escapement, combined with wheelwork and a striking mechanism, relieved the tower warden—guardian of the belfry—of his role ringing the bells with ropes and pulleys and gave rise to the origins of the *jacquemart*: a bell-side mechanical figure that springs into motion with a hammer to sound the hour.[24] "In the course of time," Silvio Bedini writes, "the jacquemart developed a family of his own."[25] Smaller bells sounding the half and quarter hours gained additional automata, and soon scores of biblical figures appeared on the faces of towers, and on clocks within the church, animated by bell-striking gear trains. Nativities, episodes of the Passion, resurrections, and ascensions were set in motion. Every history of the automaton cites the mid-fourteenth-century monumental clock in the Strasbourg Cathedral, a tower unto itself inside the cathedral's south transept, some 65 feet in height. E. R. Truitt describes it best, in the final chapter of her book, *Medieval Robots: Mechanism, Magic, Nature, and Art*:

> **Built between 1352 and 1354, [the clock] symbolized microcosm and macrocosm, demonstrated the glory of God's creation, and dramatized the timeline of salvation. Figures of the Blessed Virgin, the newborn Jesus, and the three Magi gave the clock its nick-name: the "Horloge des Trois Mages" or "Die Dreikönigsuhr." The clock also had a mechanical astrolabe, a perpetual calendar, a carillon, tables that indicated proper times for phlebotomy, and a mechanical rooster that flapped its wings and crowed every hour. The perpetual calendar**

> and mechanical astrolabe reflected the divinely ordered macrocosm. The tables for bloodletting indicated the harmonies between the macrocosm and the microcosm. The characters in the hourly pageant (Virgin, infant Jesus, Magi, and rooster) enacted a moving memorial to Jesus's sacrifice for humankind. The rooster symbolizes Peter's denial of Jesus and his repentance, as well as papal vigilance, as Peter later became head of the Church. The rooster's morning cry reminds Christians to be prepared to accept the sudden second coming of Jesus, along with the resurrection of the dead and the final judgement.
>
> ...The clock depicts cosmological, Christological, biological, diurnal, and horological time.[26]

If Chapuis and Gélis, in defining the android, ruled out bell tower automata, nonetheless automata on bell towers were the android's immediate forebears. Clockwork and "jackwork"—what the physicist and science historian Derek J. de Solla Price has called biological automata—arrived together.[27] Courtesy of the church, as Chapuis and Gélis attest, in several chapters on the rise of mechanism to animate and reify sacred narrative. "The Catholic Church, which has always excelled in the art of staging, left no stone unturned in attracting and retaining the faithful," they wrote.[28] We return to the monk and the church in part 5.

18 The Earliest Free-Walking Androids

Who made these extraordinary machines, extending the power of the clock into moving models of mythic creatures, warships, nativities, zodiac heavens—and small walking androids? Some were signed by now-famous clockmakers or goldsmiths; some carry trade guild marks; others are recognized from descriptions in imperial inventories. Surviving contracts and correspondence track yet others. Many more, including major works, remain of uncertain origin. No signatures or stamps are found in any of the eight walking androids, and no later restorers left identifying marks.

For all the variations in their character, their presumed audience, and the dates estimated for their manufacture—and amid the widely different forms of attention paid them today—the eight androids taken together offer a group portrait that widens our view of each one and demands notice in itself. Setting aside for the moment the woodcarvers' styles, the application of paint, the manner of dress, and the duet between sculpture and mechanics, we see that seven of the figures harbor an internal clockwork that follows a cognate design. Only the cittern player in Vienna differs in mechanical layout but for comparable function: a gesticulating figure on a self-propelled three-wheeled rolling chassis capable of directional change, two carved wooden feet stepping forth from the hem of a fabric robe, powered by a hidden mainspring, fusee, gear train, and an anatomy of distributed levers, cams, and return springs.

As we assembled the table of comparisons (see appendix), additional affinities appeared. Of the six androids that perform a series of turns in the direction they walk, all accomplish the turn in the same way. As the two perambulation wheels (hidden under the costume robe) drive the figure forward, feet stepping forth all the while from the robe's hem, a cam on the fusee arbor periodically pushes a vertical pivot to the table near one wheel and temporarily lifts the wheel a few millimeters off the surface of the table. This allows the other wheel to roll in an arc around the pivot. The figure continues to "walk" its feet in imitation of a decision to turn. The rear trailing wheel maintains stability. The length of the pivot and the shape of the cam that depresses it determine how long the lifted wheel stays up, and thus how acute an angle the turn will be. The Smithsonian monk and the Milan matron, though they walk at very different speeds (the monk slow, the matron fast), both execute turns at angles greater than 90 degrees, resulting in sharp changes of direction. The clockmaker need only calculate the dimensions of the cam and pivot, simple enough variables in a well-tested design for navigational mobility. When they turn, all six androids turn to their right.[1]

We noticed that all the lute players, and the cittern player, strum with their right hands, fingers facing down, and fret with their left hands, fingers facing up (which also allows the left hand to support the neck of the instrument in the crook between thumb and fingers). The saint in Budapest also possesses this orientation of hands. Further, those right hands all move rapidly, whether strumming a lute, ringing a bell, or rapping a drum, and here the Deutsches Museum monk and Milan matron join the group. Thus not only is the gear train layout similar, but in the case of a reciprocating right wrist, seven of the eight automata possess a similar offset crank and linkage, with transverse connections through a wooden shoulder and elbow to the wrist. Only the Smithsonian monk differs: his right arm moves not at the wrist but at the elbow, and only intermittently, to beat his chest. Yet this too is a back-and-forth reciprocating motion, generated from an offset crank at the far end of an arbor in the gear train. The speed of the motion depends on which arbor powers the crank: the slow fusee arbor or the faster-turning second or third arbors in the upper steps of the gear train. Left arms too were animated by offset cranks in various applications, though several of the androids have fixed or nonmoving left arms (the saint, the matron, and the cittern player).

Whether still functional today, seven of the eight androids could move their eyes. Of these, five also move(d) their mouths. Those whose heads have been opened or X-rayed (monks and matron) share a similar arrangement of levers and return springs. The method for mounting the tiny machinery behind the wooden face, and coordinating it with a hinged wooden jaw, is also comparable. The presence of a movable mouth required more space for the hardware inside the head, more adaptation in the clockwork below, and considerably more strength from wood and metal. The wire or chain linkage from gear train to mouth, passing through the neck, subjected the head and neck to more stress, to be withstood without loss of the neck's separate capability to turn and nod. Monks, saint, and matron all possess this sturdy articulation of the mouth, neck, and head—still intact and animate today. Eyes and eye-turning mechanisms have stood the test of time as well.

Significantly, these figures all possess iron, not brass, internal clockworks. Just this difference—between machines made of iron and those made of brass—allows horologists some confidence in establishing dates for unsigned clocks and automata. Here a telling difference is found among the eight androids. We have noted that the clockworks inside five of them are made of iron: the four figures named above with moving mouths and the Vienna cittern player. The remaining three, all lute or mandolin players, are of brass. The distinction is important, for the history of metallurgy helps us with chronology.

Artisans in Europe before 1600 employed brass primarily as a decorative medium, or in rare clock movements designed to be exposed to view.[2] By century's end, increased trade access to mined copper and to calamine (a mineral containing what we now know is zinc ore), along with advancements in alloying, soon made brass cheaper and more available.[3] Softer than iron and thus more easily cut and shaped with iron tools, brass was first prized for its resistance to corrosion and its similarity in color to gold. Bronze was darker in color and brittle under the hammer.

But like bronze, brass could be melted and cast into plate, replacing the labor of hammering iron ingots into plate on the anvil.[4] Clockmakers further discovered during the last quarter of the sixteenth century that brass offered a significant reduction of wear when paired with a dissimilar metal at points of sliding friction, as between mating teeth in a gear train. Brass wheels driving iron pinions suffer so little wear that such clocks have kept accurate time for hundreds of years.[5] All three of the brass-works automata boast the duet in enough key locations to likely be original (not later repairs), including brass/iron pairings among other moving parts that bear on one another—levers, cranks, hinges, and cams transmitting motion to the automata's heads and limbs. The fusee too would soon be made of brass rather than wood, as seen in later clocks, and the CNAM mandolin player contains just such a brass fusee, unique in the family of eight. Because their clockworks are primarily of brass, and exhibit innovations in metallurgy, the three brass-works musicians were likely made a generation or more after the automata with iron movements—the CNAM figure much later still. David recalls Klaus Maurice suggesting a date of roughly 1620 for the BNM and Smithsonian lute players. They are currently dated at around 1600 (BNM) and 1640–60 (Smithsonian). As for the monk and the brother and sister androids whose movements are entirely of iron: this is the first signal that enables horologists to assign them to earlier dates.

A second signal comes from the iron itself: clues on the surfaces of individual parts of the clockwork indicate how the metal was smithed and tooled. As we have noted (see chaps. 8 and 11), the labor involved in the "free-form" cutting of clockwheel teeth leaves a signature of its own on the metal, indicating the tools and methods employed by the maker. David, who has used period tools across his career and is known among colleagues at the Smithsonian for being adept at "reading" the metal, estimates that a skilled clockmaker or journeyman in the late Renaissance required two days or more to make a single clockwheel. Hammering out the plate, cutting the disc with a handsaw, truing it in a foot- or hand-operated lathe, indexing it with a dividing plate to mark equidistant teeth around the perimeter, and cutting each tooth with saw and file: David himself deliberately followed most of this sequence to make each of the seven steel wheels and pinions for his model of the monk. "The final task is the most demanding," he has written. "The upper half of both sides of each tooth must be filed to an epicycloidal curve, a shape similar to a bishop's miter, with its point directly above the center of the tooth. This will ensure the correct profile for reducing friction between the teeth of the mating wheels. Absolute concentration is imperative: one imperfectly filed tooth will ruin the wheel and halt the entire gear train" (**FIG. 18.1**).[6] Every step of this process leaves telltale marks. The wheel teeth in particular, seen under a magnifying glass, reveal slight differences one from the next, particularly in the gutters between teeth. Layout marks were often left visible, as in **FIGURE 12.6A**. Moreover, there were no standardized tools of measure in this period; each shop often made its own instruments for measuring out the dimensions of mating parts so essential in clockmaking.

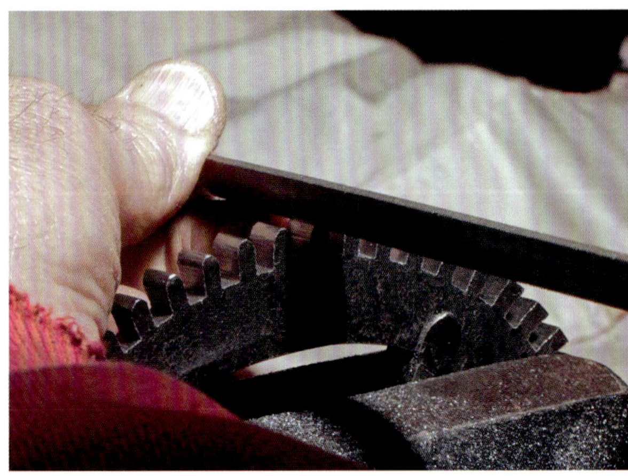

FIGURE 18.1.
David Todd files the "bishop's miter" on one tooth of a clockwheel. Though much larger than a wheel in the monk's gear train, this wheel requires the same method and skill to make. Between 2010 and 2012 David fabricated a weight-driven alarm clock patterned on sixteenth-century German technology, using only the methods of the age. Working at a coal-fired blacksmith's forge in Heathsville, Virginia, he hammered out his wheel blanks, plates, and bars from scraps of mild steel, shaped them on the swage block and anvil, measured and cut them with hand tools, and heat-riveted, mortised, and pinned parts together following early tradition. The process gave him a profound appreciation for the finesse and command of the sixteenth-century craftsman.

As David has found throughout his experience with clock conservation, it is extremely difficult to mimic the methods of the Renaissance clockmaker. The iron clocks and figure clocks that show such characteristics, and were signed, help confirm dates for those that weren't.

The small scale of these clockwheels, we should add, must be seen to be appreciated. **FIGURE 8.4** shows one of the rubbings David made of parts of the monk's wheel train: the third wheel, printed at actual size. There are sixty teeth in this wheel. Imagine cutting each tooth by hand and filing it to shape.

With the introduction of the wheel-cutting engine in the seventeenth century, many such wheels could be indexed and cut in a day's work. Juanelo Turriano himself is said to have invented a prototype engine, in order to make his great astrarium.[7] David had his colleague Stan Nelson at the museum sketch an early such engine (**FIG. 18.2**). Yet even the early wheel-cutting engines were hand-operated tools, adapted from principles of the lathe combined with the dividing plate. Power from a motor, so instinctive to us today with any tool called an "engine," would originate only with industrial production in the distant age of steam. But an engine-cut wheel can show more uniformity in the profiles of wheel teeth. David finds that the clockwheels in the three automata he has personally handled were cut freehand one by one, tooth by tooth. He estimates that this is the case for the others as well, including even the later CNAM mandolin player.

Metallurgy, process, and a "period eye" for both thus leads historians and conservators to assign most clockworks made entirely of iron to points across the sixteenth century or earlier. Those made of brass appear in the closing years of the century and begin to gain precedence in the first quarter of the seventeenth century. The androids follow suit. As more knowledge of these figures comes to light, further refinement of dates will undoubtedly continue. Comparisons

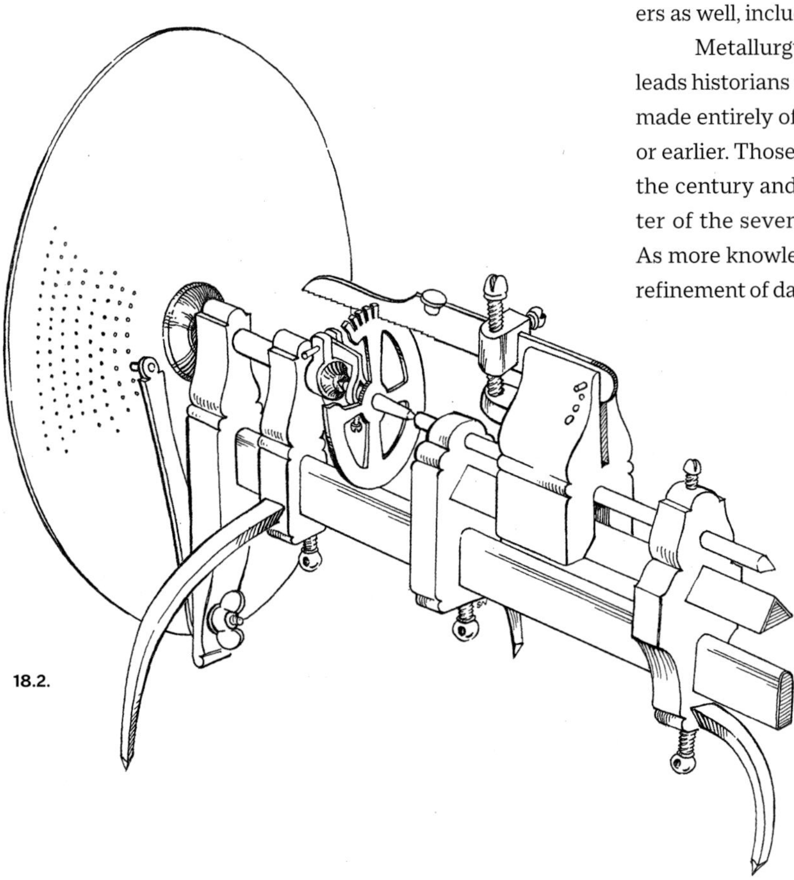

18.2.

FIGURE 18.2.
An early wheel-cutting engine might have looked like this: a lathe combined with a dividing plate and compass. The clockwheel "blank," already riveted to its arbor, is secured between centers. For each tooth, the wheel would be locked in place, and the saw blade worked by hand to make the cut. The wheel would then be rotated a fraction of a turn to its next position, based on direct calibration from the dividing plate—the disc with rings of measured holes—on the shaft at the lathe head. A compass-like arm set hole to hole prescribes each advance of the wheel, for a precise number of equidistant teeth around the wheel perimeter. Drawing by Stan Nelson, specialist in graphic arts, National Museum of American History (NMAH), from David Todd's sketches.

just among the three brass-works players reveal categorical exceptions.[8] The fact that the BNM lute player once had a movable mouth (and eyes), like the early iron-works figures, makes this a transitional machine among the eight. The outlier in the group is the CNAM mandolin player, pinned to the late eighteenth century, a date that radically extends the life span of this clockwork design. So good was the design that we see it put to use across generations, and this makes the search suddenly interesting for a prototype, a first appearance, and an evolution. At the very least, we can say that the brass mechanisms in the three musicians were derived from the earlier iron examples. The surviving wheel work in the Paris figure shows the kinds of no-frills simplification that may point to later small-batch shop production.

With so many materials and crafts combined in one object—metal, wood, cloth—the eight automata harbor multiple genealogies. Carved wood—the androids' flesh and form—gives fewer material clues for assigning dates. But in some cases the species of wood, if it can be named, allows proposals for the geographic origins for an automaton. Peter Friess made a point of determining this for the Deutsches Museum figure, identifying body, head, and hands as made of poplar—a splinter sample of which he had microscopically analyzed. European poplar grew abundantly in both Germany and Spain, and it was frequently used by sculptors in both regions. "Had linden wood been used," he wrote, "a German origin would have been more likely."[9] Microscopic sample analysis is available only for this figure and for the BNM lute player, whose remnant body shell has been identified as mountain ash (see chap. 14 n. 4). Head, hands, feet, body shell, and fusee: Are all made from the same wood in any of these figures? Michael Baxandall, in *The Limewood Sculptors of Renaissance Germany*, gives a firm geography of linden, its microscopic signature, and a palpable idea of its handling and value for the German sculptor.[10] (Linden and limewood are considered interchangeable names for the same Old World species, once abundant on German soil.) Would that we had so elegant a portrait of poplar—or of any other wood in the history of sculpture. Poplar is the wood that Smithsonian wood conservators initially indicated for the carved head and limbs of the monk, based on visual inspection, though since then other species have been considered, most recently beech. By visual inspection too the Milan matron is reported as poplar; the Iparművészeti Múzeum saint, linden. But only the analysis of cell structure under the microscope yields sure identification. Today it can be performed on a wood sample as small as a breadcrumb. Our hope is that museums will pursue this analysis in the future, for this would add an important piece to the puzzle of the origins of each automaton. Of course, wood itself, like many raw materials including metals and ores, was traded and shipped across continents.[11] Yet if we should learn that the Smithsonian monk is made of linden, not poplar, this would make Spain a little less plausible as this figure's native sphere.

Where indeed were these androids made? Their differences call not just for a chronology but for locations and makers. One possibility is that the monks, the saint, and the sister musicians emanated from the same locality, over an extended period of time. It is not unreasonable to hypothesize a single company of clockmakers with a proprietary mechanism, altering its design as demand shifted and as brass came to replace iron in shops across Europe. A likelier scenario singles out one or two figures as prototypes, later copied and adapted by other makers. Elizabeth's opportunity to view the iron clockworks of the Deutsches Museum monk and the Milan matron in particular and compare them to the Smithsonian monk was a chance to see the elegance and pride evident in all three machines. As David believes, their embellished hidden mechanisms suggest a pursuit of beauty for its own sake. Brackets, clicks, pillars, and other components were gracefully shaped, chamfered, and decoratively nicked. Much of this is absent in the later brass clockworks, whose makers streamlined a prior clockwork design to less complicated effect, shifting priority to the automaton's outward appearance, character, and dress.

A third proposal has the clockwork mechanism produced in one shop and exported to another for development. This might account for the wide variation across the eight androids in size, material, date, performance, and character. The Budapest saint's mechanism, for example, distinctive for its stationary pose, yet with a baseplate showing cutouts for perambulation wheels: Did a second shop adapt it and replace the walking mechanism with the glockenspiel?

Once again, the Vienna cittern player, with its unique gear train layout, stands alone. For the others, it is possible that all these scenarios were in play. So coherent was the clockwork design that something very close to it appeared in Japan in the eighteenth century in the form of exquisite tabletop tea-serving automata made almost entirely of wood (**FIG. 18.3**).[12]

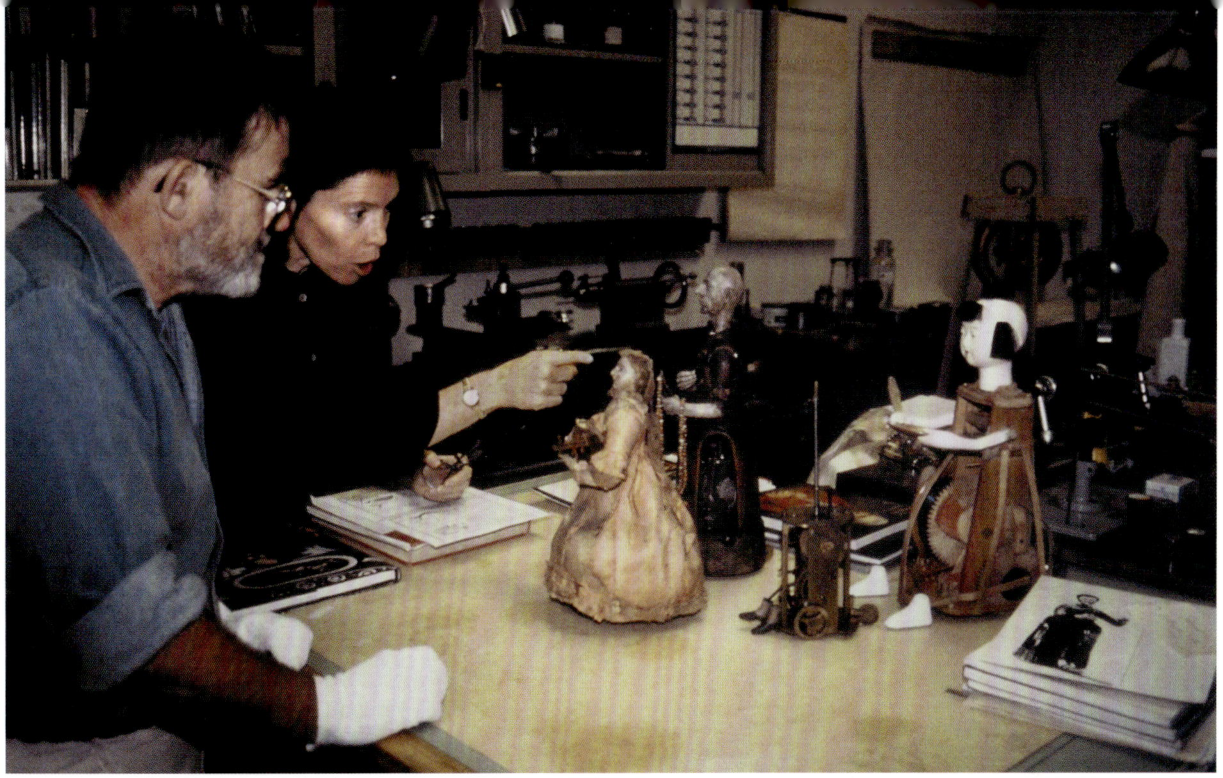

Clockwork-driven automata of all kinds—single androids, animated figures on clock striking trains, and complex composite machines—shared iterations of a mainspring, fusee, gear train, cams, and levers to effect animation. Clocks themselves, and the models of the movements of moon, sun, and stars that gave rise to clocks, drove the technology. But this was also a moment of tremendous vitality in art's capture of the moving human form in sculpture. Amid these wonders, a small doll-like figure that walks by itself may have been a humble achievement. But as the great historian George Kubler wrote in his book *The Shape of Time*, many prime objects appear first as mere mutations, only later to be recognized as the first links of a paradigmatic chain. "In other words," he wrote, "when people create new forms, they commit posterity at some remote interval to continue in the track by an involuntary act of command, mediated by works of art and only by them. Here is without doubt one of the most significant of all the mechanisms of cultural continuity, when the visible work of an extinct generation can still issue such powerful stimuli."[13] Kubler even demonstrates how two distinct crafts, each in play at the same moment in history, can combine to create a new class of forms.[14] A small wooden sculpture with a hidden clockwork was once crafted to move as if with a mind of its own. The philosophy of mind and body in the next centuries took notice. Today we are obsessed—haunted—by robots.

Modern labels identifying the eight automata still give them no common genealogy. Horology's written record fails to chart them. Period scribes looked elsewhere for the defining technologies of their time. But the shared blueprint is unmistakable. The collaboration between clockmaker and sculptor yielded a new kind of object, the first of its kind.

Among the costly composite tabletop automata mentioned in the previous chapter—mobile theaters, we called them—we know of at least one that was equipped with a probe, beneath the base and forward of the wheels, that could detect the table's edge and halt the machine to prevent it from falling.[15] Other such masterpieces surely included this protection. Not so, the walkers, alas. We are lucky that a few of them steered clear of the edges of their world.

FIGURE 18.3.
David Todd's NMAH lab at the Smithsonian, September 1999. David and Elizabeth look at the lute player (whose internal clockwork has been removed and set out for study), the monk, and a replica of an eighteenth-century Edo-period Japanese tea-serving automaton. Murakami Kazuo, expert in Japanese automata, brought the tea-serving *karakuri ningyō* to the United States and took this picture to compare it to its European forebears. The replica was made in Osaka in the early 1990s by Minesaki Sougo, renowned restorer of early Japanese automata.

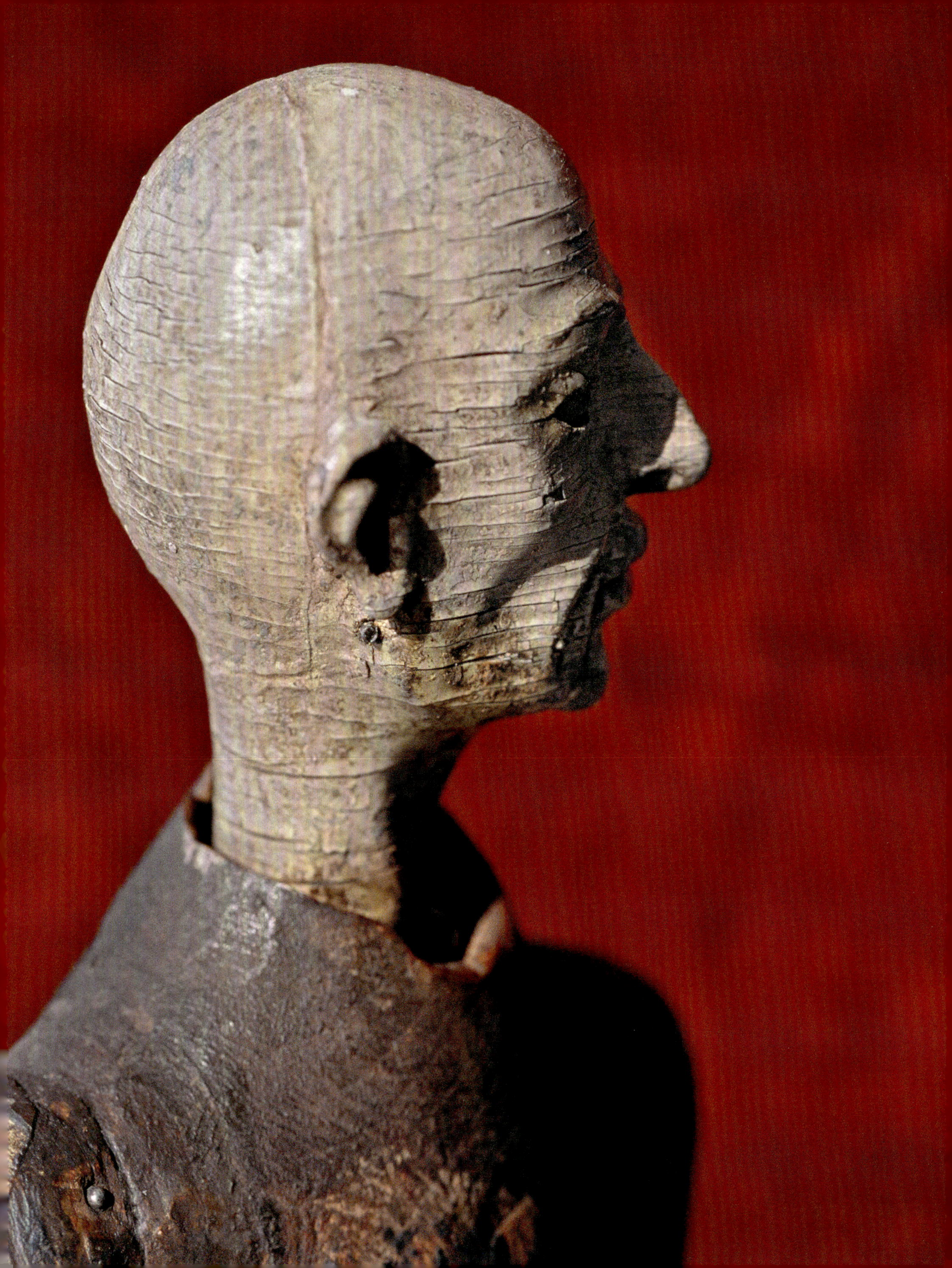

PART IV

A Search for Origins

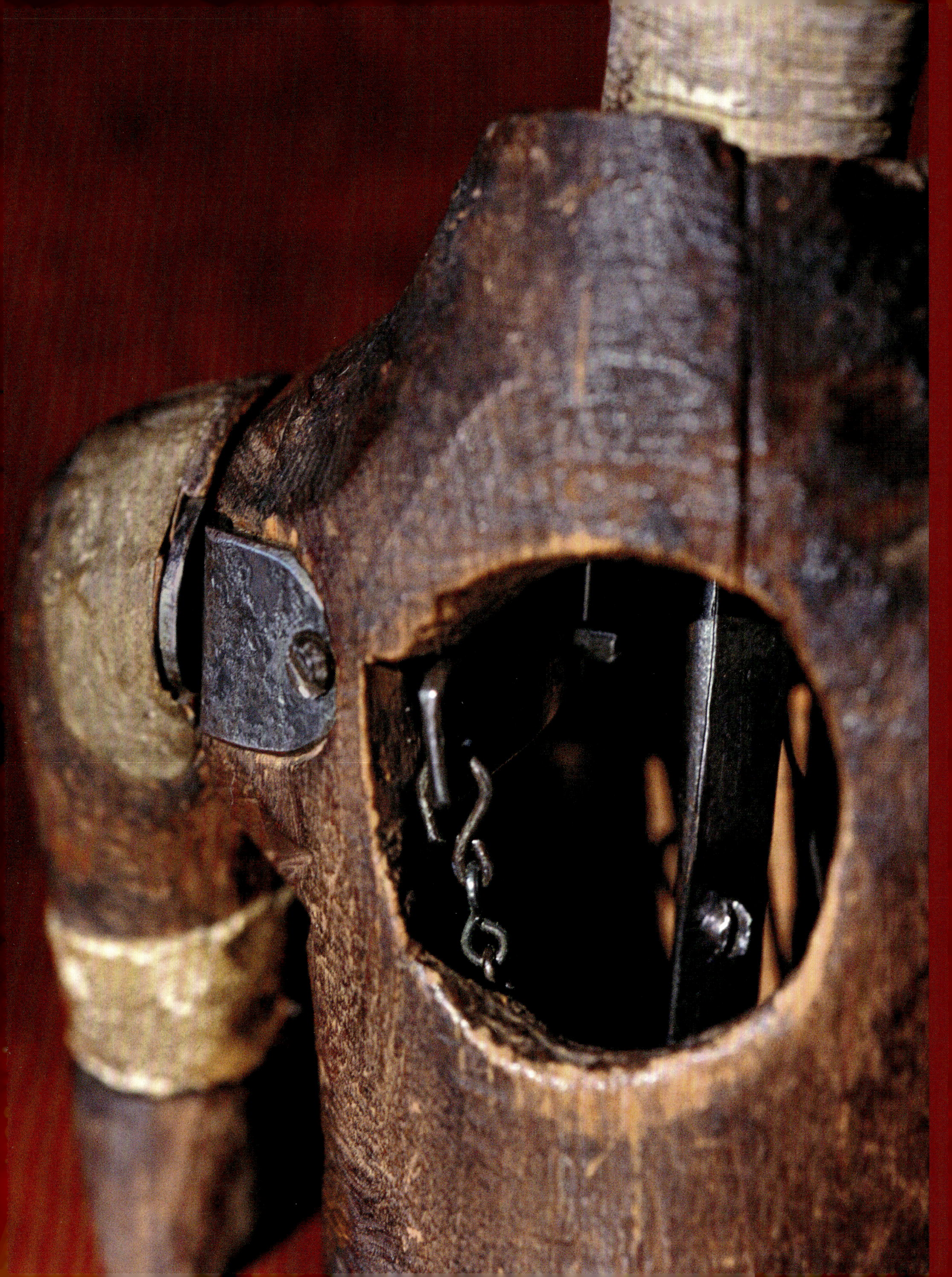

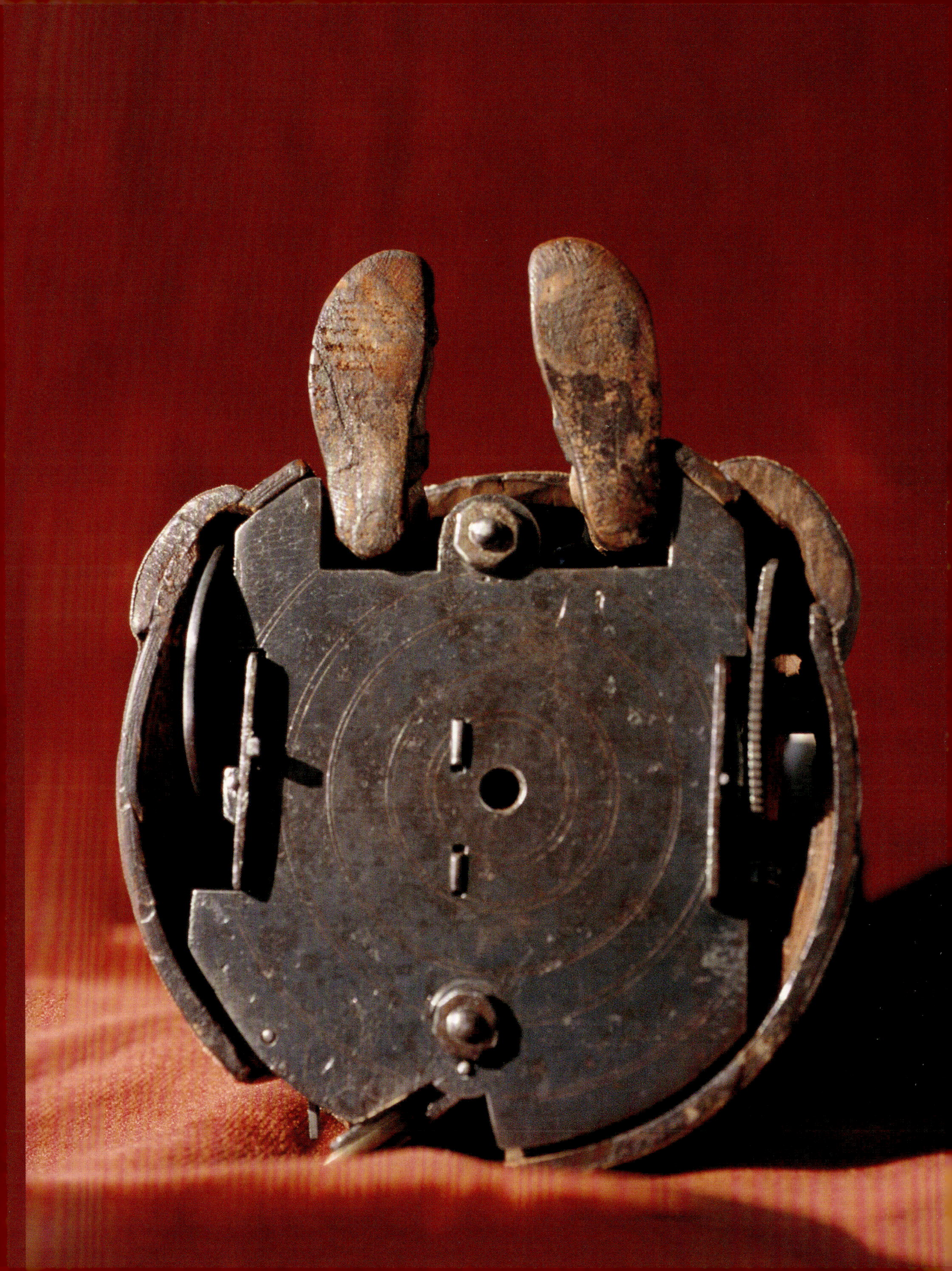

19 Craft Guild or Imperial Court?

By now, it is clear that the Smithsonian monk was made by more than one master. A clockmaker, a sculptor or woodcarver, a painter, and certainly a tailor all contributed their skills to the task. The integration and coherence between metal and wood—clockwork and sculpture—is a question of particular interest to David and Elizabeth, clockmaker and sculptor, respectively. What if this had been our assignment? Was the clockwork made first, then the wooden parts carved and pieced together around it? Or was the sculpture roughed out—head, arms, feet, and body shell—and the clockwork then measured and built to fit?

The X-ray of the monk's wooden head reveals a crisp longitudinal seam just behind the ears. As mentioned in chapter 12 (regarding the Deutsches Museum figure), the head would have been rough carved, then split or cut in half and hollowed out with a chisel, and the interior prepared to receive the small machinery to animate the eyes and jaw.[1] The sculptor worked from the outside in, then the inside out, to create eye sockets and to section the jaw. The seam between the reassembled halves of the monk's head is just visible under the paint, and straight enough that it was likely saw-cut rather than split. Because wood is more forgiving than iron, the iron components must have been prepared before the final carving of the head. The iron eyeballs, the sled and return spring for their motion, the chassis that supported these, together with the levers, arbor, and spring for the jaw (see David's diagram of the inside of the head, **FIG. 10.14**): we believe these parts were made after the sculptor roughed out the wooden head but before the final carving. Locating the chassis so the iron eyeballs would register on all axes in their wooden sockets, and rotate without scraping off their paint, required both wood and iron to be cut and teased and adjusted for the fit. The sculptor could final carve the thin outer eyelids only once the eyeball placement was certain. The same duet would be essential for installing the two arbors for the jaw. Both arbors are visible in **FIGURE 10.12**: one beneath the ear at the jaw-hinge point (for the lever to the chin), the other just below the corners of the mouth (the pivot for the wooden chin itself). The iron mechanism fits inside the wooden head as precisely as a walnut fits inside its shell. The monk's arms illustrate a similar negotiation between the competing requirements of image and movement. Surely sculptor and clockmaker worked side by side to build this figure. And when their work was finished, the painter took over.

A complex multifigure automaton like the Trompeterautomat, or the Diana and centaur, mentioned in chapter 17, required a pyramid of trades for its manufacture. The clockmaker or goldsmith who undertook such a piece depended not just on journeymen and apprentices under his supervision but on smiths and founders, toolmakers, bell casters, woodworkers, sculptors, painters, enamelers, gem cutters, and so on—some of whom depended in turn on the mining and smelting of ores and the ready supply of metals. The monk would have called for a humbler collaboration but a collaboration nonetheless. Because the clockwork shows more methodological and stylistic similarity than the sculpture and because the walking automata are above all forms of *clockwork*, they surely originate within the domain of the spring-driven clock, acquiring virtues from the other arts by necessity.

The European craft guild system presents a prime setting for a community of interdependent artisans. Guild records from sixteenth-century cities with ties to the Habsburg realm and the Holy Roman Empire show Nuremberg and Augsburg as craft centers where clockmaking flourished and expanded. Nuremberg at midcentury was a nexus of trade routes reaching across the continent. Augsburg, with its sophisticated bank and credit system, would gain even greater precedence by century's end. Iron and brass were smelted and cast in local foundries, and specialists of all kinds prospered. The metalworking trades in particular were highly regulated: adherence to guild and city council laws proved effective in holding resources and proprietary technology close, promoting reputations, and protecting livelihoods.[2]

A royal court clockmaker, in contrast, singled out by talent and called to work at the pleasure of a king, pursued a career free from guild regimentation. Producing one-of-a-kind objects to adorn a world-class collection or to symbolize the power exchange in a gift of state, such an individual was expected to exercise the very invention and innovation that the guild system often suppressed. Practitioners of many other crafts including sculpture and textiles would also be players in a court setting, on call to contribute their skills to princely commissions.

The divisions between guild and court were not impermeable. Nuremberg and Augsburg were both free imperial cities: self-ruling, independent of German territorial claims, they were bound only to the emperor in Vienna or Prague. Guild contracts record major works commissioned by Habsburg nobility, and such works are identified in inventories of imperial and patrician collections. Both guild and court artisans appear in chronicles compiled by scribes vying for history and precedent for native cities and city-states. Artisans themselves sometimes moved between roles, as did the Nuremberg smith Jakob Bulmann, called on occasion to the court of Ferdinand I in Vienna.

In either setting, the presence or absence of a maker's mark on a work was unpredictable. Guild law often required it on clocks, both to certify the originating city and to monitor the work of guild members. Automata presented a particular dilemma. The absence of signatures or guild stamps on some automata may be due to the fact that they were not subject to the limitations of a single guild. Klaus Maurice gives examples of automata produced by makers of chamber clocks, by goldsmiths, by locksmiths, and by others; the clockmaker was not always the originating contractor.[3] A goldsmith could initiate a piece, and if his work was considered the most valuable or visible, protocol might

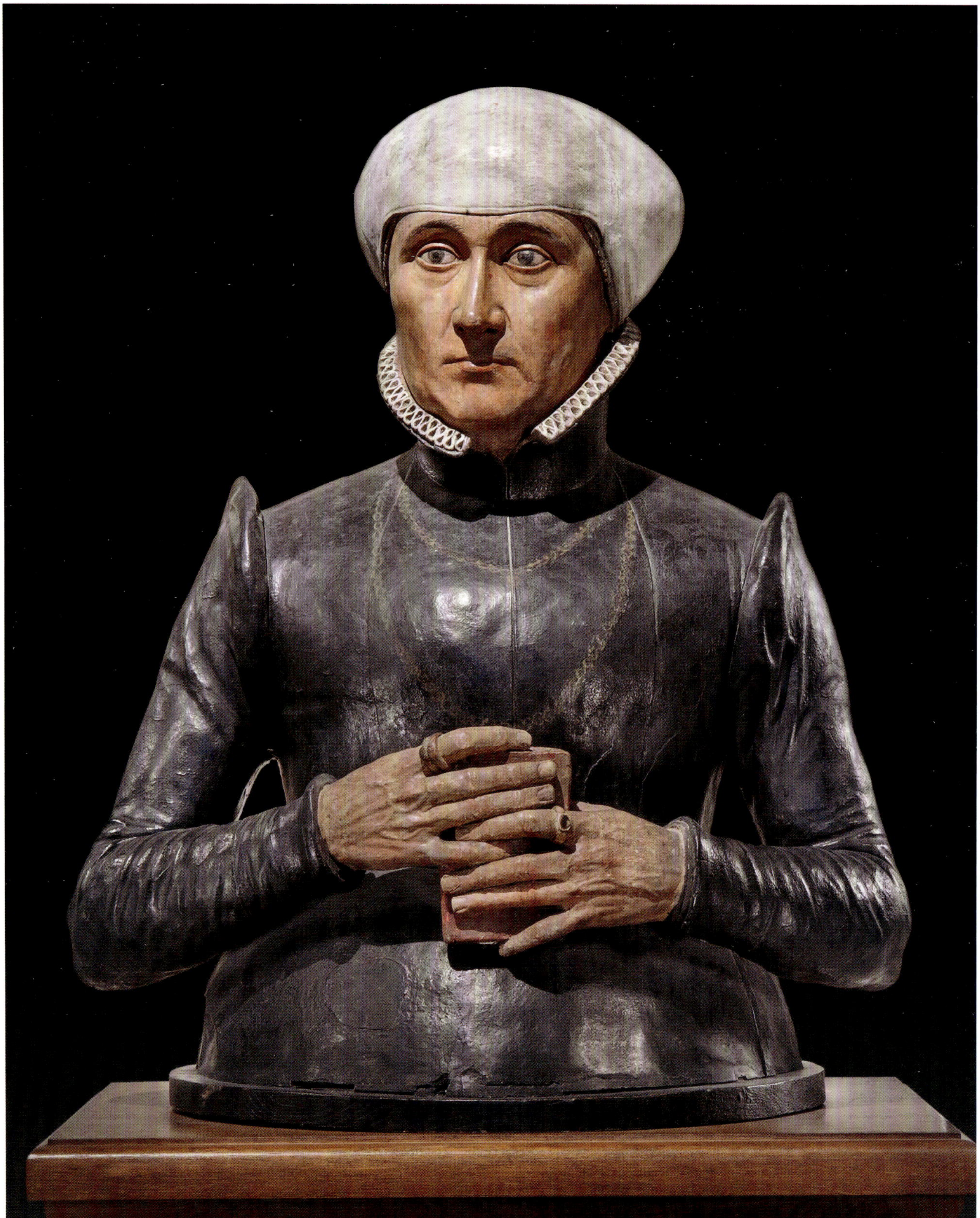

omit the clockmaker's name. Further, works that might have been considered copies may have remained unsigned for reasons of "copyright" or breach of propriety. The status and identity of the sculptors whose work was essential in the production of figure clocks is further shrouded, as Maurice shows in searching the Augsburg records for the makers of the small cast bronze and silver figures in these popular thematic clocks.[4] But the Netherlands-born sculptor Johan Gregor van der Schardt was active in Nuremberg in the late sixteenth century; he was known for small works and portraits in bronze and painted terracotta **(FIG. 19.1)** and known as well for his collaboration with the goldsmith Wenzel Jamnitzer. Maurice mentions an automaton made by Jamnitzer's nephew, Barthel Jamnitzer (ca. 1548–1596), fashioned from a small sculpture by Van der Schardt: a silver Cupid with bow and arrow.[5]

A signature on a work made within a court setting had its own mix of custom and circumstance. Schlottheim's nef, made on commission for Rudolf II, was inscribed with words celebrating the emperor, not the maker or the city where it was made.[6] On the other hand, Ambrosio de Morales shows Juanelo solemnly signing the great astrarium he made for Charles V.[7]

But the kinds of documents unique to these settings—guild, city, and court—preserve invaluable records of makers and works, and historians comb them for clues to the identity of objects of every kind. In the case of the clockwork automata, as we have seen, horological historians have found mention of solo walking figures in several kinds of archives, along with a tantalizing few names of clockmakers who made them. Juanelo's name comes to us from encomia by Ambrosio de Morales, Bernardo Sacco, Antonio Campi, Marco Girolamo Vida, and others. Jakob Bulmann's name survives in guild and city records. Christof Margraf is named in a royal inventory from Prague; a Milan inventory contains a description of an automaton cited for the Milan matron. These documents were produced for quite different reasons, and comparing them will generate a new set of category challenges. Keeping this in mind, let us see what they reveal. Two names, Jakob Bulmann and Juanelo Turriano, are mentioned most often today in the literature on the walking automata. Across regional and cultural divides, Bulmann hails from South Germany in the central European region of the empire; Turriano, Italian born, carried out his best-known work in the court of Spain.

FIGURE 19.1.
Johan Gregor van der Schardt, *Half-Length Portrait Bust of Anna Imhoff*, ca. 1580. Painted terracotta. Height: 65.5 cm (25¾ in.). Bode-Museum, Berlin (inv. 539). Anna Imhoff was the wife of the Nuremberg art collector and merchant Willibald Imhoff. Van der Schardt, known for his portrait sculpture, also had numerous contacts with Nuremberg goldsmiths and may have provided small sculptures in bronze and silver for figure clocks.

20 Two Masters, Two Cultures: Jakob Bulmann and Juanelo Turriano

The locksmith Jakob Bulmann (d. ca. 1535) is named in a 1547 manuscript account of Nuremberg's "most refined artists and craftsmen" of the century, composed by the calligrapher, mathematician, and historian Johann Neudörfer (1497–1563).[1] The entry on Bulmann, among the shortest in the manuscript, reads in full:

> This Bülmann [*sic*], though untrained in writing and reading, was nonetheless almost skilled and learned in the field of astronomy, which is why he was able to set in motion, through the application of 80 pounds of weights, the clockwork of the Theoriam planetarum, which nobody else before him had been able to accomplish. He also fashioned male and female figures for clockworks, which would walk about and strike their measure on lutes and timbals. King Ferdinand, for whom he had done much work, had him brought to Vienna in a sedan in his old age, for the sole purpose of showing His Majesty a clockwork. Moreover, he fashioned many fine items in his capacity as a locksmith, especially the scales with its two arms, allowing the goods to be weighed outside of the scale.[2]

Neudörfer's original manuscript of seventy-nine biographies received full publication and proper credit only in 1875, through the careful research of the Nuremberg archivist and historian Georg Wolfgang Karl Lochner (1798–1882).[3] In the meantime, his briefs on Bulmann and many others found their way into print under the names of later authors. In 1730, two centuries after Neudörfer's time, the mathematician and astronomer Johann Gabriel Doppelmayr (1677–1750) included Bulmann in an expanded list of Nuremberg mathematicians and artists, copying the biography from Neudörfer. Here is the key sentence, in Doppelmayr's phrasing: "Furthermore, he crafted movable figures in the shape of male and female persons with clockworks of such a fashion as to make them walk to and fro, striking a measured beat on timbals as well as on lutes."[4] He credits Neudörfer only with a footnote at this very sentence:

> This is confirmed by Johann Neudörfer the Elder in a manuscript [stating] that among the craftsmen renowned in his time, it was an altogether novel and special thing. In our time, such artificial machines are no longer so rare, as many have devoted themselves to such works in a variety of representations with great industry.[5]

These two versions of the reference to Jakob Bulmann and his automata were carefully studied by Adelheid von Herz, whose 1990 thesis, "Androids of the Sixteenth Century," we've consulted throughout our work. Singling out Neudörfer's original sentence, which reads, "[Bulmann] also fashioned male and female figures for clockworks," von Herz asks if this might refer to figures made *for* clocks. That is, did the text refer to figures incorporated into the striking mechanisms of large weight-driven clocks such as the clock in the Nuremberg Frauenkirche, called the Männleinlaufen?[6] In studying the semantics of the text, von Herz too was searching for sure historical references to independently moving androids: clockwork-driven but not themselves accessories to clocks. Doppelmayr's version could seem to be just such a reference, for he describes figures moving "with clockworks" instead of "for clockworks."[7] Yet Doppelmayr's rhetoric and later date might just as well signal, as his footnote suggests, a mistaken connection to a class of figures more common two hundred years later. Lochner adds a more secure clue, for he appends a long editorial note to Neudörfer's text, examining an array of legal documents from sixteenth-century Nuremberg archives to ascertain Bulmann's correct name and vita, and in the process he mentions a ledger document dated October 23, 1535, which notes payment received from one Hanns Retsch for an object made by Jakob Bulmann, "a walking sculpture of a maiden [*ein gehend Jungfraubild*]." Drawing special notice to this reference, von Herz concedes the plausibility that Bulmann indeed made such androids, self-propelled by a clocklike mainspring drive.[8]

Brief as it is, Neudörfer's portrait of Jakob Bulmann reveals an individual working both as an independent craftsman in Nuremberg and as a contracted court clockmaker to Ferdinand in Vienna. Ferdinand I, brother of Charles V, was king of Bohemia and Hungary at the time, a powerful patron with imperial tastes and ambitions. He would later become Holy Roman emperor on Charles's abdication in 1556, with sovereign rule over the Austrian Habsburg lands.

Further, Bulmann, not unlike Juanelo Turriano, was called upon to repair a celebrated astronomical clock. In Juanelo's case, chronicles refer to the great planetary clock made in the fourteenth century by Giovanni de' Dondi of Padua (ca. 1330–1388), its ruins still known in 1529 or 1530 when Charles V learned of it and Juanelo was brought forward to renovate it. **FIGURE 20.1** shows a modern reconstruction of de' Dondi's clock.[9] In Bulmann's case, Neudörfer's assertion that Bulmann "was able to set in motion... the clockwork of the Theoriam planetarum, which nobody else before him had been able to accomplish," leads of course to the question: What was the Theoriam planetarum? Klaus Maurice speculates that it was a planetary clock built in the previous century by Johannes Müller, known as Regiomontanus (1436–1476), a clock that had either fallen into disrepair or was left unfinished in Nuremberg at the time of Regiomontanus's early death.[10] At the very least, we learn from Neudörfer that Jakob Bulmann encountered a complex clock from an earlier time and mastered its restoration.

Bulmann's name was first paired with that of Juanelo Turriano, as candidate makers of automata, by Ernst von Bassermann-Jordan, whose 1926 essay on the Vienna cittern player identified them both. Not by accident, that essay appears in Bassermann-Jordan's encyclopedic volume, *Alte Uhren und ihre Meister*, directly following his longer and

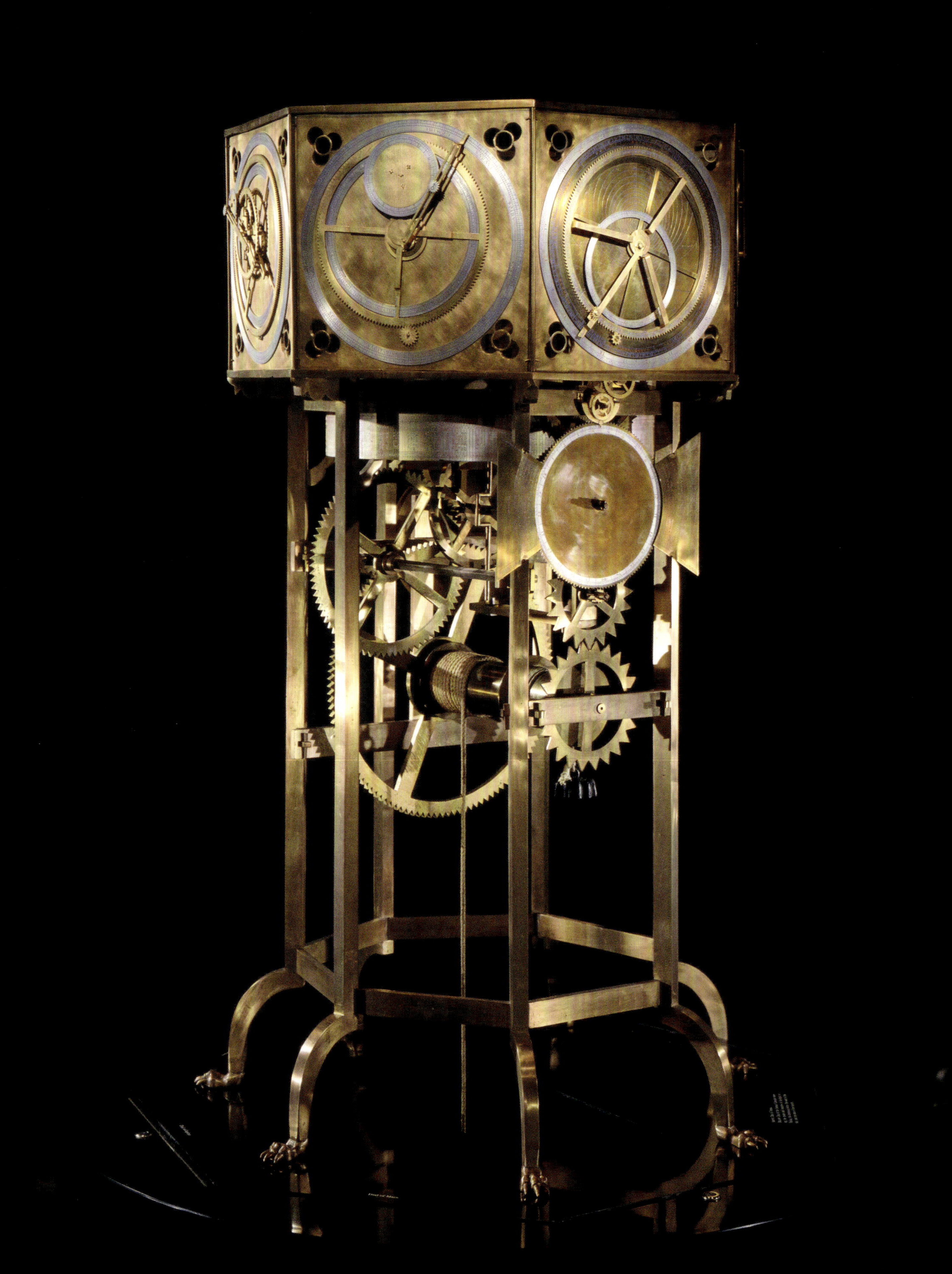

equally readable chapter on Juanelo himself, "Emperor Charles V and His Last Clockmaker." Narrated in detail is Juanelo's pride of place among the courtiers whom the emperor took with him into seclusion in the Hieronymite monastery of San Yuste in 1556. Juanelo's most celebrated works, including his great astrarium, his legendary automata, and his late, famous Artificio (the unprecedented hydraulic pump and aqueduct he built for Philip II in Toledo) are laid out and appraised along with the primary-source authors who chronicled his public life.[11] Together, the two essays—Juanelo's story, followed by that of the Vienna automaton—make a strong case, in part by pure proximity, for Juanelo as maker of that automaton. Here is what Bassermann-Jordan concludes:

> But to which cultural sphere does this [Vienna] figure belong and where did it originate? Italy as a land of origination is hardly thinkable. The craft of making small clocks was not well developed in Italy at the time and depended on imports; no Italian automata of small size are known, the artistic style of the time aimed rather for monumental effect. But we know of one Italian whose fate led him to the court of Emperor Charles V and to follow the emperor to the remote monastery of San Yuste: commonly referred to as Giovanni, this was Juanelo Torriano of Cremona, the mechanic, clockmaker, and engineer of whom I spoke at length in a previous essay. Ambrosio de Morales, who knew the master personally, informs us in his "Las Antiguedades de las ciudades de España" (Alcalá, 1575, pp. 91 ff.) that Torriano created the automatic figure of a lady that danced on the table to the beat of her tambourine.
>
> But there is another account from the time period in question that could also refer to the automaton. In his "Historical Account of the Mathematicians and Artists of Nuremberg" (Nurnberg, 1730, p. 285), Johann Gabriel Doppelmayr writes the following about Hanns Bullmann: "An artisan locksmith, though untrained in reading or writing, acquired through much contemplation prompted by his good natural intellect, such skill in various arts and crafts, that all accorded him the title of an artist.... Furthermore, he presented various figures in the shape of men and women, set in motion by clockworks so that they walked to and fro, striking a rhythm on timbals and lutes...."

FIGURE 20.1.
A modern reconstruction of the fourteenth-century astrarium of Giovanni de' Dondi. The model built in 1960 by Thwaites and Reed Ltd., England, with technical guidance by H. Alan Lloyd. Brass. Height: 132.1 cm (52 in.) Diam.: 76.2 cm (30 in.). National Museum of American History, Smithsonian Institution, Washington, DC. The machine was a clock, a calendar, and an instrument for demonstrating the motion of the sun, moon, and the five planets then known, on the Ptolemaic system. The original astrarium was constructed by de' Dondi between 1348 and 1364, historians say, in Padua, and acquired in about 1381 by Duke Gian Galeazzo Visconti, who installed it in his Castello Visconteo in Pavia. According to chronicles, its legendary fame, and its ruins after two centuries, led Emperor Charles V to call for its repair, and Juanelo Turriano undertook the task.

> **At least the priority of invention, then, must be given to the Master from Nuremberg, the city of Peter Henlein, who survived the Master Hans Bullmann. The broad shoes, which never really gained popularity in Spain, also speak for a German origin. Not so the head of the doll, which seems to derive from the nativity sculptures and polychrome sculptures of the Romance countries. In light of the rather tenuous artistic evidence, no final judgment shall be made here. However, we lean toward the assumption that this piece is to be attributed to the sphere of Spanish art, although it must be borne in mind that particularly in the field of precision mechanics, a brisk exchange of experience, masters, works, and perhaps even mechanical parts was taking place between the European states.**
>
> **In any case, the automaton discussed and depicted here deserves an excellent place as the oldest preserved work of its kind in the history of technology; moreover, the little walking doll is a first-rate exquisite antique.[12]**

Though it concedes the overall scarcity of the evidence and gives Bulmann precedent in making earlier figures, Bassermann-Jordan's 1926 study of Juanelo and the cittern player now in Vienna has provided the prime reference inherited by every sister and brother automaton that has come to light since (with or without the nod to Bassermann-Jordan's influential conjecture). Of his two candidates—one a trade locksmith in South Germany, one a court clockmaker in Spain—his final guess is a nod to both. A Nuremberg precedent for the clockwork but a Spanish sphere for the character of the figure... and a reminder that the empire covered a flexible map of exchange, craft, and trade.

Our opinion is the reverse of Bassermann-Jordan's: we would give precedence to a court setting for the innovation of the clockwork mechanism and credit a guild setting for its subsequent propagation. We wonder what Bassermann-Jordan would say with the full family of figures before him. The Vienna figure, as the solo subject of his essay, is still the exception that puts all proposals in question. Von Herz too argues for a court-produced prototype group (the male figures) and a subsequent series of guild examples (the brass-movement lute players). She posits a cooperative association between sculptor and clockmaker, both, she says, likely of South German origin. But the Vienna figure, she believes, is Spanish.[13]

The differences of opinion concerning region and setting are driven as much by the automata themselves as by the names Bulmann and Turriano, as history brings two consummate craftsmen to us via ledgers and legends. In a curious unity, both were described as prodigious mathematicians despite little education, both were associated with great astronomical clocks and works of technical engineering, and both were reputed makers of fabled android automata. They are united by the cultural trope of the gifted outsider, the mythic blacksmith. They were thinkers and makers both. Following Bassermann-Jordan's lead, Chapuis and Gélis, in *Le monde des automates*, name Bulmann and Turriano as the two sixteenth-century "premiers chercheurs" known to have made genuine androids.[14] Silvio Bedini follows suit in his 1964 essay, "The Role of Automata in the History of Technology."[15] Vienna's Kunsthistorisches Museum too, in its 1966 catalog, gives equal mention to Turriano and Bulmann in discussing its cittern player.[16] Maurice and Mayr continue the double reference in 1980 in their entry on the Smithsonian monk in *The Clockwork Universe*.[17] Finally, apropos of these two astronomer-blacksmiths, and the connection of heavenly clocks to androids, we must not forget the declaration of Derek J. de Solla Price on astronomical and biological simulacra: "These two great varieties of automata go hand-in-hand and are indissolubly wedded in all their subsequent developments."[18]

The debate between Jakob Bulmann and Juanelo Turriano, and between guild and court settings, also takes the form of a tug between two distinct geographic (today we would say, national) spheres. Although the historical record yields far fewer references to Bulmann than to Juanelo, it provides a more vivid idea of the technical milieu in which Bulmann worked, thanks to the craft guild records of the South German city-states.[19] Against these, how are we to weigh the legends surrounding Juanelo Turriano? Ledgers support one sphere, legends the other.

Here is a sketch of Juanelo in his own workshop, written in Latin in 1550 by his countryman and fellow Cremonese Marco Girolamo Vida, revered poet and bishop of Alba:

> If you observe the man, nothing will be less obvious than the acuity of his talent. His face and figure are rough, ugly and coarse, and there is no dignity, no natural grace, not a sign of diligence in his features. That you always see his face, hair and beard caked and blackened with quantities of ash and clumps of soot just adds to his vileness. His hands and gross, enormous fingers are always coated with rust. Further, he is untidy, badly and disagreeably dressed, so that you might think him a Brontes, a Steropes or some other servant of Vulcan, who shapes everything he makes at the anvil with his own hands—a man born for working at the forge. However, lest a person think that some outstanding master in mathematics has written out and solved for him the principle of the spheres, their motions, and the stars, and that this master has determined it all for him in advance because he understands nothing about these things and is only skilled in craftsmanship, let that person know that he prepares everything himself and constructs it all without any help whatsoever, using his own talent, his own research, "his own force," as they say. He is both inventor and executor at once, one of those I have mentioned who stands out among all our fellow citizens in the mastery of astrology, even though no one ever taught him to read or gave thought to a teacher for him. Not only is he versed in the workings of the heavens themselves in this way, but with his intelligence he discerns the very causes through his calculations and understands them. Moreover, he often and quickly contradicts recognized authorities on astrology—men who write books on the subject—and he does so knowledgeably and assuredly. Indeed, on no few occasions he proves them wrong with incontrovertible arguments, obliging them to admit their errors and mistakes.[20]

As García-Diego has commented, "The astronomer who at the same time worked with his hands, of whom Vida speaks, was then, and also later, a rare phenomenon."[21]

Juanelo was once recorded as saying, "Princes are deprived of three very lovely and good things worthy of being esteemed by everyone, namely: of being told the truth, of seeing the dawn, and of feeling hungry."[22] To this day, the street in Toledo where Juanelo Turriano lived is called Calle del Hombre de Palo, the street of the wooden man (**FIG. 20.2**).[23]

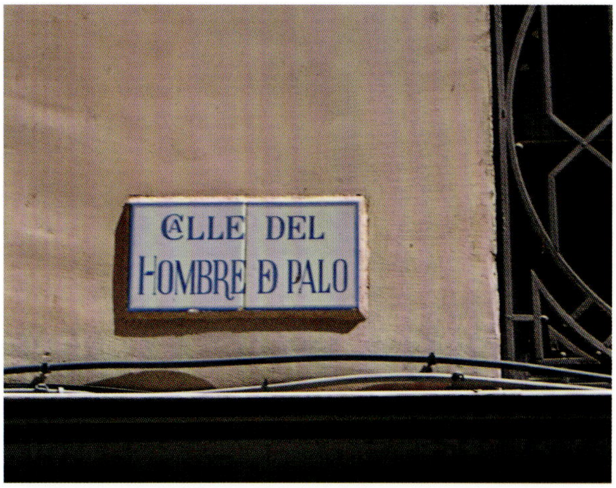

FIGURE 20.2.
Calle del Hombre de Palo, a street sign in Toledo.

21 A Nuremberg Clockmaker's Workshop

Although the mechanical clock is thought to have originated in northern Italy or England before 1300, its later development in South Germany is well documented in metalworking guild records, famous for their detail. The imperial free cities of Augsburg, Nuremberg, and Ulm and the Bavarian capital of Munich numbered among important centers of metalworking technology from the fifteenth to the eighteenth century, and many specific trades—locksmith, gunsmith, toolmaker, clockmaker—evolved in time from the blacksmith's craft. Augsburg and Nuremberg produced scientific instruments, sundials, clocks, and automata that were world renowned. Nuremberg, the city where Jakob Bulmann plied his trade in the early half of the sixteenth century, was known for its innovative figure clocks, hence our choice of this city for envisioning one possible scenario for the monk's origin. Johann Neudörfer, whose manuscript preserves the first account of Bulmann, also singled out some of the greatest names we associate today with Nuremberg: Peter Henlein (1485–1542), inventor of the watch; the great limewood sculptor Veit Stoss (ca. 1445/50–1533); the goldsmith Wenzel Jamnitzer (1507/8–1585), who would later serve in the court of Rudolf II; Neudörfer's close friend and neighbor Albrecht Dürer (1471–1528), who employed Neudörfer's script Frakturschrift in his woodcuts; and Hans Sachs (1494–1576), shoemaker, poet, and Meistersinger, about whom more below. Nuremberg was also the chosen home of Johannes Müller, better known as Regiomontanus, who set up an observatory, workshop, and printing press in 1471 and published leading modern and ancient texts on mathematics and astronomy.[1]

By the mid-sixteenth century, clockmaking practice in Nuremberg flourished within a broad network of interdependent metalworking trades. Iron ore was smelted in blast furnaces fueled with charcoal made of wood from the surrounding forests of Saint Lorenz and Saint Sebaldus; the ore in turn became the raw material for the city's master smiths, instrument makers, and clockmakers, together with those who fashioned the primary tools employed in trades of every kind.

Nuremberg also served as a nexus of established trade routes dating from the Middle Ages, carrying commerce and exchange across and beyond the Holy Roman Empire, in a network that included Genoa, Venice, Lisbon, Madrid, Paris, Toulouse, Antwerp, London, Danzig, Constantinople, Africa, the Levant, and the Orient. Nuremberg merchant families stationed relatives as trading agents in many European cities, where they had established credit and reciprocal customs treaties. Records show, for example, 232 Nuremberg merchants in Venice alone around 1500. Well into the sixteenth century, Nuremberg remained outwardly neutral in the face of social and political unrest associated with the Reformation and Counter-Reformation, peacefully doing business with all parties, while Lutheranism, with its independent spirit, quietly gained appeal among the patrician ruling families, the professions, and the skilled trades alike.[2] An old proverb celebrated the city's economic primacy: "Nürnbergs Tand / Geht durch alle Land" (Nuremberg's hand / Through every land).[3]

Among the larger cities where cognate trades were organized into systems of self-governing craft guilds, Nuremberg was unique in the direct administration of its trades by the city council, and as a result the merchant class exercised considerably more power over craft practitioners. If Nuremberg artisans plied their skills with less guild protection and more open competition, they nonetheless navigated a closely scripted set of contracts among masters, journeymen, apprentices, and oversight councils, which regulated everything from pricing, business territory, and shop size to resource and labor use, subcontracting, and client relations.[4] Above all, the city council dictated the requirements defining the "masterpiece," the technical term for work submitted by an applicant for status as a master. In its prime in the first half of the sixteenth century, this urban environment fostered an unprecedented level of standardization and virtuosity across the trades, including armorers, metalsmiths, weavers, potters, glassmakers, carpenters, and coin stampers. Type founders, printers, and bookbinders produced the books that we study now in order to learn about the age that produced them. Prized among these is Jost Amman and Hans Sach's famous *Ständebuch*, or *Book of Trades* (published in Nuremberg in 1568).[5] Urban regulation, whatever its social costs to citizens, has redeeming value for historians in search of information about a vanished era.

The sixteenth-century clock—a Nuremberg specialty—took a form very different from the routine timepiece we now take for granted. Clock time was at first incidental to an instrument that demonstrated the complex relative motions of the sun, moon, planets, and so-called fixed stars. The astronomical clock, or astrarium, as much a spatial tool as a temporal one, told location and influence in the revolving cosmic order. Such a machine naturally entailed the representation of first causes and prime movers: the figures of the constellations, of myth, religion, and imperial agency. The astrarium epitomized a new world science: a manifestation of the power of measure, a symbol of wealth, and an object of wonder. Neudörfer mentions such a clock in connection to Jakob Bulmann, as we've seen. The word itself, *astrarium*, is closely connected to the unprecedented machine made by Giovanni de' Dondi (see **FIG. 20.1**). These, and elaborate figure clocks and automata from Nuremberg and Augsburg, were commissioned as royal peace tributes to the Ottoman sultanate, as Jesuit mission gifts to China, as currency of imperial diplomacy in courts across Europe, and as the furniture of public relations in patrician households. Smallness was regarded as a measure of genius: the more actions that could be generated by a diminutive mechanism, the higher the price. However, basic wall-hung alarm clocks were also in demand, for the bell towers of monasteries, for public buildings, and for larger houses, while table clocks and animated toys, humbler marvels, found many buyers among the merchant class.[6]

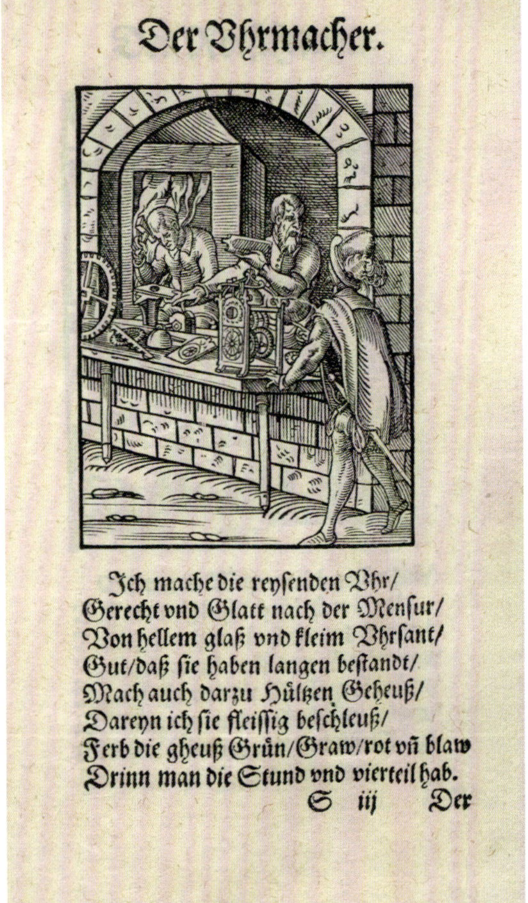
Der Vhrmacher.

Ich mache die reysenden Vhr/
Gerecht vnd Glatt nach der Mensur/
Von hellem glaß vnd kleim Vhrsant/
Gut/daß sie haben langen bestandt/
Mach auch darzu Hülßen Geheuß/
Dareyn ich sie fleissig beschleuß/
Ferb die gheuß Grün/Graw/rot vñ blaw
Drinn man die Stund vnd viertel hab.
S iij Der

FIGURE 21.1.
"The Clockmaker," from Jost Amman and Hans Sachs, *Eygentliche Beschreibung aller Stände auff Erden* [Exact description of all ranks on earth] (Nuremberg, 1568), popularly called *The Book of Trades (Ständebuch)*. Original sheet with Amman's woodcut and Sachs's verse about 20 × 15 cm (just under 8 × 6 in.), bound for a page size of about. 7 × 5 in. Amman's woodcut shows a clockmaker displaying a fine example of a period mechanical clock, though Sachs unaccountably praises a sandglass, or hourglass: "The charming clocks I make are nice, / Polished smooth and most precise. / Their glass and sand is always pure, / So that they will long endure. / Their wooden cases I make, too, / Paint them green, grey, red, and blue, / To show, when fitted round the glass, / Ev'ry hour and quarter pass." (Verse translation by Theodore K. Rabb)

A shop producing one-of-a-kind clockwork pieces, or small-batch productions of pieces, might typically comprise a master, one or more journeymen, a few apprentices, and in some cases journeymen from other trades whose skills were wanted in the clockmaking enterprise, such as coppersmiths or goldsmiths. Auxiliary crafts might include the construction of wooden clock cases, brass founding, bell casting, dial painting, and sculpture: these tasks would have been contracted to other shops or to individual artisans. Only late in the next century did a higher degree of specialization begin, as when several masters in Augsburg were permitted to take up the exclusive production of clock springs.[7] Only later too did the metallurgical development of brass as a noncorrosive and easily worked replacement for iron become commercially affordable.

A midcentury Nuremberg clockmaker's shop is illustrated by Jost Amman (1539–1591) for the famous *Ständebuch* produced with Hans Sachs in 1568 **(FIG. 21.1)**.[8] The full title of this compendium of 114 woodcuts by Amman and short verses by Sachs devoted to the trades, professions, callings, and ranks in the city of Nuremberg conveys a worthy history lesson in its own right. Translated, it reads:

> **Exact description of all ranks on earth, high and low, spiritual and secular, of all arts, crafts, and trades, etc., from the greatest to the smallest; also of their origin, invention, and usages. By the far-renowned Hans Sachs most diligently described and put into German rhymes very useful and pleasurable to read, and also with artistic illustrations the like of which has never been seen before, printed for the honor and satisfaction of all ranks included in this book and particularly for the use of all artists, such as painters, goldsmiths, etc. With the permission of the Holy Roman Emperor. Printed at Frankfurt am Main. 1568.**

The parade begins with Pope, Cardinal, Bishop, Priest, Monk, Emperor, and King…and ends with four classes of fool: the money fool, the gluttonous fool, the jester, and the natural fool. But the heart of the *Ständebuch* consists of the arts and crafts of manufacture, with the clockmaker in their midst, together with the closely allied locksmith **(FIG. 21.2)**.[9] The sculptor too is represented **(FIG. 21.3)**.[10] The book,

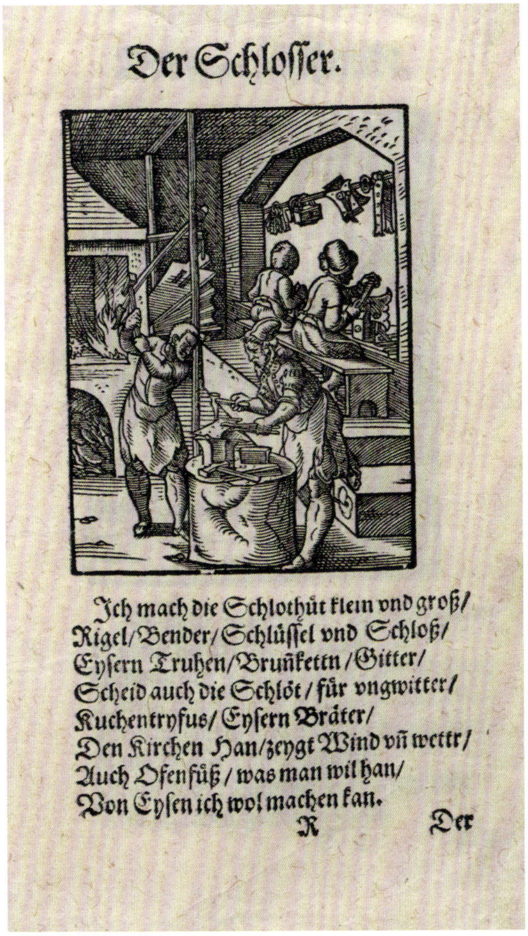
Der Schlosser.

Ich mach die Schlothůt klein vnd groß/
Rigel/Bender/Schlůssel vnd Schloß/
Eysern Truhen/Bruñkettn/Gitter/
Scheid auch die Schlöt/für vngwitter/
Kuchentryfus/Eysern Bräter/
Den Kirchen Han/zeygt Wind vñ wettr/
Auch Ofenfůß/was man wil han/
Von Eysen ich wol machen kan.
R Der

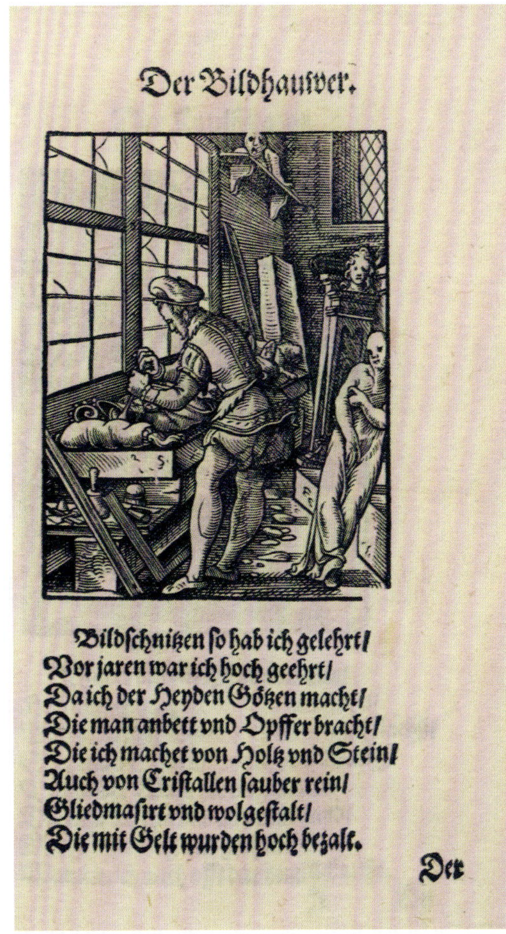
Der Bildhauwer.

Bildschnitzen so hab ich gelehrt/
Vor jaren war ich hoch geehrt/
Da ich der Heyden Götzen macht/
Die man anbett vnd Opffer bracht/
Die ich machet von Holtz vnd Stein/
Auch von Cristallen sauber rein/
Gliedmasirt vnd wolgestalt/
Die mit Gelt wurden hoch bezalt.
Der

FIGURE 21.2.
"The Locksmith" (Amman and Sachs, *Ständebuch*). Amman gives a good view of the furnace, its bellows poised for pumping air to boost the charcoal's heat. Two smiths wield tongs and hammer to shape a piece of hot iron on an anvil, the anvil mounted on a stout tree stump. Many a locksmith also made early clocks. Sachs's verse: "There's grates and bolts and straps and keys; / Locks as well, and things like these; / There's railings, chests, and iron chain; / Chimney linings for the rain; / There's pans and stands for kitchen fires, / Weathervanes on churches' spires— / If iron-ware you want, then I / Make the goods you need to buy." (Verse translatation by Theodore K. Rabb)

FIGURE 21.3.
"The Sculptor" (Amman and Sachs, *Ständebuch*). Amman depicts a remarkably well-dressed sculptor at his bench with gouges, mallets, and drill, at work on what appears to be a tomb effigy. Good light, many works in progress, and gentleman's attire confirm his status. Sachs's verse: "The art of sculpture, which I've taught, / Brought me honors when I wrought / The idols which the people so / Dearly worshipped long ago. / I carved them out of stone and wood, / Purest marble, clear and good. / The well-proportioned forms I made / Earned me wealth—I was well paid." (Verse translatation by Theodore K. Rabb)

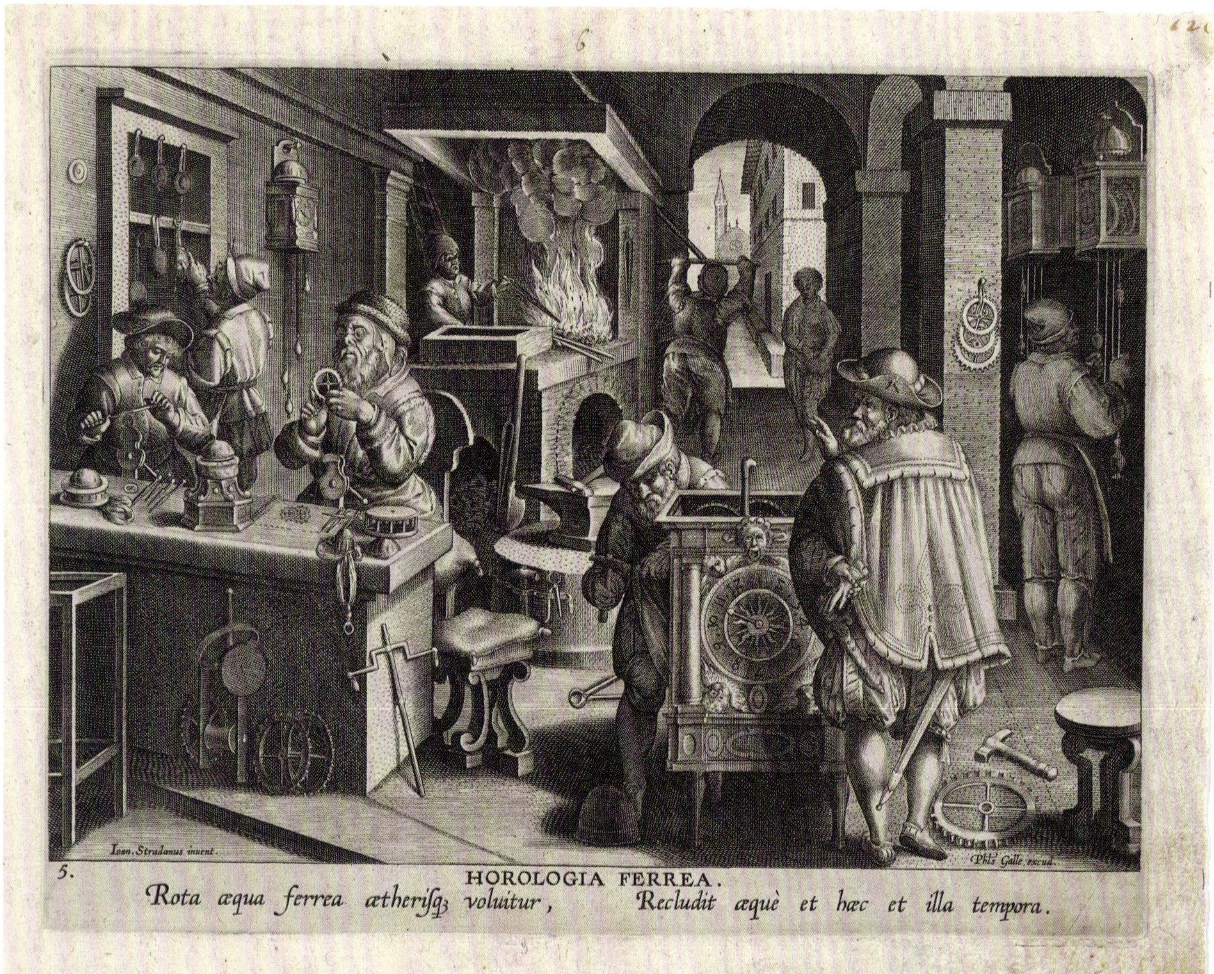
Ioan. Stradanus inuent.
Phls Galle excud.
5.
HOROLOGIA FERREA.
Rota æqua ferrea ætherisq; voluitur,
Recludit æquè et hæc et illa tempora.

celebrated for the accuracy of its detail, is also contemporary with the period of the monk's estimated origin, making it a valuable resource for our study.

Amman's woodcuts of the clockmaker's and locksmith's shops illustrate some of the tools essential to clockmaking: the forge fire with its bellows; the smith's anvil; tools of persuasion, such as the hammer, drift, and swage; cutting tools like the file and punch; and assorted holding tools, vises, tongs, and pliers. To which David would add the handsaw, the drill, the dead-center lathe, and measuring devices: calipers and rules and dividing plates. Some tools came from local specialist smiths and toolmakers, but many more were produced in the shop where they would be used. The clockmaker first makes the tool, then, by means of the tool, the clock.

These elementary tools made possible astonishing precision and artistry. In our industrial and postindustrial world, skill and labor hold profoundly different meanings, which make it difficult for us to imagine the sheer manual bravura of artisanal iron-working technology before the introduction of electricity and power tools. As a clockmaker of long standing, while making and using early-pattern hand tools, David learned much about the characteristics of iron, along with the huge advantages of "batch production." In 2010 he set out to fabricate a simple wall-hung alarm clock in the style of the mid- to late sixteenth century, using only the tools and methods of that period, and it entailed more than two years of work at forge and bench (see **FIG. 18.1**). This convinced him that even at this early date, the master could make a reasonable profit on a clock only by contracting or employing other artisans. The master would have had overall control of the assembly and fine-tuning, but the individual parts would have been produced by a number of different practitioners, each specializing in a particular skill, some in the shop, others as outworkers. A few decades after Amman and Sachs published their *Ständebuch*, another artist took a closer look inside a clockmaker's workshop (**FIG. 21.4**), this one filled with workers.[11]

Ambrosio de Morales's chronicle of Juanelo's production and Vida's description of his blackened face and hands offer but a brief glimpse of this preindustrial world. The workmen who knew about iron, as the Siennese metallurgist and engineer Vannoccio Biringuccio wrote in his celebrated 1540 text, *De la pirotechnia*, were "never able to enjoy any quiet except in the evening when they are exhausted by the long and laborious day that began for them with the first crowing of the cock. Sometimes they even fall asleep without bothering about supper."[12] A complex clock would require many different types and sizes of toothed wheels, together with a multitude of arbors, pinions, cams, and levers, as well as the driving weights or mainspring, the subsidiary springs, the components of the escapement (such as the verge and its foliot or balance wheel), and the frame and various supporting brackets. All these crucial parts, each made individually from forge anvil to final polish, had to work together flawlessly in concert.

FIGURE 21.4.
Horologia ferrea (Iron clocks), a copperplate engraving from the series *Nova Reperta* (New discoveries), printed by Philips Galle in Antwerp, ca. 1591. Sheet: 23.2 × 29.2 cm (9 × 11½ in.). Influenced by Jost Amman's clockmaker's shop in Nuremberg, Jan van der Straet, known as Johannes Stradanus (Flemish, 1523–1605, active in Antwerp, Florence, and Rome), produced this detailed scene showing the diversity of skills needed to make a clock. A master works on a virtuoso clock while a finely dressed nobleman looks on. At left, a bespectacled craftsman examines a finished clockwheel, while his bench mate files the teeth of a new wheel, held in a foot-operated vise (see **FIG. 18.1.**). Behind them, a worker inspects a set of oval neck watches (once called "Nuremberg eggs") hanging in a cupboard, while at rear, two men work the forge. At right, a worker adjusts the pendulum weights of a wall-mounted chamber clock. Through the shop's rear door, a clock tower is visible in the distance.

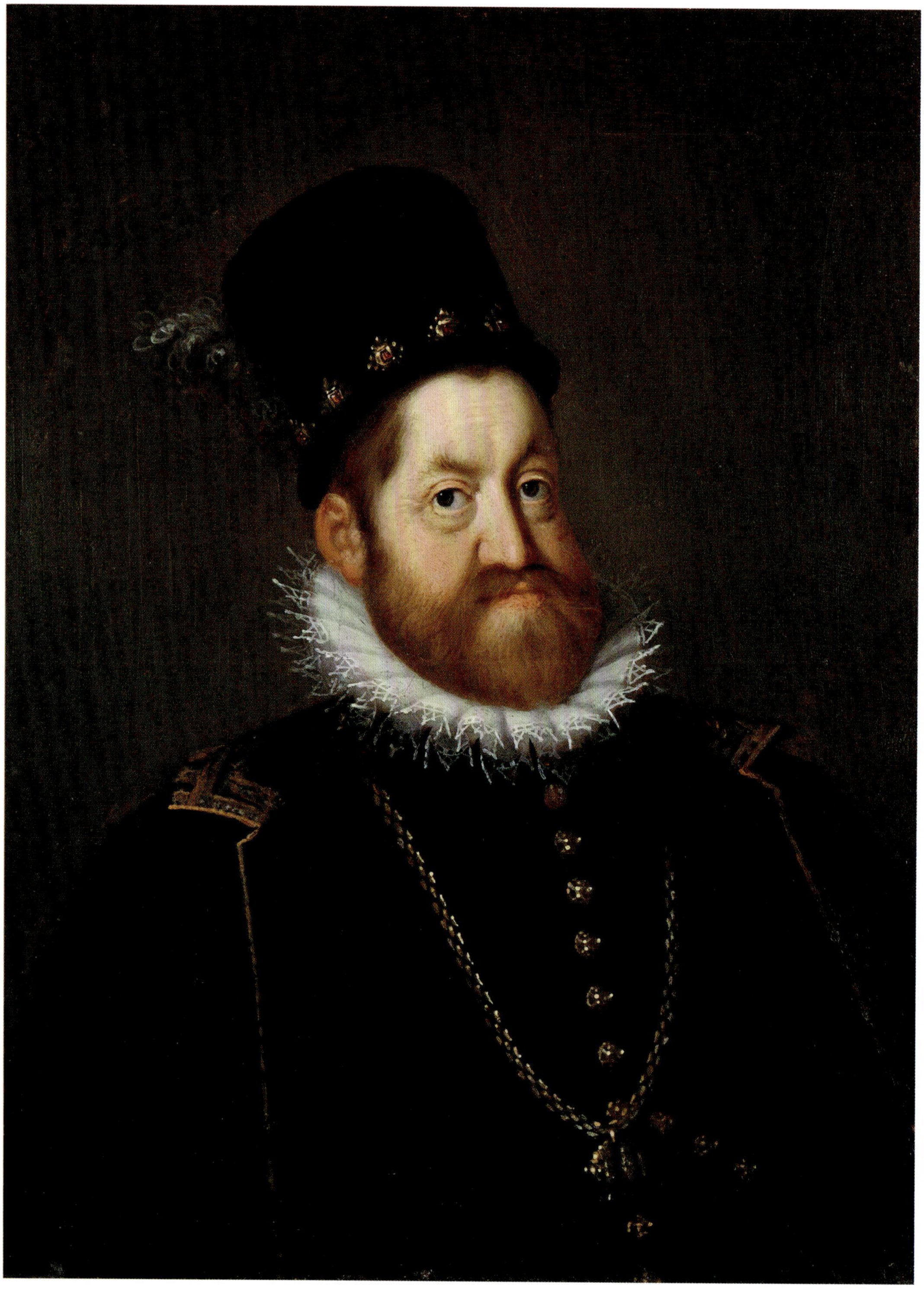

22 The Art Cabinet of Rudolf II

FIGURE 22.1.
Joseph Heintz the Elder, *Emperor Rudolf II*, ca. 1592. Oil on copper. 16.2 × 12.7 cm (6⅜ × 5 in.). Kunsthistorisches Museum, Vienna. Rudolf II sent emissaries across Europe and around the globe in search of treasure for his Prague *Kunstkammer* and amassed the most encyclopedic collection of natural and artificial wonders of the northern Renaissance. Instruments, clocks, and automata are well represented in surviving inventories.

Among the most demanding and persistent customers for objects produced by the masters of the German craft guilds were reigning members of the Habsburg dynasty and, in particular, the sovereign collectors who created *Kunstkammern*, art cabinets. One such avid collector, Ferdinand I, ruled the Austrian hereditary lands for his brother Charles V and became Holy Roman emperor when Charles withdrew to San Yuste. Like his brother, he was fascinated by clocks and instruments. His sons, Maximilian II and Archduke Ferdinand II, governor of Tyrol, built eminent collections of their own. But the collections in Prague amassed by Emperor Rudolf II, grandson of Ferdinand I and firstborn son of Maximilian II, are justly celebrated as the most famous and extensive treasury of the northern Renaissance (**FIG. 22.1**). As the Czech art historian Eliška Fučíková has asserted, "Even though Rudolf's collection and the Emperor's court had their predecessors—one need mention only Lorenzo de' Medici and his collections—in magnitude, quality and diversity Rudolf's collections were without equal, not only in their time but for centuries to come."[1]

Kunstkammern, and kindred cabinets referred to as *Wunderkammern*, *Schatzkammern*, or cabinets of curiosities, depending on their contents, were compendia of natural and artificial marvels containing samples of all the world's knowledge and of all the phenomena of Creation. Art and wonder cabinets arose from the will to capture and to categorize all that man and nature could make. Precursor of the museum as it is instituted today, encyclopedic cabinets initiated the marshaling and division of knowledge itself. R. J. W. Evans, in *Rudolf II and His World*, writes, "The assembling of many and various items reflected the essential variety in the world, which could nevertheless be converted into unity by a mind which brought them together and divined their internal relations one with the other.... [W]e see the analogy from microcosm to macrocosm and the search for a 'key' to the harmony of the created universe."[2] It is no accident that imperial cabinets expanded at the moment when Europe was encountering for the first time the wonders brought back by explorers and conquerors of what was called the New World. A collection was a world in parallel: physical encyclopedia, art museum, repository of dynastic regalia, of ethnographic plunder, scientific laboratory, and philosophical archive. Not least, it was an instrument of imperial prestige, diplomacy, and power. "A prince expresses his *virtus*, his worth, in his collections," observes Thomas DaCosta Kaufmann.[3]

Rudolf II (1552–1612) ruled the Habsburg lands of Central Europe from 1576 to 1612. His two grandfathers were brothers: Ferdinand I (through his father) and Charles V of Spain (through his mother); his parents, Maria and Maximilian II, were first cousins. He was educated and came of age in the court of his uncle, Philip II in Madrid, sent there by his mother from Vienna for a good Catholic upbringing.[4] Only seven years younger than his cousin Don Carlos, he would have arrived in Madrid at the age of eleven, the year after Don Carlos's illness and cure. On the early death of his father, he assumed rule at age twenty-three as archduke of Austria, king of Hungary, Croatia, and Bohemia, and (by age twenty-four) Holy Roman emperor. Such were the imperatives of dynasty.

Rudolf's extensive collections, including treasures secured from his Habsburg forebears, were housed in three great halls of the Prague castle, with a fourth vast chamber given to the *Kunstkammer* proper and its library.[5] Objects in the *Kunstkammer* comprised *naturalia* (natural history, zoology, botany, and mineralogy), *scientifica* (clocks, astronomical instruments, terrestrial and celestial globes, and compasses), and *artificialia* (weapons, textiles, coins and medals, prints, and artworks of surpassing craftsmanship).[6] Works by the finest goldsmiths and clockmakers in Europe were represented: Anton Schweinberger, Wenzel Jamnitzer, Gerhard Emmoser, Jost Bürgi, Christoph Margraf, and Hans Schlottheim. Court craft workshops flourished, such as that of the incomparable gem-cutter Ottavio Miseroni and his family. The library contained treatises by Paracelsus, Giordano Bruno, Galileo, Giovanni Battista della Porta, Tycho Brahe, and Johannes Kepler, as well as archives of codices on astronomy, mechanics, mathematics, alchemy, and the search for perpetual motion. The *naturalia* featured precious stones, medicinal herbs, bezoars, ambers, fossils, dinosaur bones, preserved creatures of air, land, and sea, monsters and mandrakes, marvelous fakes, and the products of the occult crafts.[7] Beyond the halls of the *Kunstkammer* itself, Rudolf's collections boasted artworks by Titian, Correggio, Veronese, Tintoretto, Dürer, the elder Brueghel, Giambologna, and the court artists Giuseppe Arcimboldo, Bartholomeus Spranger, Hans von Aachen, Adriaen de Vries, and Joris Hoefnagel, among others.

The *Kunstkammer*, Fučíková recounts,

> was not merely a static display to please the eye of the onlooker; it was a living organism and a fascinating resource. Various ores, natural gemstones, shells, horns, rare nuts, and the like were deposited in it not merely because of their unusual shape, the rarity of their material or their origin in distant lands: they represented a store serving the craft workshops of the court, to which exotic or precious raw materials were dispatched only to return to the Kunstkammer in the form of artefacts.[8]

Rudolf famously kept his *Kunstkammer* secluded and private, yet it was also available as an open laboratory to appropriately credentialed visiting and court alchemists, astronomers, artisans, and artists.[9] The collection reflected Rudolf's secular passions. Though educated in Philip's court, he resisted the orthodox pressures of the Spanish sphere in his patronage and his reign.

Rudolf's *Kunstkammer* did not survive. At Rudolf's death, his brother Matthias succeeded in moving only fractions of the huge collection to Vienna; other parts were sold in the next decade by Rudolf's successors in order to finance military campaigns; and the remainder was pillaged in 1648 by Swedish forces in the sack of Prague at the close of the

Thirty Years' War.[10] The posthumous inventories composed in 1619 and in 1621 reflected a collection much reduced and its order lost. Scholars searched in vain for an inventory said to have been produced during Rudolf's lifetime. Piecemeal reports from letters by court visitors give a mythic sense of its size and diversity, but few visitors privileged with an invitation had time to absorb Rudolf's method of organization. In the absence of a direct record, the nature and arrangement of the *Kunstkammer* remained elusive, contributing to the reputation of the emperor as a recluse secluded among his fathomless treasures, neglecting affairs of state.

Just after World War II a manuscript was found in the Bibliothek des Regierenden Fürsten in the tiny nation of Liechtenstein, the German-speaking Alpine principality dating to the late medieval era, whose nobles had once served Habsburg rulers. Archived in the library only as an "inventory of an unknown collection," it was discovered by Gustav Wilhelm, director of art collections for the Principality of Liechtenstein, who thought he recognized its identity. Knowing there were at least some surviving original *Kunstkammer* objects that had passed into the collections of the Kunsthistorisches Museum in Vienna, Wilhelm produced a transcription of the manuscript in 1947 and gave it to Erwin Neumann, curator and later director of the Collection of Sculpture and Decorative Arts at the museum. Neumann, confirming what Wilhelm had suspected, published an account of the find in 1966, announcing, "The author of the present text…finds himself in the agreeable position of being able to report the unexpected discovery of the original inventory, long believed lost or even nonexistent, of the Rudolfian Art chamber." (**FIG. 22.2**).[11]

Out of one war came the key to an earlier war's ruins. Neumann foresaw that besides furnishing a provenance for works of art that had survived the dispersal, the inventory would lead to a broad reappraisal of Rudolf and his controversial reign. Scholars transcribed its full text in the next decade, working from a bound parchment codex of 415 sheets, listing almost three thousand entries. Their assessment lays out the method, coherence, and scope of the collection, as well as Rudolf's unflinching promotion of the arts and the natural sciences in the face of inquisitorial Rome and the western empire under Catholic Spain.[12]

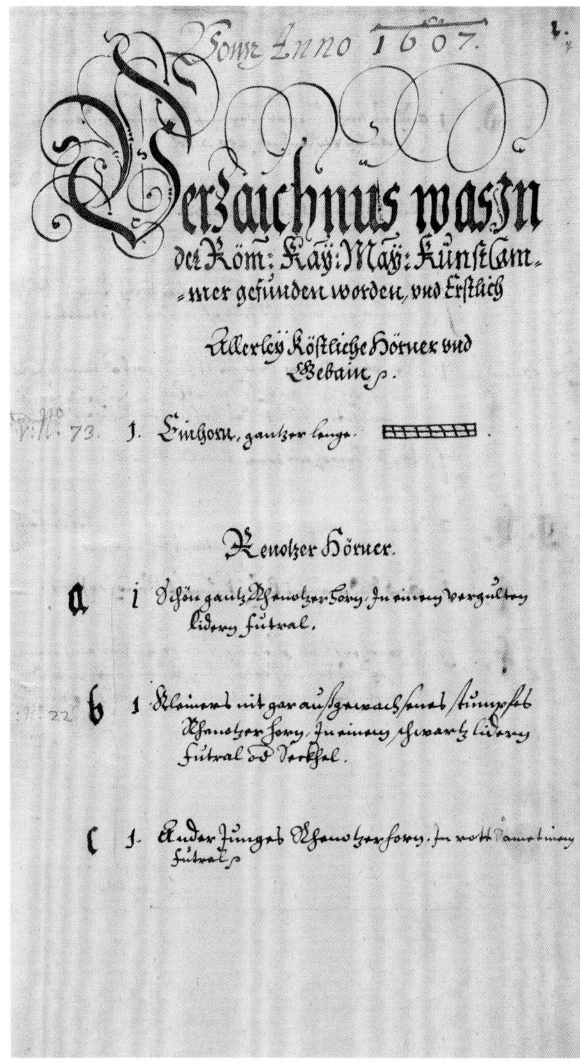

FIGURE 22.2.
Title page of the *Kunstkammer* inventory of Rudolf II, 1607–11. Kunsthistorisches Museum, Vienna. The title reads in part, "VON ANNO 1607. VERZAICHNUS, WAS IN DER RÖM: KAY: MAY: KUNSTCAMMER GEFUNDEN WORDEN" (From the year 1607. Directory of what has been found in the *Kunstkammer*).

The inventory itself is unsigned, but its author is believed to be Daniel Fröschl (1563–1613), painter, miniaturist, and, from 1607 until the emperor's death, Rudolf's trusted court antiquarian. Clearly, Fröschl was intimately familiar with the collection, and his many notes on the sheet margins document the active use of the collection over time.[13]

Are automata mentioned in the inventory? In that golden age of the mechanical clock, indeed they are. Many were small figurative embellishments on the striking trains of clocks, but others are listed as complex performing clockwork machines, not timepieces. A few entries name makers. Augsburg and Nuremberg loom large as manufacturing locales. A number of the finest machines listed in the inventory still exist today, preserved in the Kunsthistorisches Museum in Vienna. Surviving works include the gilded silver ship made for Rudolf by the Augsburg master Hans Schlottheim, which we described in chapter 17.[14]

But are any self-propelled androids listed in the inventory? Perhaps fashioned of more humble materials, dressed in real cloth? In fact, there are two such entries, tantalizingly brief. Listed under the heading "Clocks and clocklike works [*Von uhrn und dergleichen rederwercken*]" is entry 2145.h/379:

> Another of C. Marggraven's works is a waxen likeness/image, covered in violet velvet, [which] strums a cittern, moving its head and its hand. Wound up like a clockwork, it strums a fantasia.
>
> [Ein ander werckh vom C. Marggraven mit veyelbraunem samet überzogen, ist ein waxin bild, schlegt uff der cittern, mit dem kopf und hand sich bewegendt, man zeuchts auf wie ein uhrwerckh, schlecht ein fantasia.][15]

Later in the inventory, under a different heading, "Clockworks in pictures and other diversions [*Uhrwerck von bildern und andern kurtzweyligen sachen*]," we find entry 2194/426:

> A young boy dressed in a long coat, wound up like a clock, beats the drum with natural movements and walks about.
>
> [Ein jung knäblin in einem langen röckhlin beklaidet, welches man auch wie ein uhr aufzeucht, schlegt natürlich die trummel und geht umbher.][16]

Precious few words, and much hangs on these tacit phrases! The name "Marggraven" leaps out, for this is Christoph Margraf, whose name was added to those of Bulmann and Turriano as makers of mechanically moving figures in the description of the Vienna cittern player in the 1966 catalog of the Kunsthistorisches Museum.[17] Erwin Neumann himself wrote that catalog description, as he was by then head of the Collection of Sculpture and Decorative Arts at the museum. He does not cite a source for his reference to Margraf, but with the transcript from Liechtenstein in his hands, he must surely have spotted the entry describing the velvet-clad cittern player. But does the entry describe a walking figure? And is it a female figure? The German phrase that concludes the entry, "schlecht ein fantasia"—strums a fantasia—might equally apply to a music box supporting a partly movable figure rather than a walking android.[18] And how did Fröschl know this was a Margraf piece? Was it signed, or did Fröschl know the court clockmakers so well that it was familiar to him? If only those two Rudolfian inventory entries were *combined* into one marvelous automaton, we would have an account of an android as complete as that of Ambrosio de Morales's *dama* by Juanelo.[19] Both entries mention the figures' clothes: "covered in velvet" and "dressed in a long coat." Robes or dresses would hide the clockwork, an important clue, as we have mentioned. None of the descriptions identify the material of the clockwork itself, if of brass or of iron.

But Neumann's addition of Margraf to the list of clockmakers known to make walking automata is important. From 1595 to 1612, Margraf was a celebrated clockmaker in the court of Rudolf II. A signed and dated clock by Margraf, today displayed in the collection of the Kunsthistorisches Museum in Vienna, is his 1596 *Kugellaufuhr* (lit., "rolling ball clock") based on Galileo's discovery that a rolling ball responds to gravity in a way that can measure time. An elaborate machine, the clock takes the form of a box: when its lid is tilted open a mirror on the underside of the lid allows a view of the clockwork inside. There, on an inclined glass plate, along wires strung in a zigzag pattern, a small white ball rolls back to front and top to bottom. Viewed through the mirror, the ball appears to float in space, resetting itself and continuing its motion as long as a hidden mainspring and fusee run. All the while the ball transmits its motion to the clock face on the front of the box and to a striking bell within. With its illusion of perpetual motion, it was said to be the first of its kind—and it won him special privileges in Rudolf's court.[20] Such an object exemplifies the kind of innovative machine that a court clockmaker had the opportunity to develop. Future research on automata must include a search for works by Margraf.

The much-awaited publication of the Rudolfian inventory in 1976 allowed historians to at last find their way to individual entries in that great list.[21] The news traveled fast, for by 1980, when Klaus Maurice and Otto Mayr presented the Smithsonian monk in *The Clockwork Universe*, they added a new reference to its modest bibliography. To the sources naming Bulmann and Turriano, they appended one more: "In Emperor Rudolf II's collection of art treasures there were a number of spring-driven clothed dolls that beat upon drums and walked about."[22]

This is the landscape behind the circumspect bibliography Maurice and Mayr assembled for the Smithsonian monk. "Spring-driven clothed dolls that beat upon drums and walked about"—the simple but telling paraphrase Maurice and Mayr borrowed from the inventory in indicating cousins to a small, dressed, mobile, self-propelled automaton beating, not a drum, but its own chest. Morales in Toledo, Neudörfer in Nuremberg, and Fröschl in Prague: three eyewitnesses observed such automata.[23] The texts they wrote served different aims. Morales, on royal assignment for Philip II, tours cities of Spain and reports Juanelo's accomplishments in Toledo. Neudörfer and Doppelmayr include Bulmann in biographical chronicles of outstanding citizens of sovereign Nuremberg. Fröschl names Margraf for an object in an imperial cabinet inventory. But all three confirm the existence of walking clockwork figures and allow us to posit plausible settings for their manufacture and destination. It is just possible that one or two of the objects described in these brief passages parade before us now. The texts chronicle a family, a genus, to which the figures belong. Although it is difficult to envision a monkish android in Rudolf's secular *Kunstkammer*, we might picture a figure not unlike it among wonders kept hidden in the emperor's private theater, set before a select audience on the right occasion. Automata took their place among clockwork marvels and demonstrated both the power to fashion a lifelike being—and the power to possess it.

To the Rudolfian inventory we should now add another list from a later collection: this one in Italy, assembled by the seventeenth-century nobleman Manfredo Settala (1600–1680). As mentioned in chapter 16, the clockmaker Giorgio Gregato, who restored the Milan automaton, has argued that an entry in the 1666 Settala inventory identifies this automaton. Settala was a visionary Milanese cleric famous both as a collector and as an inventor: he called himself an artisan-scholar. "Settala created a museum filled with natural and ethnographic curios, that arrived from the far reaches of the formidable Catholic missionary networks, as well as mechanical and optical inventions of his own device," the historian Paula Findlen writes.[24] Among independent patrician collectors in the seventeenth century, Settala vastly expanded a collection of medicinal and scientific matter begun by his father and pursued a vigorous effort of classification that advanced the meaning and definition of collecting well beyond the by then typical *Wunderkammer*. Colleague of Athanasius Kircher in Rome, and of a wide circle of thinkers, makers, princes, and cardinals, he had his inventory published in both Latin and Italian, for broadest notice, and won the support of powerful patrons, including Federico Borromeo, archbishop of Milan, who had founded the Biblioteca Ambrosiana in 1607. At Settala's death, the collection passed through subsequent generations of the Settala family and ultimately to the Ambrosiana in 1751.

In citing the 1666 inventory, Giorgio Gregato adds a welcome source to the small list of period records that refer to walking automata.

23 Searching the Lost and Found

The 1607–11 inventory of Rudolf's *Kunstkammer*, emerging from oblivion, gives us a glimpse of how automata might have been seen, and by whom. A lost inventory, rediscovered, sets a cascade of revisions in motion. Just as the monk too has appeared only recently, further discoveries will surely come to light and change what we can know. With each discovery, we marvel first at the odds of survival. The Rudolfian inventory lay neglected for 350 years; the monk, far longer. What lost news will reach us yet?

The search in the historical record for citations of self-propelled clockwork figures remains quite new. The most defining characteristics of the monk—that he walks by himself, that he interacts with onlookers in a direct way, that he is *a monk* and not, say, a lute player—would seem to be distinctive features ripe for comment by period witnesses. Morales, Neudörfer, and Fröschl single out an automaton's walking capability, definitive for a figure that moves under its own power. But would not further detail about mobility—especially the movement of mouth and eyes—be equally defining features of such a machine? Above all, wouldn't a *praying* monk warrant special notice?

The automata that have survived to our time have become, in retrospect, small archetypes: prime objects that launched the evolution of the great androids that followed. Contemporary interest in the history of artificial life turns a spotlight on a class of early automata that may have been perceived as only incidental novelties at the time they were made. The more complex composite machines like Schlottheim's nef were seen as outstanding accomplishments of their day, worthy of preservation even as technology moved on and patronage changed. The clockwork astrarium retains its high status even today, rightfully so. A solo figure, in contrast, with wooden body and perishable dress, was perhaps only "a toy and fit for mirth," as Morales wrote, to be discarded with other trifles as a culture outgrew them. What purpose could the monk have served? Now that a number of kindred figures have emerged from history's shadows, one can hope and expect that the discovery of additional neglected records, and even newfound automata, will follow. A new understanding of the significance of these figures has already brought more of them to light.

These small figures may have been lost in time because they are protean and mutable. The metal clockwork outlasts the wood sculpture—and even damages it: recall the small felt pad added to the monk's chest to dampen the splintering blows of his own hand. All the figures we examine in this study sooner or later received new costumes, for the cloth dress was the first to deteriorate. The mandolin player in the Conservatoire national des arts et métiers in Paris was identified as a nineteenth-century figure, then later understood to be an older machine under a newer dress. Only the brass mechanism itself inside the lute player

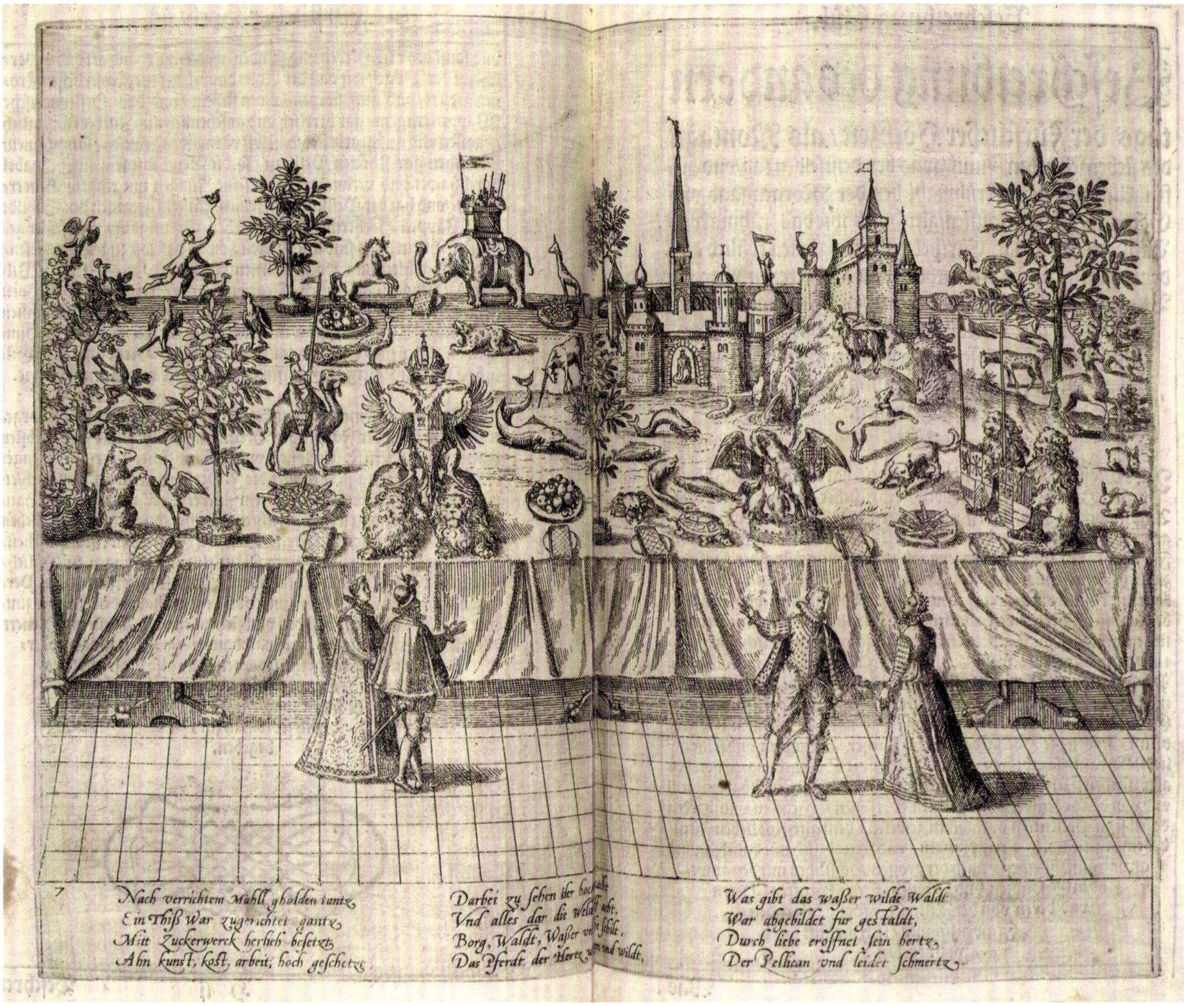

in the Bavarian National Museum in Munich (and remnants of the wooden body shell) is original. With such similarity of clockwork design across the family of automata, is it possible that these durable machines were always intended to be finished on demand? Perhaps they were meant to animate relatively inexpensive, replaceable, even expendable characters. Or perhaps they were showpieces to be categorized with other (to us, equally marvelous) table ornaments meant for onetime festivals and court celebrations **(FIG. 23.1)**.[1] Such utilitarian or ornamental purposes would distinguish them from wonders of a different order, with only a few rare figures finding their way to the *Kunstkammer*. Could this explain why little apparent value was placed on the figures we study here—that is, if the scarcity of contemporary descriptions of them should be so interpreted?

FIGURE 23.1.
From *Fürstliche Hochzeit* (A princely wedding) by Dietrich Graminaeus, engraving by Franz Hogenberg, Cologne, 1587, a book that commemorates the extravagant wedding of Johann Wilhelm, duke of Jülich-Cleves-Berg (son of Maria of Austria), and Jakobea von Baden, 1585, Düsseldorf. The engraving depicts a tabletop garden menagerie, sculpted in sugar (Zuckerbäckerei). Klaus Maurice writes that automata too were created for decorative banquet settings like this one, a world in miniature.

Talking Heads

A subtle moment in the monk's performance, as David parsed and timed the sequence of actions in chapter 10, bears a second look.

> **The kiss done, the monk raises his head and continues his advance across the table, and now his eyes again move slowly to the left, giving the uncanny appearance of a sustained sight line fixed on a single point as the monk walks past it. Brief as this is, if an onlooker happens to occupy this spot, he will likely feel as if the monk has singled him out for special attention, in effect keeping an eye on him. The surveillance lasts just so long, then the eyes dart forward, as if in sudden dismissal. For each revolution of the fusee, the eyes drift left and then spring forward three times. The third time, as the monk pivots to march in a new direction, his eyes stay momentarily locked on the person toward whom he had been headed, and he holds the gaze as he turns away.**

This remarkable bit of clockwork required the tight coordination of separate parts of the gear train. We watch the automaton, and it watches us back. One thinks of the ancient phenomenon of the "following illusion" in the Egyptian Old Kingdom reserve heads, with eyes ingeniously cut from rock crystal in such a way that they seem to move with the motion of the viewer. Or of the all-seeing eyes of the frescoed medieval Pantocrator. The artist's depiction of the eye and its gaze is a subject unto itself. But representing the act of looking performed by a moving sculpture was something altogether new to art and to mechanics. Even today, David notes, sustained sight line is one of the biggest challenges in the technology of human/robot interaction.

Adelheid von Herz's research emphasizes the mobility of the *faces* of the three male figures—the Smithsonian and Munich monks and the Budapest saint (and to these we now add the Milan matron). Referring to them as "die cephalen androiden" (lit., "the cephalic androids"), she suggests that the motions of their heads, eyes, and mouths represent the enactment of perception and response. A head turns, its eyes make contact with an onlooker, its mouth opens to speak.[2] Especially the movement of the mouth, von Herz says, links these automata to the literature of talking heads in medieval and Renaissance lore.[3] Although no actual voice is produced, the repeated opening and closing of the mouth in the Smithsonian monk might correspond to uttered language associated with the Catholic liturgy, especially the repentant "mea culpas" of the beating of the breast. But von Herz asserts further that the sequenced actions of the head and eyes show a clear intent to represent a gaze that connects with an onlooker in a performance of admonishment and warning. The mobile sensory organs of the head imitate not merely acts of perception but feedback structures of cognition and reaction as well. "These mimetic qualities involve a visualization of invisible processes," she writes. "Driven by a concealed mechanism, the figure performs visible movements simulating human behavior that constitute a reflection of inner processes. That is to say, sensory, noetic, and moral human faculties are being simulated."[4]

This representation of inner process goes beyond the simple simulation of motor activities in the female musician figures, she continues. The period writers seem to describe not the male figures but the lady musicians with their less complex, more programmed movements (Morales, Neudörfer, and the Margraf figure in Rudolf's inventory). "It would appear," von Herz comments, "that the 'female musician' was a topos, that is to say, a characteristic and widespread form of android that was a popular motif for ornate toys designed for use in a sophisticated entertainment."[5] She argues that the kinetic image of the female musician does not go beyond a simulation of human motor actions. The three male figures, however, mimic the paired actions of looking and of speaking; they give the impression that they observe their surroundings, that they are aware of the things they hold in their hands; they "speak" in response to what they see, in a pattern of syntactic perceptions and reactions. We could add that the monk, seen in performance, even mimics human spontaneity. He seems to turn not on cue, not on a mechanical pattern, but as if abruptly making up his mind.

Automata with head and face mobility would surely have been noted by annalists, even if at the time they were considered comparatively minor pieces in terms of their clockwork. Von Herz comes to the same conclusion we have reached, reviewing the historical record: "None of the texts describes a figure possessing as complex a program of movements, especially involving the head, as is found in the surviving male figures. Since the sources [we do have] repeatedly emphasize the naturalistic qualities of the movements, it seems to me that they would hardly have overlooked or failed to mention such a complex program of movements."[6]

Later Sightings?

Equally surprising is the absence of sure images of such automata in the many late Renaissance paintings and illustrations depicting collections, studios, and *Kunstkammern*. For all the pleasure of the search—for example, in the encyclopedic suite of paintings titled *The Five Senses* (1617–18) by Jan Brueghel the Elder and Peter Paul Rubens in the Museo del Prado, Madrid, replete with clocks, instruments, workers at the forge, and the bounty of art—no automata can be found.[7] Not until the eighteenth century do we alight on an engraving that clearly pictures a small walking automaton: *La charmante catin*, 1742 **(FIG. 23.2)**. The print bears an inscription in French:

> *Sans badiner, malgré ce que nous sommes,*
> *Tel est souvent nôtre destin;*
> *Les Grands quand il leur plaît font à peu près*
> *des hommes*
> *Ce qu'un ressort caché fait icy de Catin*
>
> [Without jest, despite what we are,
> Such is often our fate;
> The Great Ones, when it pleases them make of men
> the same as a hidden spring here makes of the doll][8]

Horst Bredekamp opens his groundbreaking study, *The Lure of Antiquity and the Cult of the Machine*, with this print.[9] Image and inscription are credited to Charles-Nicolas Cochin, and the ironic message is clear, for by then the clockwork automaton had begun to assume its now-familiar pejorative metaphor. "The doll," Bredekamp writes, "bathed in a diffuse and flickering light, dominates the scene with the power of a supernatural apparition. Facing the magic of this autonomous movement, the plainly-dressed woman in the foreground falls to her knees in a praying gesture and the other, more reserved witnesses also seem deeply affected." *Without jest*, the image jests, depicting astonishment where wisdom should be. The engraving, however ironic its message, also captures the genuine astonishment that must once have greeted such an automaton. We know better now, the inscription implies, and the magician at far right with his magic lantern, who has just wound the doll up, must be enjoying the show.

What the engraving also demonstrates is that Mssr Cochin had in fact seen such an android. Its size is correct, full dress covers the clockwork, and the figure moves under its own power, sharing the floor with the onlookers. The

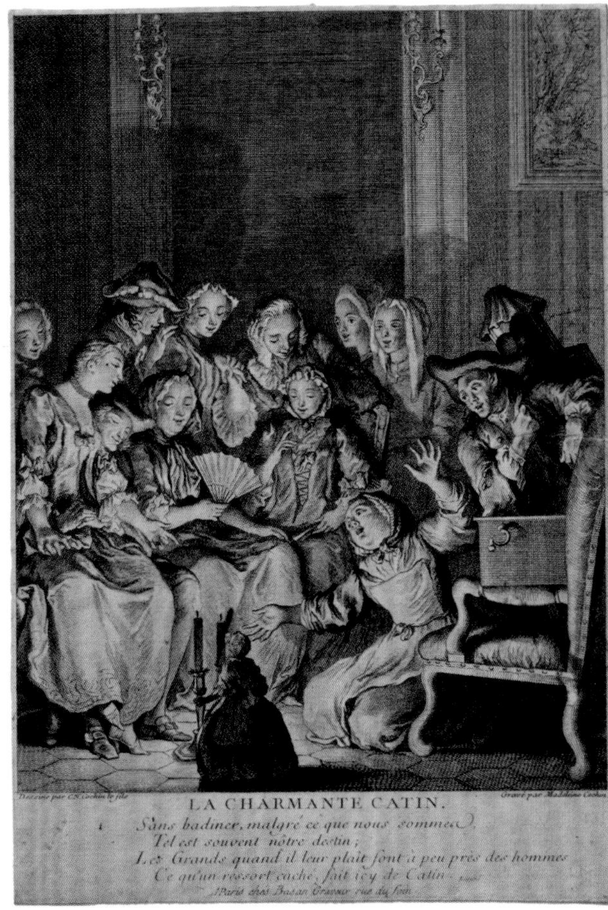

FIGURE 23.2.
Louise-Magdalene Hortemels, after Charles-Nicolas Cochin, the Younger, *La charmante catin*, 1742. Engraving. Sheet: 27.5 × 19 cm (10¾ × 7½ in.). A small walking automaton performs for spectators in a candle-lit room.

doll's feet and wheels may be faintly visible in the shadow under the hem of the dress. Another print with the same title, dated 1737, also French, in an album of etchings of working-class Parisiens, features a peasant girl staring at a doll in fancy dress (**FIG. 23.3**). At least one wheel is clearly visible beneath the dress hem.

Here we can make a connection to the comment of Johann Gabriel Doppelmayr, quoted above (see chap. 20 n. 5), who wrote in 1730 that such artificial machines were no longer rare in his time, citing three contemporary descriptions. Of these, one by Kaspar Schott (1608–1666), a German Jesuit mathematician known for his interest in mechanical instruments, is invaluable, though we note it dates not from Doppelmayr's time but from the previous century. In a section titled "Statua automata" in his treatise, *Magia universalis naturae et artis*, 1657, Schott describes two tabletop walking maidens, both of them tantalizingly close to those we have studied:

> Moreover, automata statues in human or animal form ought to be told. In his *Mathesis Polemica*, Adrianus Romanus says he saw a statue in the size and form of a young girl, who was marching forth, bending down, rising up, extending her hand, and performing other things. She could hardly be distinguished from a real girl. What's more, in Würzburg, the most illustrious lord Konrad Friedrich von Thüngen, canon of the Würzburg Cathedral, showed Romanus a silver drinking cup made in the form of a girl and filled with wine. It proceeded across the table, even to the far end to whomever it was ordered to approach. When it reached the end of the table, it remained there without any movement, as if it knew the boundary and precipice. After the wine was drunk and the cup was refilled, it was set free and dispatched to someone else in the same way, and its motion continued for another half hour. We also saw in Palermo in Sicily at the house of the most reverend lord Abbot Gieze, among other rare things and artifacts, a statue in the form of a girl, who played the *citharam* beautifully and harmoniously and danced to the rhythm. It roamed across the table and when the song ended, it turned to those around it and greeted them with a bow. There won't be enough time and paper to go over everything I have seen at the house of the Abbot in Palermo and at many collections, or "galleries" as the Italians call them, of princes and cardinals, in Rome and various cities in Germany.[10]

This wonderful account singles out two automata marching across a tabletop, making sequential stops before guests, one even bowing to finish the performance. Could these automata have been made in Schott's time, or were they earlier works, treasures in noble collections? The silver drinking cup maiden calls to mind the small early seventeenth-century Joueuse de théorbe now in the Musée d'Horlogerie du Locle in Switzerland, described in chapter 17. The dancing *citharam* player is even more intriguing, so close is it to the lute players illustrated in part 3. Schott admired these figures among "rare things" in that house in Palermo, belying Doppelmayr's later claim that they had become common. And rare are such period descriptions yet, among the few we have found.

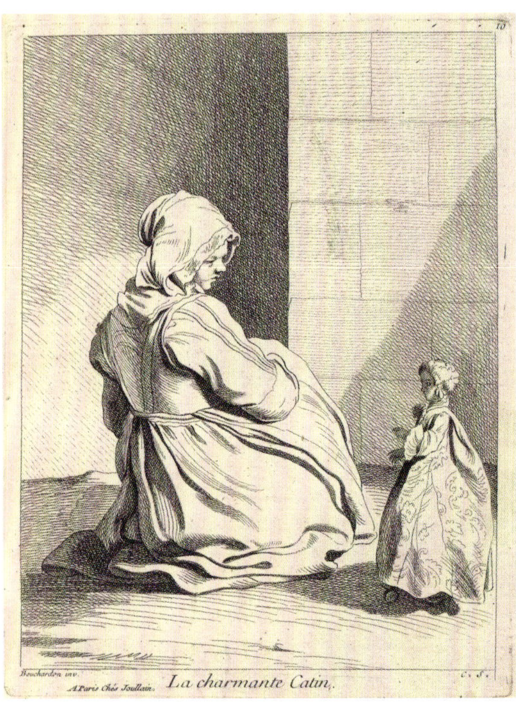

FIGURE 23.3.
Anne Claude Philippe de Tubières, comte de Caylus (etching), Étienne Fessard (engraving), after Edmé Bouchardon, *La charmante catin*, in *Études prises dans le bas peuple ou les Cris de Paris*, 1737. 23.8 × 18.6 cm (9½ × 7¼ in.). British Museum, London.

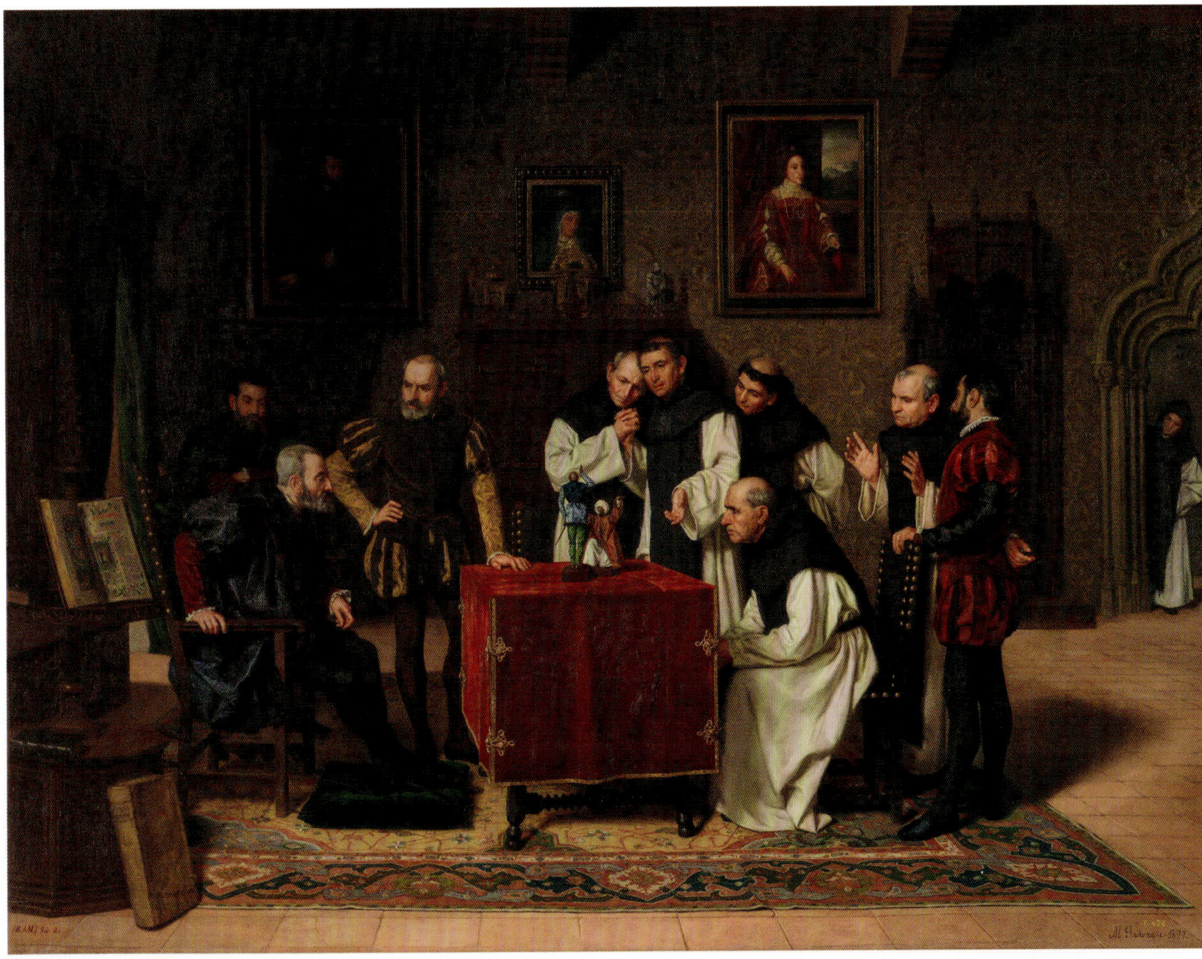

A more startling image of a group of spectators in the presence of an alleged automaton, although even more distant from the sixteenth century, is a canvas by the Spanish painter Miguel Jadraque y Sánchez Ocaña titled *Carlos V en Yuste*, finished in 1877 (**FIG. 23.4**). The painting shows the seated emperor, surrounded by friars and courtiers, leaning forward to study two diminutive figures on a cloth-covered table. One, in caftan and turban, poses on one knee with arms flung up; the other is in European dress with one hand on his sheathed sword and the other hand raised in the sign of the cross. Each is mounted on a low plinth. Their colorful garments appear to be of real cloth, with the folds of the caftan falling over the plinth to the table. Nearly every gaze of the assembled company converges on the pair, tokens of

FIGURE 23.4.
Miguel Jadraque y Sánchez Ocaña, *Carlos V en Yuste*, 1877. Oil on canvas. 147 × 193 cm (58 × 76 in.). Cámara Oficial de Comercio e Industria in Salamanca (La Casa de la Tierra), long-term loan from the Museo Nacional del Prado, Madrid. Jadraque's painting shows the aged emperor, his courtiers, the friars of San Yuste, and perhaps Juanelo himself standing behind the monarch. What holds their attention? The painting surely shows that the legends of Juanelo's automata remained vivid for nineteenth-century viewers.

the Habsburg-Ottoman struggle that dominated Charles's reign. Could they represent the much-fabled figures that Juanelo Turriano is said to have fabricated for the emperor's diversion at the monastery of San Yuste? Other objects in the painting reveal historically accurate knowledge of the objects the emperor took with him into retirement: at right is Titian's portrait *La emperatriz doña Isabel de Portugal*, Charles's much-loved wife; next to it is a smaller panel, perhaps the Titian *Dolorosa* known to have been in Yuste.[11] Both works now hang in the Prado, since 1821 and 1839, respectively. At left is a large portrait in the shadows, difficult to identify but not unsimilar to another canvas by Titian titled *El caballero del reloj* (The gentleman of the clock), once thought to be a portrait of Juanelo Turriano (**FIG. 23.5**). This too hangs in the Prado.[12] Two clocks stand atop the cabinet in the background, at least one of which is close in form to a clock by Juanelo, listed among objects at San Yuste at the time of the emperor's death.[13] In the foreground of the painting, at far left, just under the emperor's right hand, with a large book propped against it, is an octagonal cabinet made of dark wood. It is not inconceivable that this is the wood case Juanelo made to contain his great astrarium, "the great clock, made by Master Juanelo, with its case," as documents from San Yuste identify it. Size and shape are exact.[14] Juanelo himself is surely represented in the shadows, if not in the portrait hanging just above him on the wall, on call behind his monarch.

Peter Friess, former conservator at the Deutsches Museum in Munich and restorer of that museum's automaton (see chap. 12), coauthored an essay about this picture with the art historian Reinhard Steiner in 2003.[15] While the small figures on the table are the focus of the painting, Friess and Steiner draw attention instead to the small hooded monk in the background, on the cabinet next to the clocks. They compare it, first, with the famous alabaster mourners of near-identical small size found in the great tomb sculptures of the dukes of Burgundy in the late fourteenth and early fifteenth centuries, monuments familiar to Charles, who inherited rule of the Burgundian territories in 1506. They go on to observe that it is close in size to the monk automata in Munich and Washington. Citing legends of Charles's preoccupation with the details of his funeral, they propose that an ensemble of automata might have constituted a formal lamentation group commissioned by the melancholic emperor.[16]

The tiny figure of the monk in the painting is indeed very close in scale and in pose to the alabaster mourners from Burgundy, and it likewise shares the proportions of the automata. Does the painter represent here a sculpture or an automaton? The argument for the latter depends on the presence of other clues to Juanelo Turriano in the picture. As a nineteenth-century view of a legendary moment in Spanish history, the painting reflects the rich lore surrounding the emperor's final years at San Yuste. And as Friess and Steiner point out, it shows that stories of the emperor's love of clocks and automata remained alive in the popular imagination, as did legends of Juanelo himself.

The monastery of San Yuste was destroyed in 1809 by Napoleon's forces during the Peninsular War. The treasures and belongings of the emperor were recalled to Madrid long before that, at the time of his death, though the monastery itself remained active following the departure of its royal guest. It has since been restored, and a sundial said to be made by Juanelo is still there.

FIGURE 23.5.
Titian, *El caballero del reloj* (The gentleman of the clock), ca. 1550. Oil on canvas. 122 × 101 cm (48 × 40 in.). Museo Nacional del Prado, Madrid. Once considered by historians to be a portrait of Juanelo Turriano.

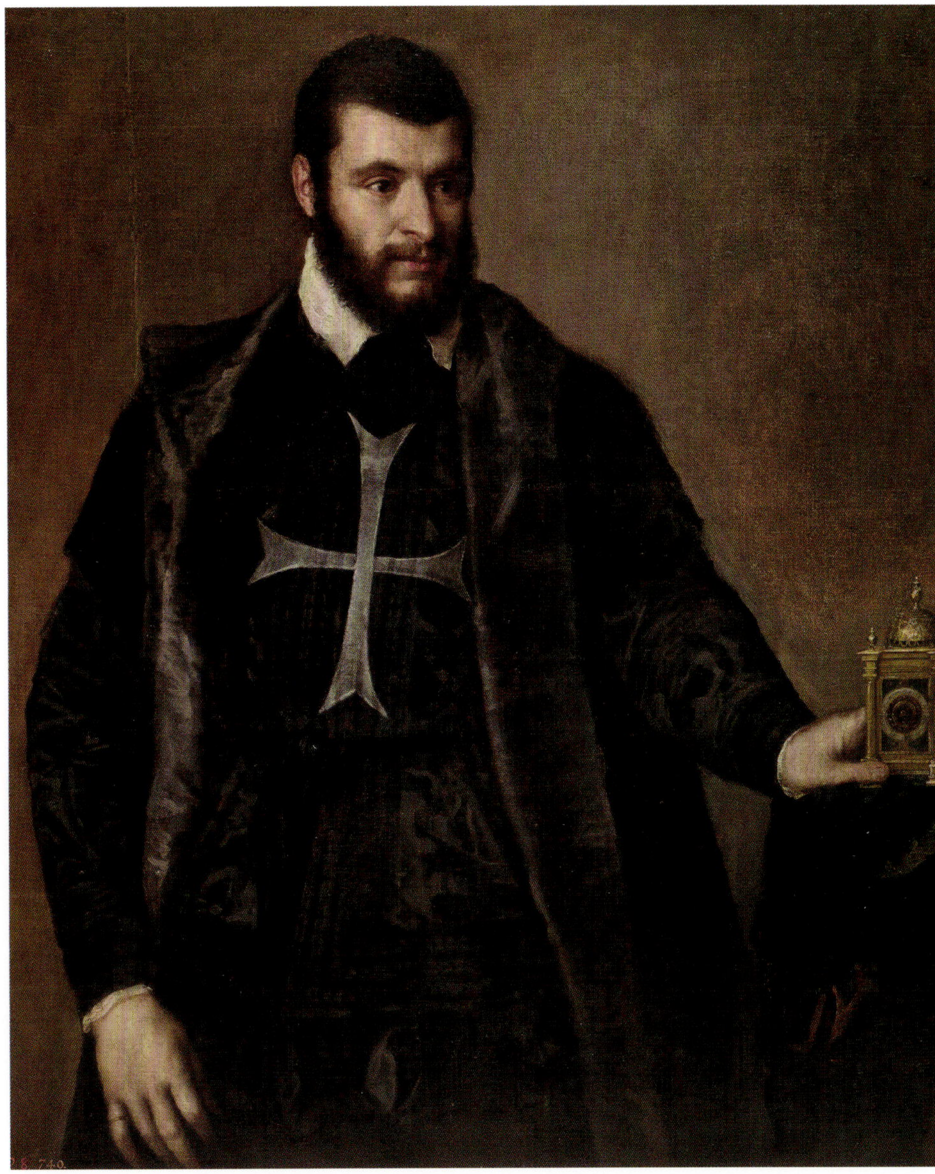

24 A Monk's Opinion, Continued

Another inventory—the catalog of objects at San Yuste, compiled after the emperor's death— together with the wealth of correspondence from the final years of his life were ultimately transferred to the royal archives in Simancas, a stronghold initiated by Charles himself and institutionalized in 1588 by Phillip II. The archives suffered disruption and partial transfer to France in the Peninsular War (1807–14) but were reclaimed after Napoleon's defeat. The task of restoring order fell to the archivist Don Tomás González, canon of Plasencia, whose labor was elegantly acknowledged by William Stirling-Maxwell in *The Cloister Life of the Emperor Charles V.*[1] González had begun his own narrative account of the San Yuste years, but he died before it was finished, and his manuscript passed through several hands and ended up, ironically, back in France in the archives of the French Foreign Office. Stirling-Maxwell, Louis-Prosper Gachard, William H. Prescott, and others were able to study not just the freshly reconstituted archives in Simancas but, on discovering it, the González manuscript in Paris for an intimate understanding of the final years of Charles V.[2] Juanelo's role in the daily round at the monastery, and his move to Toledo following the emperor's death, come from these sources. Newly available to scholars and readers in the nineteenth century, this literature surely gave the painter Miguel Jadraque y Sánchez Ocaña the details for his portrait of the emperor. The painting, with its accurate depiction of persons, clocks, and portraits at San Yuste, was shown in Paris in 1878.[3] Awareness of Spanish history and precedent was in the air. Only a few decades later, Ernst von Bassermann-Jordan saw the cittern-playing automaton in Austria and seized on the legends of Juanelo's automata.[4]

Though our search has yielded no period witness of the monk, we have been able to assemble a handful of "bones" leading to other kinds of recovery. We have argued that the family of walking clockwork androids to which the monk belongs forms a now-significant class in its own right. We can establish dates for them in the absence of makers' marks. Though their fabrication would have required the skills of more than one craft, a few individual clockmakers were said to have made movable figures that "beat upon drums and walked about." Guild contracts, royal inventories, and biographical briefs, though scarce, suggest a shared technology, a mercantile context, a court audience. We have retraced the steps of scholars who have combed the historical record, and we have observed how the repetition of a surmise from one century to the next can generate its own kind of truth. The chronology of voices citing and reciting Morales to attribute the cittern player and then the monk to Juanelo: this we understand is history's reflex. We ourselves play it forward, in the recounting. Even as other documents and proposed makers—and recovered automata—have come to the fore, Morales's "dama que tañe y dança" remains an indisputable primary source, just detailed enough to hold its own in the void.

And with Juanelo ever in the wings, the connection between the automaton monk and the miracle of fray Diego also comes to the fore. What is the backstory of this connection? The monk came to public notice in the 1970s with the legend intact.

The Legend Reprised

A review of the events is in order. Following the death of Charles V in San Yuste in 1558, the imperial clockmaker Juanelo Turriano assumed the role of engineer to Philip II, in Madrid and later Toledo. The near-fatal injury of Crown Prince Don Carlos occurred in 1562 in the university town of Alcalá de Henares, near Madrid. The prince's unexpected recovery, following the intercession of the body of the Franciscan brother Diego de Alcalá, compelled Philip and his bishops to mount a petition to Rome to confirm a miracle. In 1575 Ambrosio de Morales—court annalist to Philip II, professor at the University of Alcalá de Henares, and author of Diego's vitae sanctorum—published *Las antigüedades de las ciudades de España*, an encyclopedia of the material history of Spain, which included a detailed chapter on Juanelo Turriano and his Toledo workshop. Morales, who knew Juanelo well, inventoried marvels from Juanelo's hand and gave the clearest period description we have found of a self-propelled solo automaton. Juanelo died in Toledo in 1585. Diego was made a saint in 1588, twenty-six years after the miracle, the first Counter-Reformation saint in the Roman Catholic Church (and the first Iberian saint in more than a century).

From these well-documented facts, the legend takes its shape. When the figure of the monk first appeared in the 1970s, attempts to identify it followed several leads that converged on Juanelo and the Spanish court. The ecclesiastical nature of the automaton, the role of Charles V as patron in the golden age of clockwork, and the precedent of crediting similar machines to Juanelo: all coalesced in the Spanish sphere. The story that braided the strands together, and reinforced them by adding one more strand, is the episode of Don Carlos and the miracle. Without that added dramatic strand, an evidence-based argument could still be made for a Spanish origin for the automaton, as Bassermann-Jordan in particular had provided. With the story of the miracle, however, we have a legend that has come to have a life of its own. The legend has attached itself so irrevocably to the monk that the monk all but tells it himself, raising his cross and beating his chest. The person who accomplished this, without realizing the spark he would ignite, was the modern Franciscan historian and scholar Servus Gieben.

Gieben's 1975 letter, a response to an inquiry from a broker, is preserved in the Smithsonian archives among the monk's acquisition papers.[5] As Elizabeth would later learn, the museum contacted Servus Gieben directly to request the letter and received it from him. Gieben's proposal—that the monk was a votive figure made in the aftermath of the famous miracle—is more than reasonable, given the prevalence and popularity of votive artifacts in the church and their role in seeking or commemorating miracles.[6] The miracle of Brother Diego carries its own exhaustive church-ordained proof by evidence, and it provides the momentum for an icon in the form of a "miracle" from the technical sphere: a portrait of the saint to be, on its feet and moving, made after the prince himself gained his feet. An unacknowledged resonance with the story of the cure is set in play by Gieben's proposal.

Servus Gieben

Widely recognized as scholar and archivist, Servus Gieben, OFM Cap. (Ordo Fratrum Minorum Capuccinorum), served two terms as president of the Istituto Storico dei Cappuccini in Rome, where he was also curator and then director of the institute's Museo Francescano. The museum's collection of paintings, drawings, prints, sculpture, manuscripts, seals, medals, and functional objects spans the full eight-hundred-year existence of the Franciscan order and is among the largest of its kind.[7] For some forty years, Gieben developed, cataloged, exhibited, and wrote about its collection of more than 40,000 engravings of Franciscan subjects. Born Harrie Gieben in the Netherlands in 1924, he acquired the name "Servus"—a name he took as a call to a life of service—when he entered the Capuchin novitiate in 1942. He was ordained as priest in 1949 and completed a PhD at the Università Gregoriana in Rome in 1953. As a scholar of medieval theology, he researched the life and works of the thirteenth-century English philosopher Robert Grosseteste, contemporary of Saint Francis, leading to the definitive critical edition of Grosseteste's *Hexaëmeron* (1982) with coauthor, Richard C. Dales. "Whoever confronts this text," a commentator remarked, "can touch at first hand the scrupulousness of Father Gieben and the richness of his scriptural and theological knowledge."[8] Among Gieben's published volumes are many editions of primary manuscripts from the history of Franciscan thought.[9] He was fluent in Dutch, German, French, Italian, English, and Latin. Servus Gieben died in 2014 at the age of eighty-nine, in Tilburg in the Netherlands, not far from his native village of Sint Anthonis.[10]

Gieben himself published the story of how he came to perform his research on the automaton monk. It appears in the journal *Desipientia*, a periodical of the Catholic University of Nijmegen (now Radboud University) in the Netherlands, in an issue that transcribed papers given at a 1995 university conference on Christian iconography. Gieben's paper, "Christelijke iconografie: Problemen en oplossingen" (Christian iconography: Problems and solutions), discussed the unique use of printed images of saints—images rather than texts—for advancing iconographic understanding, and he included some unusual cases that had had real-world consequences outside the task of interpretation. "In the summer of 1975," he wrote,

> I was approached by the antique dealer Georges Sedlmajer from Geneva. He was in possession of a curious automaton, in the figure of a monk, which he wished to identify. Because he thought that it was a Jesuit preacher, he had turned for help to the Jesuits in Rome. But they pointed out to him that the doll was wearing sandals: not the habit of the Jesuits, but of the Franciscans. Thus they redirected him to our Museo Francescano. The automaton doll could perform the gestures of preaching, moving its head and rolling its eyes, and was indeed dressed as a Franciscan. The doll had a very intense facial expression, which pointed in the direction of a portrait. The only other clue was an approximate date of circa 1560.[11]

Sedlmajer had provided good photographs of the automaton. Gieben, then director of the print archives of the Museo Francescano, describes his search through the collection and his study of images of Didacus van Alcalá, which he felt offered a "more than sufficient" likeness to the automaton. The automaton's simple cross and rosary as iconographic identifiers and the shaved head (identifying a lay brother rather than a priest) further confirmed a representation of Diego. From there Gieben singled out, among the prints' depictions of the saint's miracles, the sickroom scene.

> In despair they brought the body of the holy Didacus to the royal palace and placed it on top of the sick boy. Upon this touch the boy awoke from the sleep of death and urned to health within a few days. It is understandable that Philip II now put full effort in speeding up the process of canonization of Friar Didacus in Rome. And to motivate the young Carlos into a more structured life (apart from being sickly and without grace, he was—according to the chroniclers—also extremely cruel and

> **prone to all sorts of foolishness), his father gave him the admonitory automaton with the appearance of his benefactor, the holy Didacus of Alcalá. The author-maker of this complicated mechanism could only have been the clockmaker of the Spanish court, Gionello Torriano.[12]**

Gieben then related that the Smithsonian's National Museum of American History had contacted him a few years after his exchange with Sedlmajer for information about an automaton they presumed was in his own museum's collection, saying they themselves had a similar specimen on display, bought from a Swiss dealer.

> **We quickly solved the misunderstanding about the assumed specimen in our museum when I sent them a copy of my correspondence with Mr. Sedlmajer in Geneva. The specimen in their museum turned out to be the same one I had identified for the Swiss antiquities dealer. They informed me that my information about the holy Didacus had also been cited in a recent Spanish book on the clocks and automata of Gionello Torriano ([by] Jose A. Garcia-Diego of Madrid...). Without my knowing about it, my historical interpretation of an automaton had had far-reaching consequences.[13]**

Unaware of this article, Elizabeth wrote to Servus Gieben in 1999 at the Istituto Storico dei Cappuccini in Rome, and he kindly responded, providing a reference to his *Desipientia* essay. He added that the impetus for his identification of San Diego, in response to Sedlmajer's query in 1975, came on November 12 of that year, the feast day of San Diego, as he was reading about the saint's life.[14] Indeed, his now-essential letter was dated November 16, just a few days later. Remarkably, his study did not touch directly on the writing of Ambrosio de Morales, and he was not aware of any other automaton similar to the monk (such as the Vienna cittern player) or attributed to Juanelo Turriano. As he explained to Elizabeth, both image and text archives pointed to the right setting for a votive object and an admonition to the unruly Don Carlos, and the emperor's clockmaker had the skill to make it.[15] "There is no tradition for this kind of automaton in the Franciscan order, and I would exclude that the monk could have been made for the friars," he affirmed.[16]

Later in their correspondence Servus Gieben mentioned his lifelong interest in clocks and his awareness of the emperor's famous clockmaker. But since he is a scholar of the lives of the saints, his opinion comes from a different sphere than does that of the horological historian Ernst von Bassermann-Jordan, who marshals a separate set of sources in support of Juanelo. They arrive at Juanelo from different points of the compass. The monk belongs equally to the history of technology and to the history of the church. This is why the texts by Ambrosio de Morales, witness to prince, saint, and clockmaker all, are so invaluable. While Otto Mayr and Klaus Maurice introduce us to the clockwork monk, Servus Gieben's remarkable letter gives the monk its identity as votive object, icon, and cautionary figure. Father Servus is our missing art historian—our historian of the spirit—with his eye for art's depiction of saints and his resources as curator and director of one of the world's significant collections of Franciscan images. Such a convergence of expertise across distant bodies of knowledge is precisely what it will take to discuss this automaton, now and in the future.

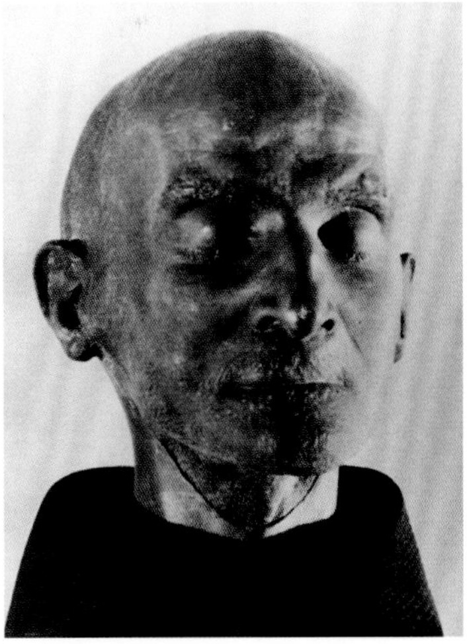

FIGURE 24.1.
Death mask, wax, identified as that of San Giacomo della Marca (1393–1476). Collection of Convento S. Maria La Nova, Naples. Portraits of the saints were often based on such death masks, prized for true likeness. Servus Gieben speculated that such a mask, should one have been made of Diego, might have informed later representations.

"Shaved-Head Saints"

Even a concession to Juanelo opens new questions. Given Charles's well-known love of clocks, would it not be more likely that Juanelo, if he made the monk, made it for the emperor rather than for the emperor's wayward grandson? In this case, since the emperor's death took place four years before Carlos's ordeal, a votive portrait of Brother Diego and the life-granting miracle would be quite ruled out. As for King Philip, formidable as was his belief in the agency of Diego, there are few indications in the historical record that he shared his father's passion for clockworks. To connect the automaton, the clockmaker, and the court of Spain, three alternative scenarios might be tested: (1) an automaton but not a portrait created for Charles V in San Yuste between 1556 and 1558; (2) an automaton made in Toledo for Don Carlos after the 1562 miracle; or (3) an automaton made by another clockmaker, perhaps an associate or former apprentice of Juanelo, on the occasion of Diego's canonization in 1588. It is also conceivable that Juanelo made automata at another time, before his arrival in Spain. He was famous first in Lombardy and worked in Milan. Morales described Juanelo's dancing lady in Toledo, late in Juanelo's life, but gave no actual date for it. Was the automaton there in the workshop, and did Morales actually see it?

Finally, to retain a connection to Spain, there is the real possibility of an automaton made under Juanelo's supervision but not by his hand, for he indeed had assistants. Bassermann-Jordan names the clockmaker Jean Valin, together with an assistant, Jorge de Diana, working with Juanelo at San Yuste.[17] Stirling-Maxwell also mentions one "Jorge de Diano, a boy employed in the workshop of Torriano" at San Yuste.[18] The following comment about Juanelo Turriano and Jean Valin appears in Chapuis and Gélis, *Le monde des automates*: "We are not unmindful that Gianello was assisted by a skilled craftsman, clockmaker Jean Valin. Although we can't be certain what was done by whom in this work, we believe that here, as in many other similar cases, the learned Gianello della Torre was only the designer, while the practitioner Jean Valin executed the work."[19] Otto Mayr himself spoke of assistants to Juanelo in a letter to Elizabeth in 2000. If there is substance to the San Diego hypothesis, he wrote, it would be likely that Juanelo too would be involved. If that were the case, this would put Juanelo in his fifties or sixties, by which time he would have been busy with grander projects. It is easier to imagine him as supervisor of the work of a younger man.[20]

Servus Gieben's instinct that the automaton may represent a specific historical figure was central to his conclusions. "I think it was a question of physiognomy that led me to identify the monk as a Spanish figure and finally as a portrait of Diego," he wrote to Elizabeth.[21] He even entertained the idea that a death mask (**FIG. 24.1**), not an unusual funerary practice at the time, may have served as a model for the likenesses of Diego seen in Cornelius Galle's engraving (see **FIG. 6.1**) and in the automaton itself.[22]

Gieben responded patiently to Elizabeth's questions about his theory, across a correspondence that began in 1999 and lasted until 2007. She asked, for example, about the significance of the shaved head rather than the tonsure. "The tonsure was reserved only for clerics, not for lay brothers like Diego," he replied. She asked, "How many Franciscan lay brothers ultimately became saints? If there were several, even many, then the shaved head becomes only one of a number of reasons for choosing San Diego." Gieben responded by email, with the subject line "Shaved head saints":

> Dear Prof. King,
> The following lay brothers of the Franciscan Order have been canonized:
> Diego of Alcala † 1463
> Felix of Cantalice † 1587
> Benedict the Moor of San Fratello † 1589
> Paschal Baylon † 1592
> Seraphin of Montegranaro † 1604
> Crispin of Viterbo † 1750
> Ignace of Laconi † 1781
> Francis Mary of Camporosso † 1866
> Conrad of Parzham † 1894
>
> I have not counted the lay brothers who died in groups, like Accursius and Adiutus (Protomartyrs of Marocco † 1220), Donnolus of Castrovillari (7 Martyrs of Ceuta † 1227), etc.
>
> Yours faithfully,
> Servus Gieben[23]

If Servus Gieben thought the monk bore a resemblance to San Diego, Elizabeth could rest assured that he had the pantheon of saints in view.

Servus Gieben's 1975 letter to Georges Sedlmajer marked a turning point for all who encountered the automaton, before and after the figure was acquired by the museum. Spain, in the absence of stronger evidence to the contrary, will continue to exert its magnetic compass as the culture of origin for the mysterious monk.

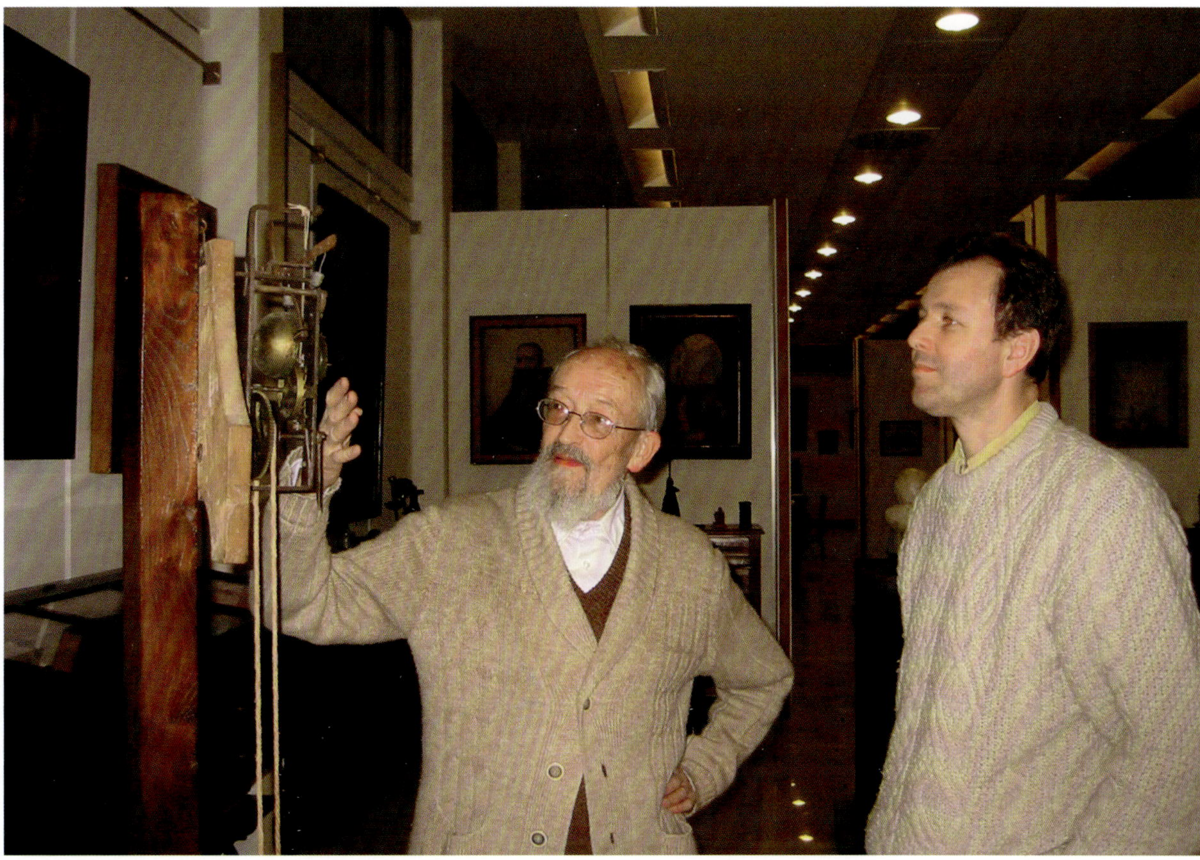

FIGURE 24.2.
Servus Gieben, OFM Cap. (at left), showing an antique meditation clock to a visitor in the Museo Francescano, Istituto Storico dei Cappuccini, Rome, 2005. The clock (made in 1757), adjustable for the timing of varying durations of prayer, was used in early Capuchin devotion. Servus Gieben was director of the Museo Francescano from 1981 to 2010.

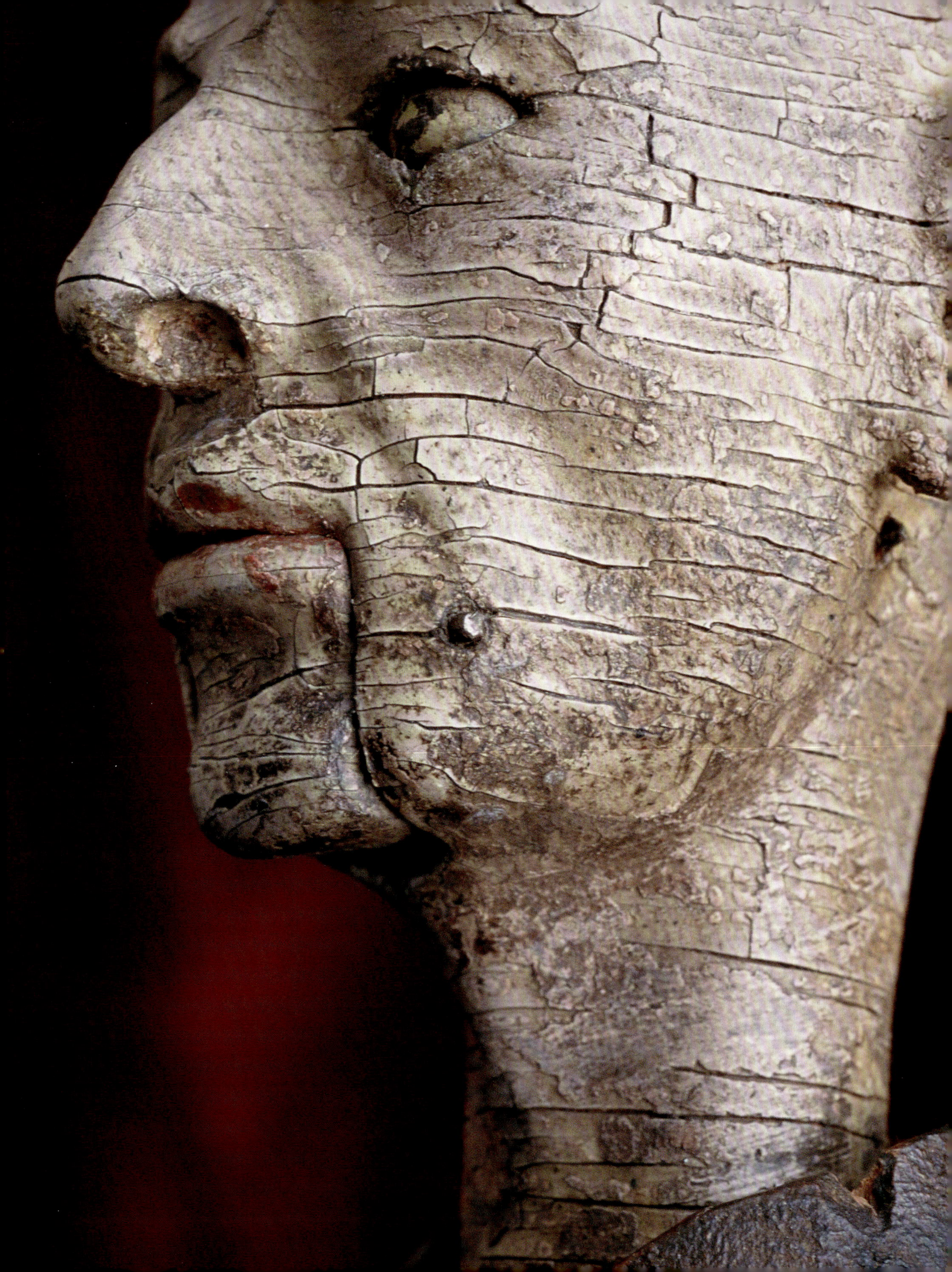

PART V

A Machine That Prays

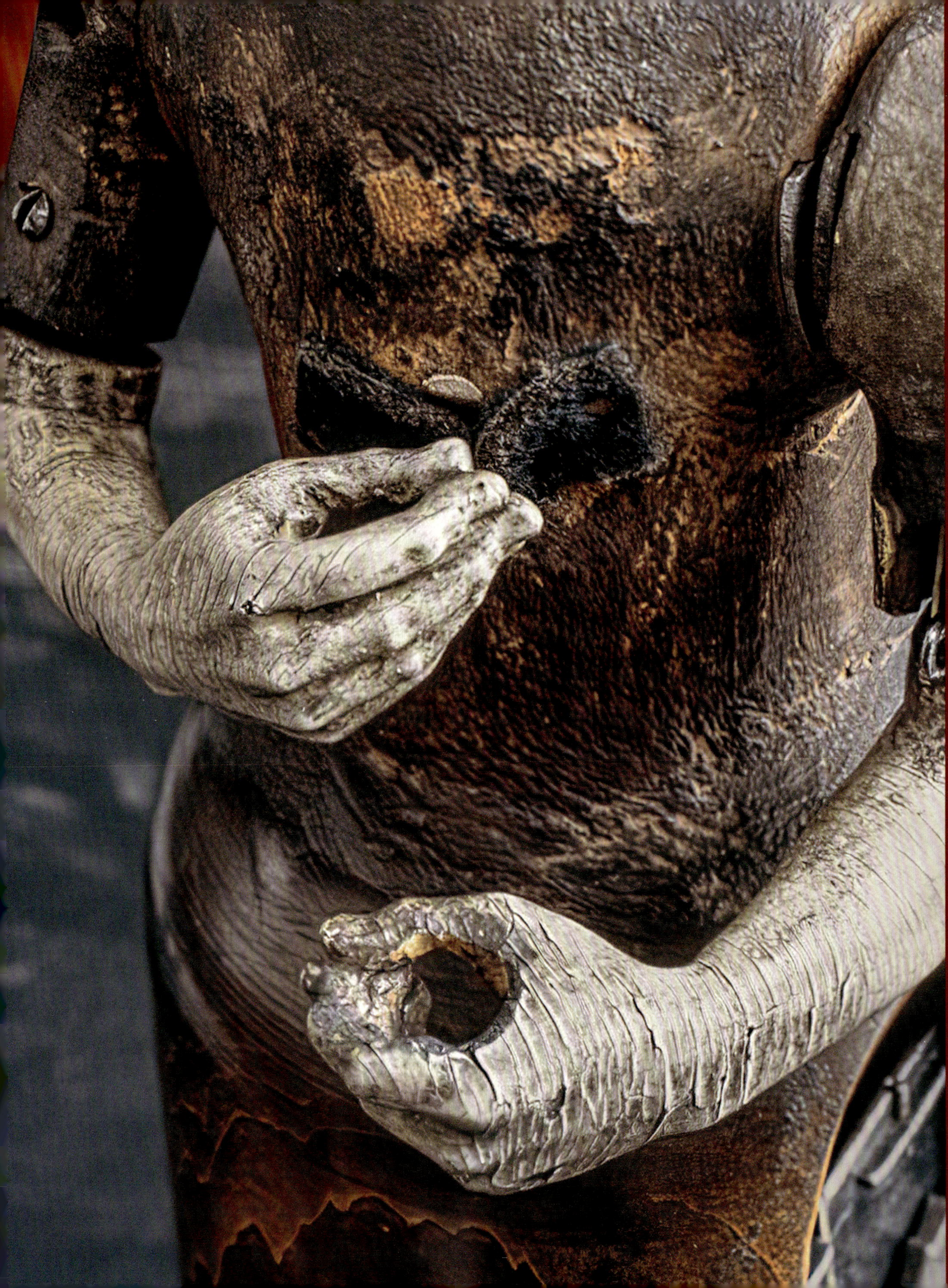

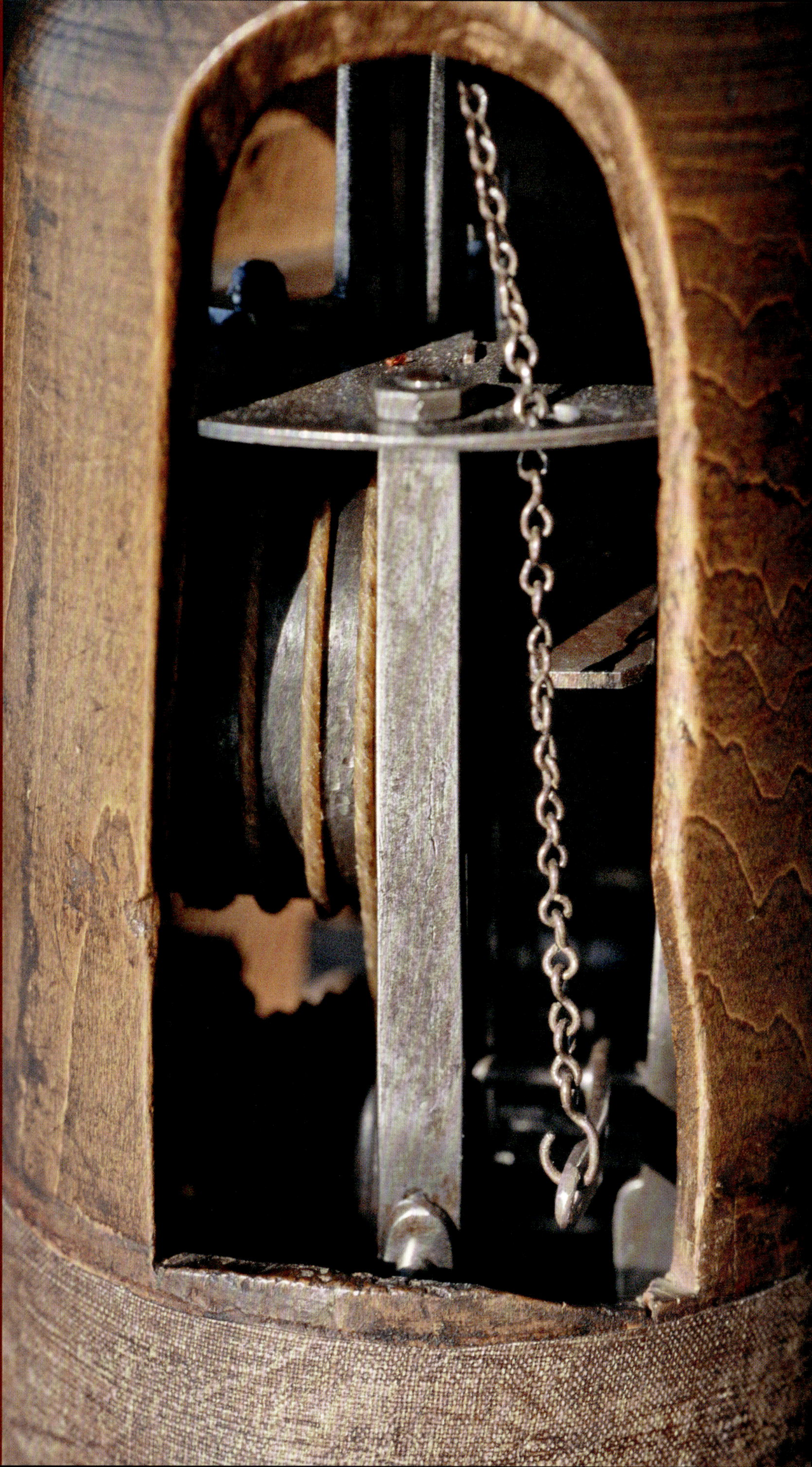

25 Miracles and Walking Sculptures

There is more to say about the automaton and the miracle. Even as we concede that the link is only an elegant hypothesis, the hypothesis refuses to disappear. Why does the story of Don Carlos and San Diego continue to grip us when we look at this figure? Indeed, *looking* at this figure is something we've not finished doing yet. The monk wins his motion from the clockmaker, but he gains his identity and a great part of his meaning from the sculptor.

Without hesitation, everyone who looks at the monk accords him agency and gender. In spite of his cracked paint, and even disrobed, clockwork revealed, he demands our interaction. We give it without thinking. His authority is confirmed for us through the character of his face and his bearing. The sculptor has endowed him with gravitas, with decorum. His features are theatrical, but even in comparison to the classic composure of the Munich and Budapest brothers, this is not the anatomy of caricature. The architecture of the brow and bridge of the nose, the rise of jaw and cheekbone from the hollows of the face, the proportions of the upper lip, the earlobe, and the delicate folds flanking nostril and mouth: these draw us into a subtle regard. The hands are masterfully carved: the fingers of the right hand press hard against the thumb for the strike to the chest. Form itself, from the sculptor's hand, instructs our response. We might say the monk is uncanny—he appears familiar, yet strangely disquieting—or that he has *duende*, the dark spirit conjured and defined by the poet Federico García Lorca.[1] Or we might call him an ur-robot, a conceit from another time and another world. But these are modern ways of speaking. What did a sixteenth-century viewer see and believe when the monk was new? His color and eyes would have possessed their original gleam, his arms and body in motion beneath the tunic, cloth folding and creasing as he moved.

Does the monk—or any work of art—possess qualities that transcend historical context and call forth responses common to generations or cultures across time? At least some of our perceptions go beyond culture and come from a primal human nerve. The monk's head-on motion seizes our instincts: a small, convincingly sentient thing heads our way on a table, and our animal reflexes stir.

The cultural historian Peter Burke cautions those who look back in time: "Our modern distinctions (between rational and irrational, natural and supernatural, religious and superstitious) were in process of formation during the period; to apply them to the years before 1650, in particular, is to invite misunderstanding."[2] So drilled are we against "presentism" in interpreting works from history, we hardly know what to do with our emotional reactions to the objects we have inherited. Yet is this not precisely how works of art connect us to the past? How can we hope to understand how representational objects were seen at the time they were made if we repress our own responses to these works? This is the question that the art historian David Freedberg asks in his milestone book, *The Power of Images*, a study of the history of human response to works of art.[3] Many classes of object come under his scrutiny: votive images; religious relics; objects of pilgrimage; *acheiropoietic* images (those not made by human hands); images said to bleed, or weep, or whose eyes moved; effigies of sorcery; wax-museum figures—works from across the canon of Western painting and sculpture. His discussion of verisimilitude in objects associated with miracles offers insights that suggest how an early walking automaton friar might have been received.

A strong devotional image may invoke what Freedberg calls an answering act of *reconstitution* performed by the beholder. Using the terminology of semiotics, he proposes that the gap between sign and signified may remain unacknowledged in the devout response. In this case, an image is seen not just as a metaphor or stand-in for a saint. It is instead a metonym: a piece of the real cloth. The power of an image, together with the desire of the beholder, collapses the distinction between representation and presence.

> The sign has become the living embodiment of what it signifies. Perhaps it will be suggested that a strong or ingrown belief in the powers of the Virgin easily inclines the believer to see her present, disencumbered of everything that makes her dead representation. Perhaps the suggestion will run that one cannot believe that the Virgin is in the picture—or *is* the picture—unless one believes, to begin with, in the Virgin. Then, wanting her to be there, to exist (because of the love we bear her), we willingly concentrate on the image, and what is represented on it becomes present again. She is, quite literally, re-presented. The slip from representation to presentation is crucial, from seeing a token of the Virgin to seeing her there. What happens? How do we proceed without engaging in the analysis of the propositional status of statements about belief, or the relations between the nature of beliefs and reports about them?[4]

The slip from *representation* to *presentation*, Freedberg argues, is something all of us experience every time we respond emotionally to an image, and such a slip accounts for the unacknowledged traces of animism in our own time. "Not necessarily 'animism' in the nineteenth-century ethnographic sense of the transference of spirits to inanimate objects, but rather in the sense of the degree of life or liveliness believed to inhere in an image," he writes.[5]

The collapse of distinctions—confounding the image of Mary with Mary herself—was among the accusations that launched the Protestant Reformation. "Thou shalt not make unto thee any graven image, or any likeness of any thing that is in heaven above, or that is in the earth beneath, or that is in the water under the earth. Thou shalt not bow down thyself to them, nor serve them: for I the Lord thy God am a jealous God."[6] Martin Luther in Wittenberg in northern Germany, Ulrich Zwingli in Zurich to the south, and later John Calvin in Geneva singled out as particularly egregious those idolatrous images that played a role in the commerce surrounding indulgences—the buying of time from damnation.

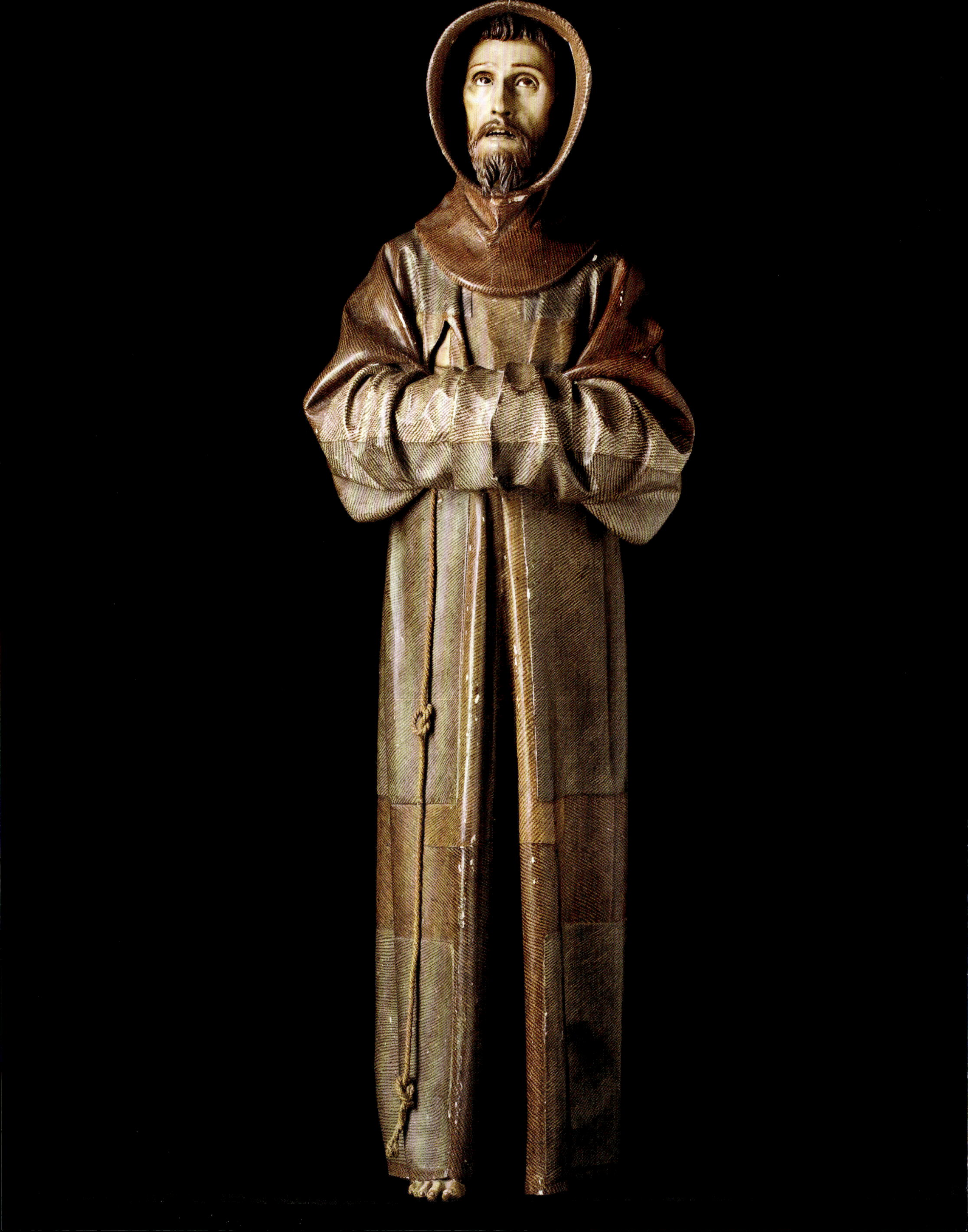

But arguments for the usefulness of images had existed for a thousand years. Pope Gregory in the sixth century declared, "The picture is exhibited in church, so that those who cannot read may, by looking at the walls, at least read there what they are unable to read in books."[7] The fourteenth-century Franciscan teacher Nicholas of Lyra argued that "the prohibition [of images] belonged to a period when God himself had not yet become Man and so could have no image: after the Incarnation, however, and under the New Law, the image was not forbidden."[8]

Luther himself softened his line on images even before the iconoclasts' violent destruction of works of art in Zurich, Basel, and Strasbourg in the 1520s. Remove them, do not destroy them, he said. The Council of Trent (1545–63) laid out Counter-Reformation support for the role of images, albeit with renewed censure for those suspected of misuse. Priests and mystics argued the fine points of the efficacy or corruption of an icon. The key distinction was between honor and worship: honoring an image for what it represented versus adoring or worshipping it as a divine presence.[9] Catholic doctrine has spent centuries of ink to try to determine the difference. Art, meanwhile, had its own designs.

Inhabited

Among well-loved works of art in Spain is the small and haunting *Saint Francis* carved in wood by the Andalusian sculptor Pedro de Mena (1628–1688). Mena created it in 1663 for the great Santa Iglesia Catedral Primada in Toledo, where it stands to this day (**FIG. 25.1**). The sculpture is said to represent not the saint but the apparition of the saint, standing alive in his crypt deep under the basilica in Assisi. The apparition was seen by Pope Nicholas V in 1449, searching for the tomb by torchlight, two hundred years after the saint's death. A cardinal accompanying the pope witnessed the figure too:

> It was a strange thing, that a human body, dead so long before, should be in that manner in which it was: for it stood straight up upon the feet; without leaning, or being supported on one side or the other. The eyes were open, as of a living man, and moderately lifted up to heaven. The whole body was entire without any corruption, white and red, as if it had been alive. It held the hands covered with the sleeves of the habit before its breast, as the Friars Minorites are wont to hold them. The Pope seeing him in this posture, fell upon his knees with great reverence and devotion: and lifted up the habit from above one of his feet, and he saw, and we also that were there present saw, that in that holy foot there was a wound, with the blood as fresh and red as if it had been then made with a nail in some living body.[10]

Reading the cardinal's words, the term "apparition" for this figure may be misleading in modern parlance, where we assume a threshold between the real and the imagined. The tactile vividness of the cardinal's description carries the lived, sensory experience of witness: the erect figure stands by itself without support, eyes open, "as of a living man," the colors red and white of the flesh, the pope lifting the fabric of the habit with his fingers, fresh blood visible in the wound.[11] Pedro de Mena, among Spain's greatest masters of polychrome realism, created an *encarnación* of what the visitors saw in the crypt.

At half life-size, the *Saint Francis* stands poised on an ebony plinth, face suspended in the dark recess of the cowl, glass eyes raised under real eyelashes, mouth open to reveal two uneven rows of ivory teeth (some missing), the teeth parted, the interior of the mouth dark. The tunic, carved and finely textured in relief to represent the coarse patched cloth of poverty, is cinched at the waist by a knotted cord made of real fiber. Hair and tunic are finished with a low-gloss polychrome; the face takes its hue from matte pigment. A sculpted tear in the tunic reveals the side wound.

Earlier artworks—sculpture by Gregorio Fernández, paintings by Francisco de Zurbarán and the Guatemalan Antonio de Montúfar, educated in Spain—depict the saint in near-identical posture, and many copies in sculpture were made after Mena. Richard Ford, nineteenth-century British traveler and Hispanophile, extolling the precedent and glory of Spanish sculpture, described it as "a masterpiece of cadaverous extatic sentiment."[12]

FIGURE 25.1.
Pedro de Mena, *San Francisco de Asís*, 1663. Polychromed wood, glass eyes, eyelashes of human hair, ivory teeth, cincture of real cord. 97 × 33 × 31 cm with plinth ($38\frac{3}{16}$ × 13 × $12\frac{3}{16}$ in.). Figure itself 83 cm in height, or just under 33 in. Toledo Cathedral. A small masterwork of Spanish polychrome realism, its pose connected to the story of a miracle.

Made in Toledo in the century after the automaton monk, the *Saint Francis* is a prime work in the Spanish polychrome idiom. Skillfully carved and painted, this is a small-scale icon intended for close-up viewing. The wooden head, like the monk's head, would have been halved and carved from both inside and outside, to open the eyes and mouth. Color amplifies its realism, and accessory materials extend it: glass eyes, eyelashes of human hair, ivory teeth, braided cord. The saint's preternatural suspense, its motionless *nunc stans*, holds our attention with its own astonishing performance of locked attention. The sculpture's hush may be the antithesis of the moving monk, though the monk exacts an equal claim on our attention, interrupting the fabric of ever-moving time with his own urgent "now." The topos of a sculpture representing an apparition touches the automaton too.

Spanish sculpture is scarcely the only arena for a discussion of Catholic saints in carved wood in any search for correlatives to the sculptural character of the automaton monk. Elizabeth will readily admit that when she first saw Mena's *Saint Francis* in Toledo, it delivered the same jolt she experienced on first seeing the automaton. Different in kind as they are, with Mena among the first-rank sculptors of Spain's artistic golden age, generations later than the unknown sculptor who carved the monk, the two works nonetheless harbor a common temper, she felt. Searching for the *Saint Francis* on a visit to see it in Toledo in spring 2000, she circumnavigated the shadowy cathedral nave for a long time before finally spotting the small figure in the sacristy behind a closed balustrade. "Where was my torchlight?" she wrote to David. A decade later came the chance to see it again, in Washington, DC, where it was included in the landmark exhibition *The Sacred Made Real: Spanish Painting and Sculpture 1600–1700*, at the National Gallery of Art.[13] As she wrote then:

> In Washington, you could step up to within inches of the vitrine. The sculpture was lit with subdued but focused precision.... One tooth caught a tiny highlight and glinted from within the mouth. You see this and catch your breath—then realize the figure, too, is inhaling. No mistaking it: the contracted lift of the head, beard just sprung free of the cowl, indrawn gulp at the mouth and jaw, taut chest: an inspired pressure at the back of the throat is inferred from every part visible around it. All who looked at this sculpture held their breath.[14]

The exhibition, curated by Xavier Bray, had traveled to Washington in 2010 from the National Gallery in London, giving viewers in the United States a rare chance to see masterpieces from Spanish church and museum collections, some of which, like the *Saint Francis*, were being borrowed from religious institutions for the first time. Bray paired works by both painters and sculptors to reveal the unique ways they borrowed from one another and, in the case of sculpture itself, to show the combination of highly specialized skills requiring both painter and sculptor to work in collaboration. (Guild regulation separated the two arts; only very few sculptors were allowed to paint their own works, Mena among them.) The show went a long way to reset the asymmetry of notice between painters and sculptors in sixteenth- and seventeenth-century Spain, especially for viewers in the United States. El Greco, Zurbarán, Velázquez, and Murillo need no first names, whereas (until recently) some of the greatest sculptors of the same period have won little mainstream recognition outside their native sphere: Gregorio Fernández, Juan Martínez Montañés and his student Juan de Mesa, Alonso Cano (who was also a painter and an architect), Pedro de Mena, Andrea de Mena, Pedro Roldán, Luisa Roldán, José de Mora, Francisco Antonio Gijón, and many others whose works graced cathedrals, convents, and private residences in Iberia and Hispanic America. These sculptors worked largely for the church rather than the court. Their depictions of the holy family, the saints, and the sacred mysteries attained an unsurpassed visceral and emotional realism. Masterpieces by Martínez Montañés, Fernández, Mesa, Cano, Mena, Mora, and Gijón were on display in *The Sacred Made Real*. It was extraordinary to see these sculptures in the flesh, and almost equally extraordinary to see their effect on visitors to the show. Works that had been borrowed from the devotional settings for which they had been made carried the weight of four centuries of worship. Conversation died away in front of them. People just stood and looked, transfixed.

Indeed, when we look at these works we are aware of the invisible presence of the generations of people through time who have looked at them before us, prayed to them, studied them. Humans leave their traces on the objects that win their regard. In the show, Mena's small sculpture was paired with Zurbarán's painting of the saint, both identified by the same title: *Saint Francis Standing in Ecstasy*. Since then, newer publications have adjusted the titles, and both are now often referred to as *San Francisco de Asís y la visión del papa Nicolás V* (Saint Francis of Assisi according to the vision of Pope Nicholas V).[15] The shift in title, from the saint's ecstasy to the pope's vision, emphasizes the images' origins in the story of the apparition. The newer title moves our attention to the crypt and puts us in the place of the pope, and in turn it moves the saint from a seer to a thing seen. This is a complicated shift, pushing the sense of sight front and center as the artists' subject. We become, not viewers, but witnesses.

And we must remember, the visit to the crypt, according to fable, took place two hundred years after Francis's death (d. 1226, canonized 1228). Further still, the cardinal's report of the apparition—the story itself—did not surface for almost another two hundred years. Not apparently until the mid-sixteenth century was the story disseminated, by the Franciscan traveler and historian Marcos de Lisboa, who compiled the Chronicles of the Friars Minor, *Las crónicas de la orden de los frayles menores*, and published it in 1562.[16] The news soon gained wider notice, among other places, in Pedro de Ribadeneira's 1599–1601 *Libro de las vidas de los Santos*. Thus launched, the story spread, and was granted its form by Fernández, Zurbarán, Mena, and many other artists after them.[17]

Diego too gained his most recognized form from a story, though not the story of the cure of Don Carlos. Pedro de Mena himself produced not one but four sculptures of San Diego, by current count. Another, made in collaboration with his mentor, Alonso Cano (1601–1667), came first, in the mid-seventeenth century. All of them followed a small model by Cano, representing Diego and the miracle of the flowers, by then the traditional portrayal of the saint (see **FIG. 3.1**; see also **FIG. 6.1**, at lower left in the cycle of panels depicting Diego's miracles).[18] Looking down, a youthful brother is about to discover to his own surprise that the bread he had stolen for the poor, hidden in his tunic, will turn to flowers when the monastery guardian confronts him. Such images, and others of Diego, were in great demand following the canonization in 1588. The earliest engravings and paintings show Diego in old age, grasping and lifting a tall wooden cross, as he did on his deathbed, where witnesses were stunned to hear his last words: "Dulce lignum, dulces clavos, dulcia ferens pondera" (Sweet wood, sweet nails, that bore such a sacred weight).[19] Although the miracle of the flowers had also been reported in Diego's late years when he was gatekeeper in the monastery of Santa María de Jesús, artists in the seventeenth century began depicting the saint as a youthful innocent, perhaps in part to emphasize his look of surprise. Diego's illiteracy and lay status even at the end of his life contributed to the reputation of one who never seemed to age.[20] In 2013 a small sculpture of Diego was purchased by the San Diego Museum of Art. The scholar and curator José Luis Romero Torres, who has written in depth about Spanish seventeenth-century sculpture, has attributed the work to Pedro de Mena (**FIG. 25.2**).[21] The figure is portrayed in midwalk, mouth open, tunic clutched in both hands, the rough cloth carved in a cataract of bunched folds held open to reveal the purloined bread. As with other renditions of the miracle, neither bread nor flowers are depicted, for in the devotional setting intended by the artist, participants placed real flowers in the open folds. In Alonso Cano's earlier depiction too, the saint holds the tunic open and empty so the onlooker may enact the miracle. But whereas Cano's saint looks down at what he holds, Mena's saint glances up and brings his guilt and surprise directly to the viewer. Glass eyes, catching the flicker of candlelight, would have intensified the gaze. Mena often intended his smaller devotional sculptures to stare directly at the onlooker.[22] To capture the viewer's eye, and build a reciprocal response, the sculpture initiates the contact, and its own gaze had to be as lifelike as the artist could make it.

The earliest images of Diego, closer to the always sought "true" image of a saint, showed him at the age he would have been at the end of his life, in a character type not unfamiliar to canonical iconography: the gaunt, ascetic, penitential hermit. The automaton monk belongs to this type too in look and in act. In a detailed history of artists' depictions of San Diego, José Luis Romero Torres finds scant record of visual markers before the campaign for canonization, and by the time of the canonization itself, over a century had passed since the friar's death in 1463.[23] No record of a death mask has appeared (apropos of Servus Gieben's surmise in correspondence with Elizabeth).[24] Unlike with Francis, canonized almost immediately after

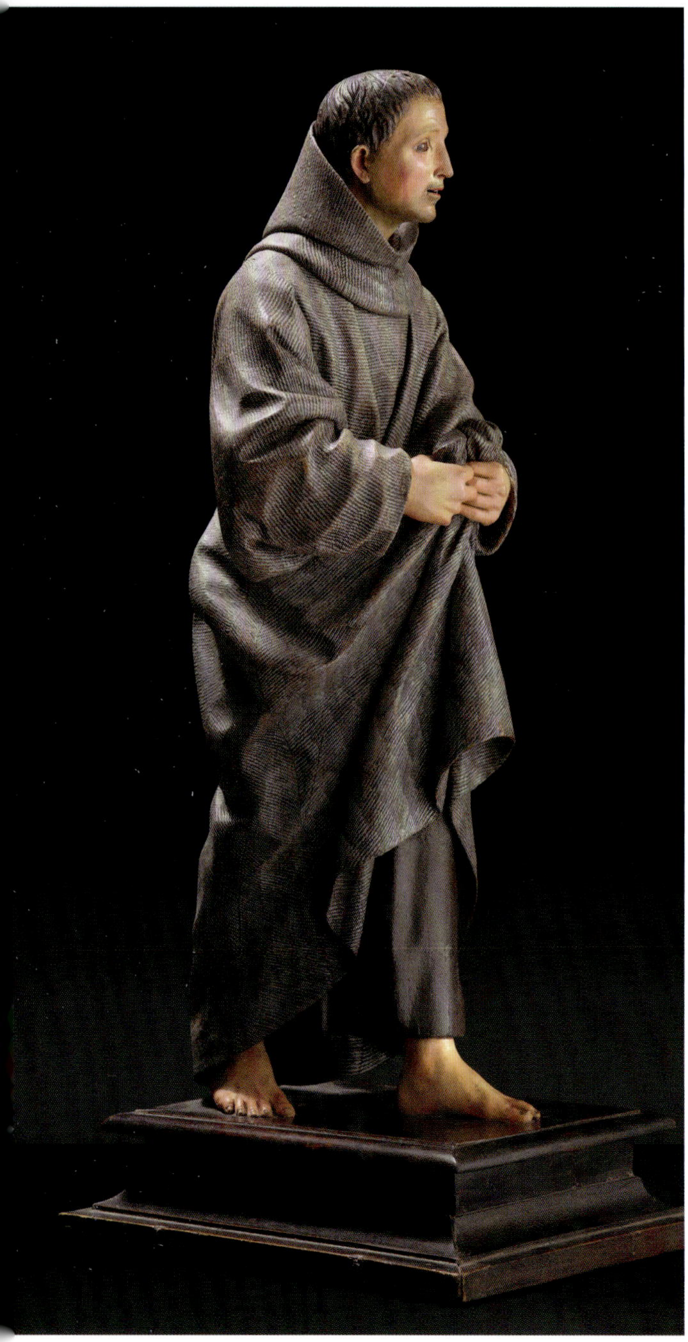

his death and of whom many images were made long before the pope's visit to the crypt (and many since), artists more or less invented an image of Diego from whole cloth. But like the depictions of Francis in the crypt, a story about Diego captured the popular imagination and came to assume an iconographic key.

Another Franciscan saint, Pedro de Alcántara (1499–1562, canonized 1669), in life far more proximate to the Spanish court than was Diego, was also a figure much represented by Spanish sculptors (**FIG. 25.3**). Mystic, reformer, and mentor to Teresa of Ávila, he was active in Extremadura during the final years of Charles V in San Yuste. The once-powerful emperor had asked the friar to be his confessor, and Pedro had demurred. Another gaunt, ascetic, penitential hermit.

Sculptures That Move

Spanish sculpture, with its pursuit of closely observed emotion and its embrace of mixed materials and rich color, flourished in the Counter-Reformation. The Catholic rebuttal to Protestant rage was to reconfirm, emphasize, and administer the role of images, however complicated a task this became. Spanish theologians declared that pictures were better than books at teaching the lessons of scripture. The more lifelike the depictions of sacred persons and stories, the more immediate to daily life they would be, and the more persuasively they would serve the Christian mission. The Carmelite mystic Saint John of the Cross (1542–1591, canonized 1726) wrote:

> The Church ordained the use of images for two principal ends; these being, to revere the saints in them, and to move the will and awaken the devotion in them.... And when they serve this [end] they are of great benefit, and [their] use is necessary; and because of this, those that are created most precise and lifelike, and most move the will to devotion, must be selected.[25]

FIGURE 25.2.
Pedro de Mena, *San Diego de Alcalá*, 1665–70. Polychromed wood, glass eyes, teeth of ivory or horn. 61.5 × 24 × 27 cm (24 3/16 × 9 7/16 × 10 5/8 in.). San Diego Museum of Art. Acquired in 2013, the small sculpture of the saint arrived in the city named for him.

FIGURE 25.3.
Pedro de Mena, *San Pedro de Alcántara*, ca. 1670s (detail). Polychromed wood. Height: 176 cm (69 1/4 in.). Iglesia del Convento de Madre de Dios, Lucena, Córdoba.

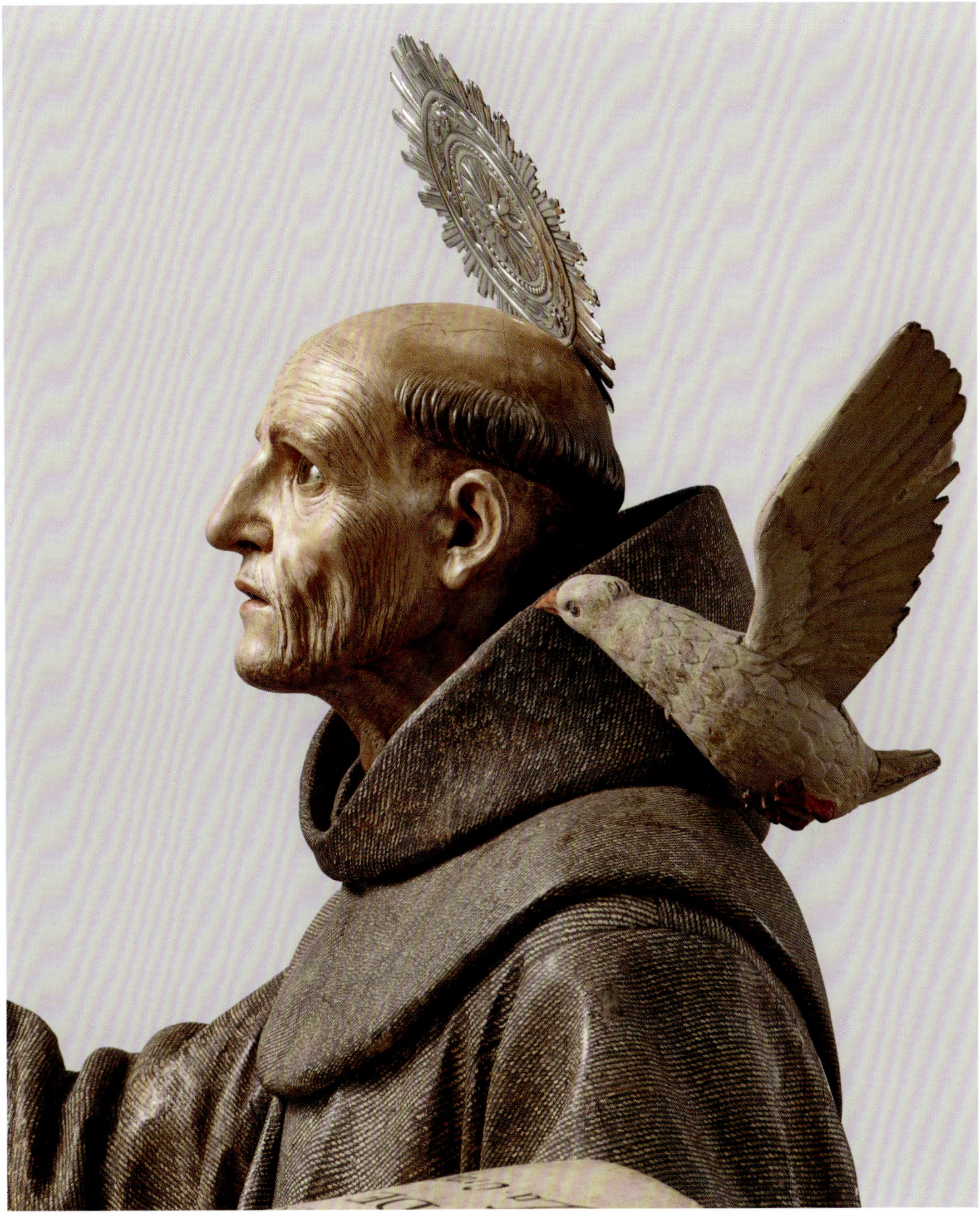

Together with Pedro de Mena's *Saint Francis*, a number of such works, made to move the will, were brought together in London and Washington for the exhibition *The Sacred Made Real* (**FIGS. 25.4–25.6**).

Yet a further step in the direction of realism appeared in the stunning processional sculptures made for the annual festivals of Holy Week and Corpus Christi in cities across Spain. Many of the sculptors who carved works for churches and monasteries also created sculptures to be brought out into the streets in theatrical reenactments of episodes of the life of Christ. A tradition with medieval roots in Christian lands in Europe and beyond, featuring both live actors and puppets, it gained a reputation in Spain for the drama of the sculpture. Carved and polychromed wood figures—many with movable limbs animated by hidden means—represented Christ and the Virgin, together with patron saints, apostles, and martyrs, in a class of devotional sculpture that entered, on a monumental scale, the real space and real time of the spectator's world. Gregorio Fernández (1576–1636) in Valladolid and Juan Martínez Montañés (1568–1649) in Seville produced some of the most prized figures and figure groups for procession (**FIGS. 25.7, 25.8**).[26]

Mounted on ornate platforms called *pasos*, so named for their depictions of the Passion, sculptures for Holy Week were born aloft in the streets by members of lay confraternities and paraded on winding routes through village or city in imitation of the procession to Calvary. The festival reached its height between 1550 and 1700, but processions continue to this day (see **FIG. 25.11**). The annual Semana Santa in Seville draws spectators from around the world. Among its *pasos* are well-conserved original sculptures, some dating to the sixteenth century.[27] The sculptures command the center of the pageant. *Imágenes de vestir*—images with clothes—are dressed in real fabric; their robes flutter in the wind, their arms and heads move, and crowns and glass eyes (and glass tears) flash light from banks of flickering candles carried by attendant marchers (brothers of light). The sculptures appear to walk, as the *pasos* heave and shift with the stepping of the teams of bearers under their load. The bearers themselves remain unseen within valances of fabric skirting the perimeters of the platforms, which are flanked by hooded flagellants (brothers of blood), many on their knees.[28] Every aspect of this public ritual was meant

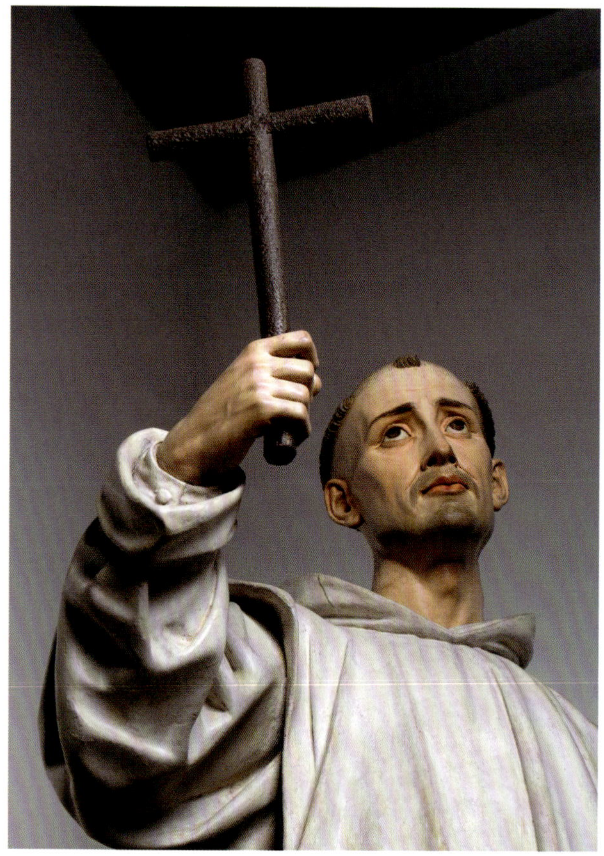

FIGURE 25.4.
Juan Martínez Montañés and unknown polychromer, *San Bruno*, 1634 (detail). Polychromed wood. 160 × 73 × 62 cm (63 × 28¾ × 24⅜ in.). Museo de Bellas Artes, Seville. Martínez Montañés was called by his contemporaries "el Dios de la Madera," the God of Wood. Commissioned for the monastery of Santa María de la Cuevas in Seville. San Bruno (ca. 1030–1101), founder of the monastic Carthusian order, here shown in fervent meditation on the cross.

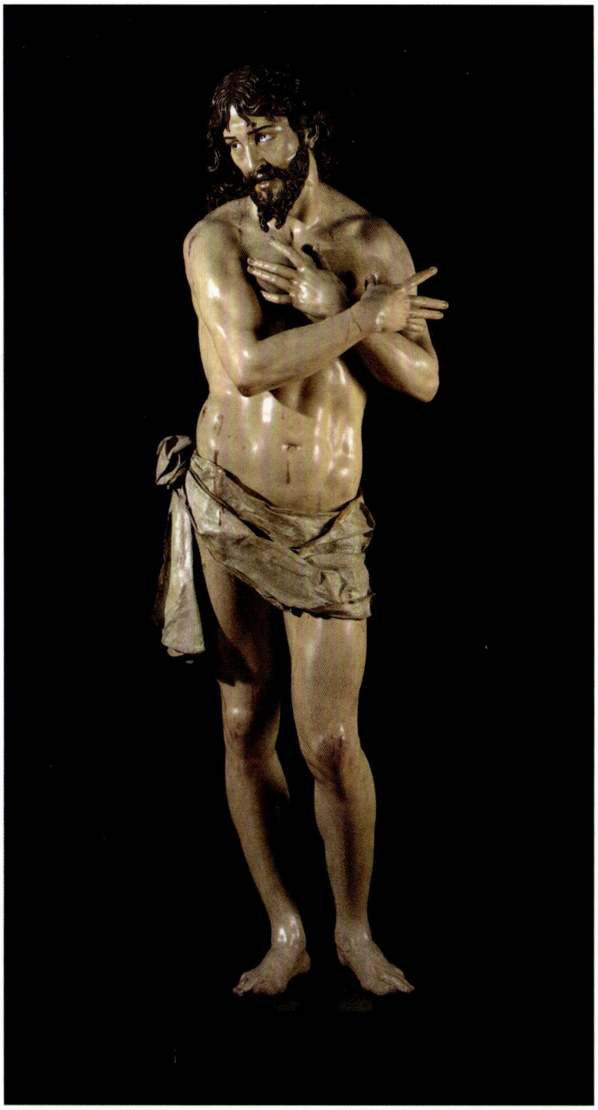

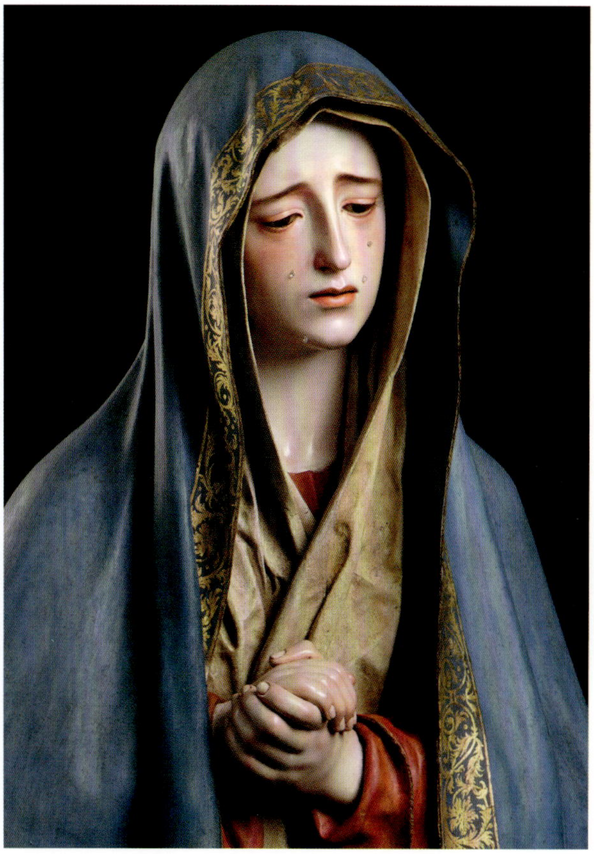

FIGURE 25.5.
Gregorio Fernández and unknown polychromer, *Ecce Homo*, before 1621. Polychromed wood, glass eyes, cloth. 182 × 55 × 38 cm ($71\frac{5}{8} \times 21\frac{5}{8} \times 14\frac{15}{16}$ in.). Museo Diocesano y Catedralicio, Valladolid. Fernández and Martínez Montañés set the stage for the charged emotional realism that reached its height in works by seventeenth-century Spanish sculptors. Both came of age in the decade following Diego's 1588 canonization.

FIGURE 25.6.
Pedro de Mena, *The Virgin of Sorrows (Mater Dolorosa)*, ca. 1670–75 (detail). Polychromed wood, ivory teeth, glass eyes, eyelashes of real hair, glass tears (now gone). 62.2 × 49 × 29 cm ($24\frac{1}{2} \times 19\frac{1}{4} \times 11\frac{3}{8}$ in.). Church and Sanctuary of Santa María de la Victoria, Málaga.

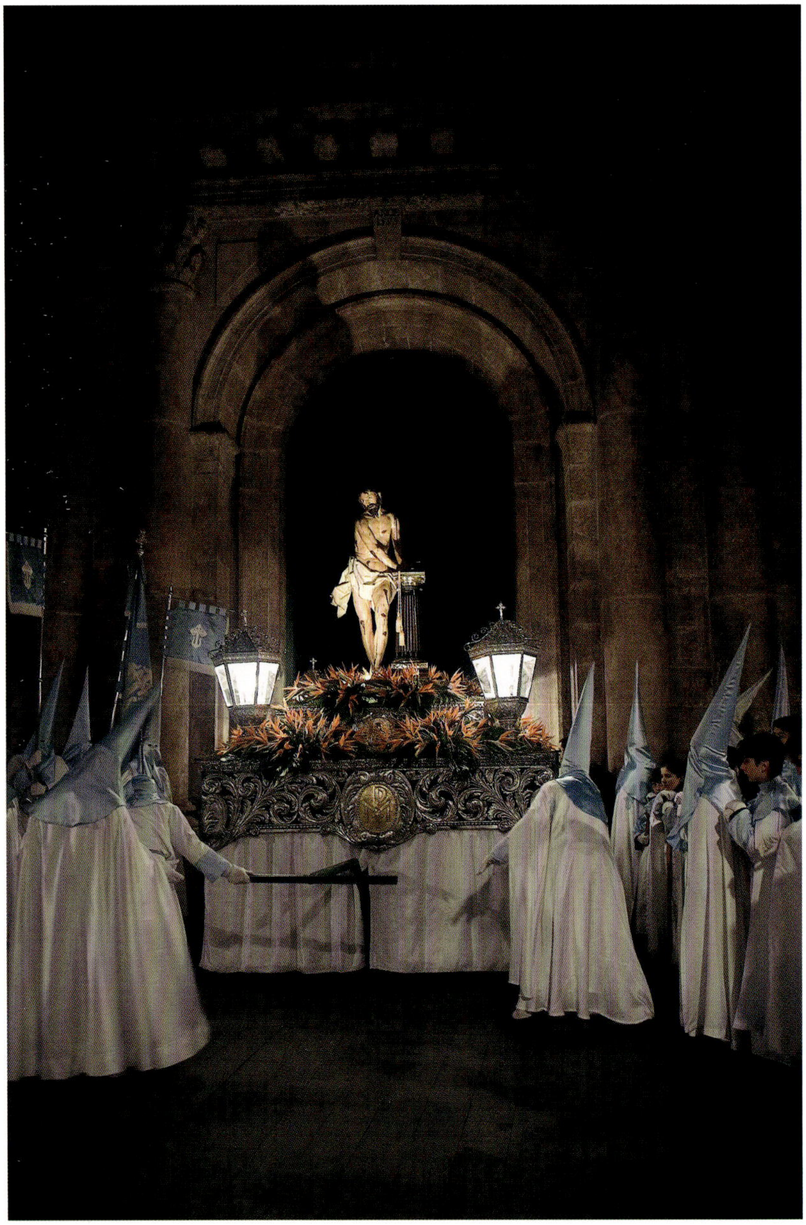

FIGURE 25.7.
Gregorio Fernández, *El Señor Atado a la Columna* (Christ at the Column), ca. 1619. Polychromed wood, glass eyes. Life-size. Iglesia Penitencial de la Santa Vera Cruz, Valladolid. Commissioned by the Cofradía Penitencial de la Santa Vera Cruz (Penitential Brotherhood of the Holy Cross), the sculpture is still taken out in procession during Holy Week. Seen here emerging from the central door: out of the church and into the street. Holy Wednesday, March 19, 2008.

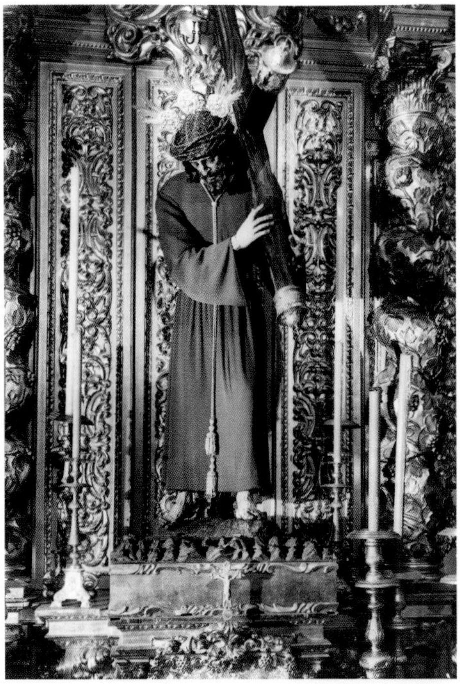

25.8A.

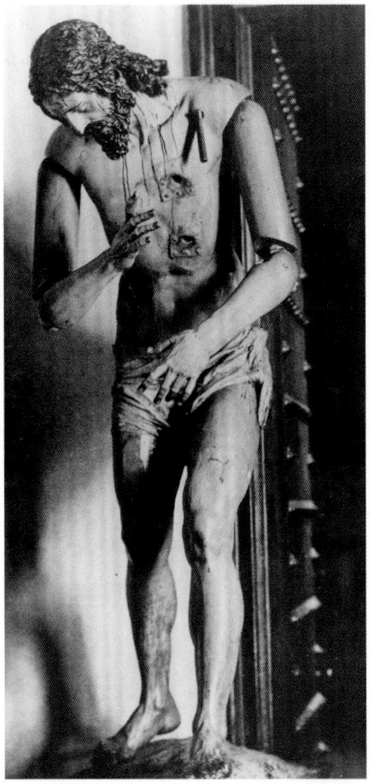

25.8B.

to inflame the passions of the devout, crowded on roadsides, balconies, roofs, and trees to see the sculptures pass. Susan Verdi Webster, in *Art and Ritual in Golden-Age Spain*, singles out the "at large" character of the sculptures.

> The spatial and temporal status of the sculptures in procession significantly enhance their mimetic effects, and their unique kinesthetic character allows them dramatic entrance into the realm of human experience. They are able to move both physically (through articulated limbs) and spatially (through the streets of the city). Furthermore, the incorporation of sculpture within a processional context acts to change a most fundamental aspect: the sense of time. No longer the static, eternal images of altars and *retablos*, their temporal state is extended so that they merge with the spectators' own experience of space and time.[29]

Like the monk and his fellow miniatures in an intimate setting, the processional sculptures in a public setting perform in the round in the onlooker's world. Both are *imágenes de vestir*—sculptures wearing clothes.

Who commissioned the sculptures for the *pasos*? Confraternities, many associated with trade guilds dating to the medieval era, were organized brotherhoods formed at all levels of the citizenry, from laborers to artisans to merchants.[30] They operated independently of both state and church, though they often maintained side chapels in local monasteries to display their sculptures when not in procession. Central to confraternity life, the processions combined religious practice, status within the community, trade relationships, and family structure. Competition among confraternities was intense. Each group commissioned sculptors, painters, carpenters, costumers, and metalsmiths to create *pasos* ever greater in size, grandeur, and artistry. Multifigure

FIGURE 25.8A, B.
Juan Martínez Montañés, *Jesús de la Pasión*, with (A) and without (B) garments, 1619. Polychromy attributed to Francisco Pacheco. Polychromed wood, cloth. Life-size. Cofradía del Cristo de la Pasión, Church of El Salvador, Seville. This sculpture, a true *imágen de vestir*, is still carried in procession through the streets of Seville during Holy Week. Arms are articulated for alternate poses and for dressing.

pasos often included smaller figures, half-size angels, for example, carrying attributes of the Passion.[31] One very small jointed figure, thought to represent Simon of Cyrene, proposed as part of a smaller-scale *paso* ensemble, can be found in the collection of the Hispanic Society of America in New York (**FIG. 25.9**).[32] Simon, charged with helping Christ carry the cross, serves as a classic processional icon.

The fabrication of the sculpture often called for a piecework process of carving only the visible heads, hands, and feet, each part then finely finished and polychromed. Torsos and legs, hidden under actual garments, were either rough carved (often in articulated sections for ease of dressing, posing, and animating) or built as schematic armatures to support and give shape to the dress (**FIG. 25.10A, B**). A few Virgins came supplied with spare heads, one face happy and one face sad, as occasion required.[33] Over time, increased attention went into elaborate dress and ornament, especially for the Virgin. Carved wooden crowns gave way to gilded crowns of tin, then silver and gold, with imperial domes augmented with halos of rays and stars.

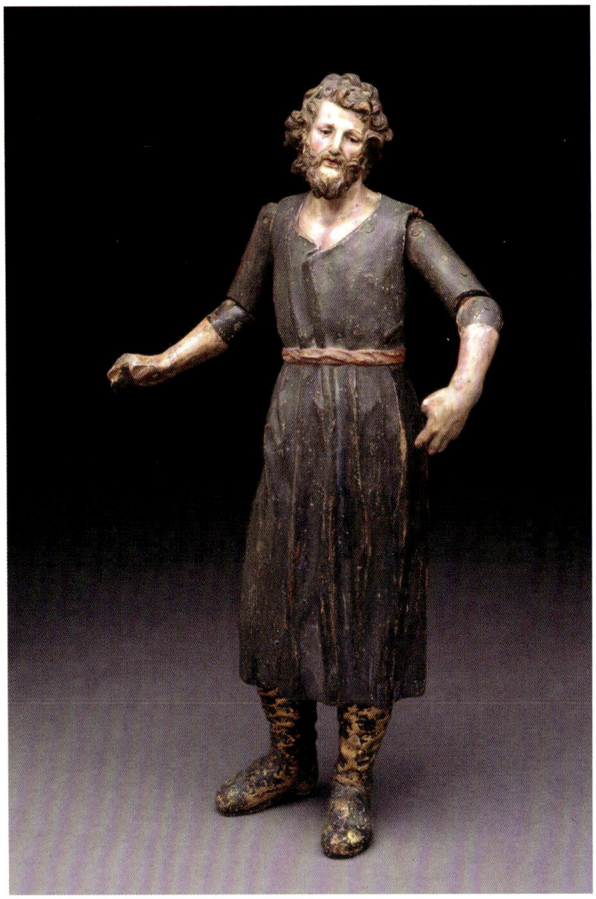

FIGURE 25.9.
Follower of Juan Martínez Montañés, *Simon of Cyrene* (?), seventeenth century. Polychromed wood, glass eyes. *Imagen de vestir*, statuette with movable shoulders and elbows. Height: 32.9 cm (13 in.). The Hispanic Society of America, New York. This small Simon, a quintessential processional icon, is close in scale to the walking monk.

Brotherhoods and the citizenry of the south of Spain routinely ignored repeated decrees by church authorities to limit the pageantry of these annual spectacles. Was the emotion generated by the sculptures a devout response, or was it a profane kind of carnival? (Jugglers, acrobats, and magicians often joined the processions.)[34] Visitors from abroad were scandalized by the Virgin's luxurious clothing and jewelry. Critics reviled the use of wigs, glass eyes, and excessively realistic wounds—granulated cork, soaked in red pigments, representing abraded flesh—as travesties of the decorum of the sacrament. Such criticisms held sway over time. The German art historian Carl Justi (1832–1912) declared, "The apparatus for moving the head, the eyes, and the mouth, the wooden dolls with real hair and real dresses, in which the head and hands alone are carved, mark the lowest level of the plastic art."[35] "A sculptor who puts glass eyes or a wig on the Virgin Mary is 'at risk of massacring her face,'" wrote the French archaeologist Marcel Dieulafoy in the early twentieth century.[36] The Reformation had already halted the flowering of a great northern tradition of polychrome wood sculpture in Germany, dubbed the early "Florid" style by Michael Baxandall.[37] By the late 1520s commissions for painted works began to fade, and a chastened form of sculpture emerged, replacing color with a subtle

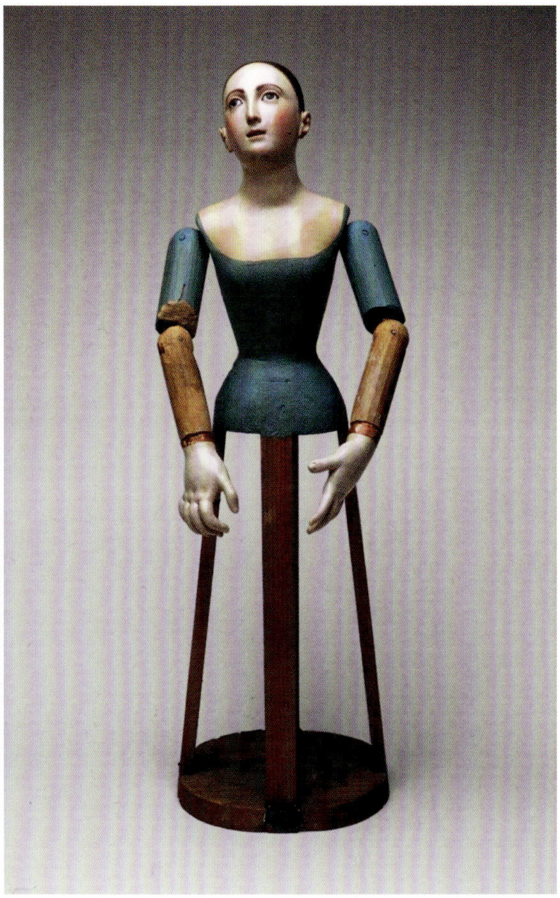

25.10A.

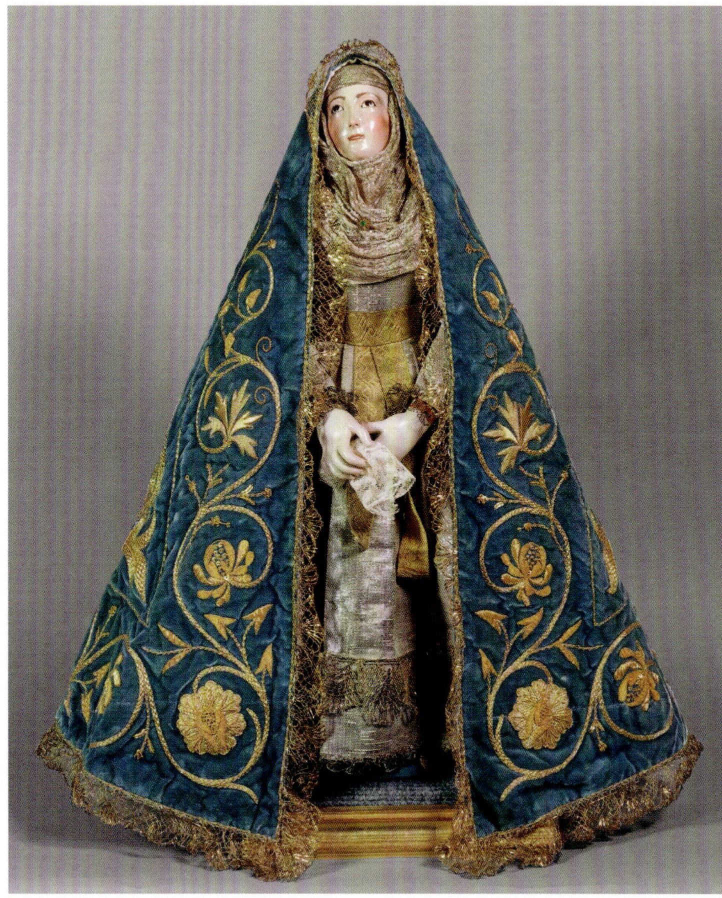

25.10B.

FIGURE 25.10A, B.
Unknown artist, Virgin (*imágen de vestir*, undressed [A] and dressed [B]), ca. 1825. Wood, polychrome, embroidery, gilt foil thread, sequins; gilt metal nimbus and silver gilt crown. Height without crown and nimbus, 54 cm (21¼ in.). With jointed arms, a traditional armature for female processional figures.

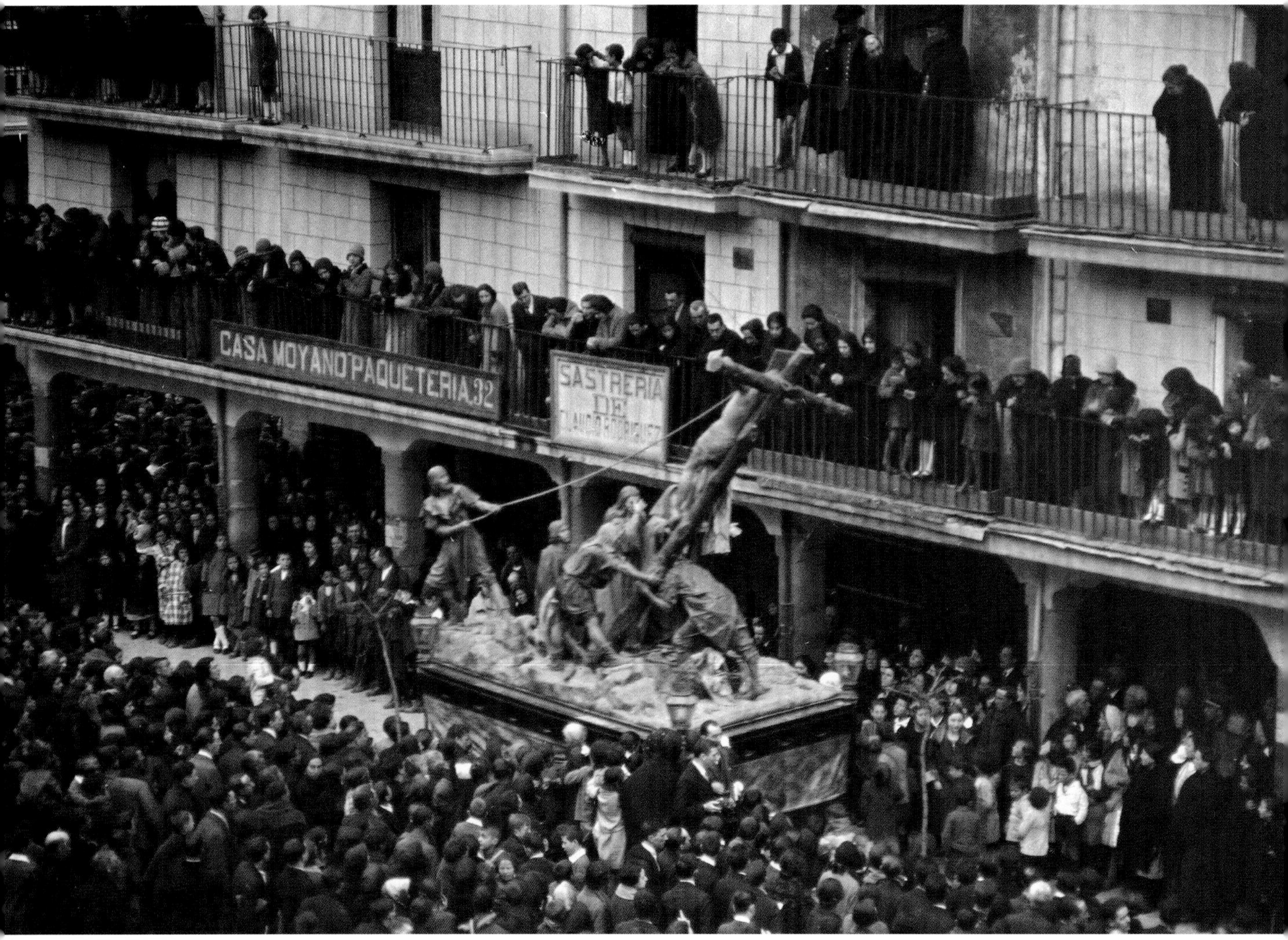
CASA MOYANO PAQUETERIA.32
SASTRERIA
DE

virtuosity of carved surface in relation to light, best exemplified in the works of Tilman Riemenschneider (1460–1531) or Veit Stoss (1445/50–1533).[38]

In Spain, by contrast, the same preachers who warned that the corporeal body was an ally of the devil gave vivid sermons appealing to the senses to inspire spiritual feeling and pious obedience. Ignatius of Loyola (1491–1556) enlisted all five senses to alarm the would-be sinner.

> The first point will be to see great fires with the eyes of the imagination, and the souls as though in fiery bodies. The second, to hear with the ears the crying, howling, voices, blasphemies.... The third, to smell with the nose smoke, sulphurous stone, sewage and putrid things. The fourth to taste with the mouth bitter things.... The fifth to feel with the sense of touch, this means, how the flames touch and sear the souls.[39]

Little wonder that Philip II so fervently collected the works of Hieronymus Bosch, who gave image to such words.

Webster gathers a wealth of reports on how the sculptures were received by onlookers and communities. Realism, from material to motion, was precisely what provoked an emotional catharsis.

> Every aspect of the procession—sculpture, clothing and adornment, conveyance, and accompaniment—can be shown to develop from a transformational paradigm, according to which the "thing represented" became "the thing-in-itself." Once established, this context became an arena in which the brotherhoods competed to reach ever-higher stature in the eyes of the public and the Almighty. The desire to create as realistic a scene as possible, and the extraordinary lengths to which the brotherhoods went to achieve their goal, resulted in the merging of physical and spiritual realities during the processions.[40]

FIGURE 25.11.
A multifigure *paso*, "The Raising of the Cross," moves slowly through the crowd, borne on the shoulders of a team of *costaleros*: men in heaving lockstep, hidden by the damask platform skirt. Holes in the perimeter frieze allow the carriers to breathe and see where they are going. Photograph by Ruth Anderson, Good Friday, April 2, 1926, Zamora, Spain.

We see David Freedberg's fusion enacted here, and at scale. "Perception elides representation with reality," he wrote.[41]

Everyone is seen to play a role in the exchange: God, church, brotherhood, sculptor, performer, spectator (**FIG. 25.11**). The sculpture too is active, referred to repeatedly by name: "The Virgin comes out of the church..."; "Our Lady stopped and waited..."; "The Lady Santa Ana came out to the bridge..." Witnesses refer to the sculptures as the agents of their own movement and behavior.[42] The whole party brings a wooden figure to life.

The processional sculptures for Holy Week enact the mystery of the resurrection. The moment they emerge from the church they come alive, in imitation of Christ from the tomb, and alive they must look and must act (see **FIG. 25.7**). For the spectator, the distinction between art and life is an open negotiation. Pedro de Mena's *Saint Francis* is also a resurrection image, showing the saint risen in the crypt. The same mystery is inherent in the clockwork monk at the moment that the stopwork lever is released and the figure comes to life. As long as the story of Diego de Alcalá hovers behind the automaton, it too is a resurrection image, recalling the prince's midnight vision: "This night the holy fray Diego appeared, standing at my bed, with a reed cross in his hand..."

Though scale and context profoundly separate them, the automaton monk and the processional sculpture share the hallmark agency of a moving figure, at large on the viewer's own turf: figures in the round, moving around us. Both entail the production of gestural movement by mechanical means. As they move through space—the tiny monk on hidden wheels, the towering sculpture on brigades of hidden human shoulders—they *do* things. The compound motion of the monk turning his head and raising his arms at the moment that he executes a walking turn: this is a pairing of two kinds of motion that amplifies, because it fuses, the movements of part and whole. From this the performance gains its irreducible three-dimensionality. One suddenly understands that the monk himself is a processional entity, the table a miniature *paso*. He too is carved and dressed in real clothing. Hands, head, and feet finished elegantly and polychromed, hidden body roughed out to support the tunic: a miniature *imágen de vestir*. He too performs a circumambulation, station to station, onlooker to onlooker, making eye contact and moving his mouth. Is this not, then, part of his family tree? His small scale and the lifelike movement produced by his greater mechanical complexity render him simultaneously more intimate—and more personal. Some quieter measure of the emotion and wonder felt by those who beheld the processional sculptures must have been felt by anyone who saw this figure move.

In 2019, the monk, on loan from the Smithsonian Institution, was displayed at the Metropolitan Museum of Art in New York. The show, *Making Marvels: Science and Splendor at the Courts of Europe*, was curated by Wolfram Koeppe and Ana Matisse Donefer-Hickie.[43] Borrowing priceless treasures from across oceans, they gathered some of the greatest objects made in early modern Europe: a *Kunstkammer* of clocks, orreries, automata, scientific instruments, optics, tools of alchemy, early metalworking lathes, sculpture, gemstones, and natural marvels. The monk, beautifully lit, was posed in a vitrine, allowing viewers to inspect it from inches away. Nearby, projected on a wall at floor-to-ceiling scale, ran the film the Smithsonian made in 1978 of the monk in performance **(FIG. 25.12)**.[44] Elizabeth, on several visits to the show, couldn't help but notice other visitors encountering the automaton and film. Conversation fell silent. People just stood and looked, transfixed.

FIGURE 25.12.
November 8, 2019: on loan from the Smithsonian Institution, the monk is being installed for the exhibition *Making Marvels: Science and Splendor at the Courts of Europe* at the Metropolitan Museum of Art, New York. The monk was on view from November 25, 2019, to March 1, 2020.

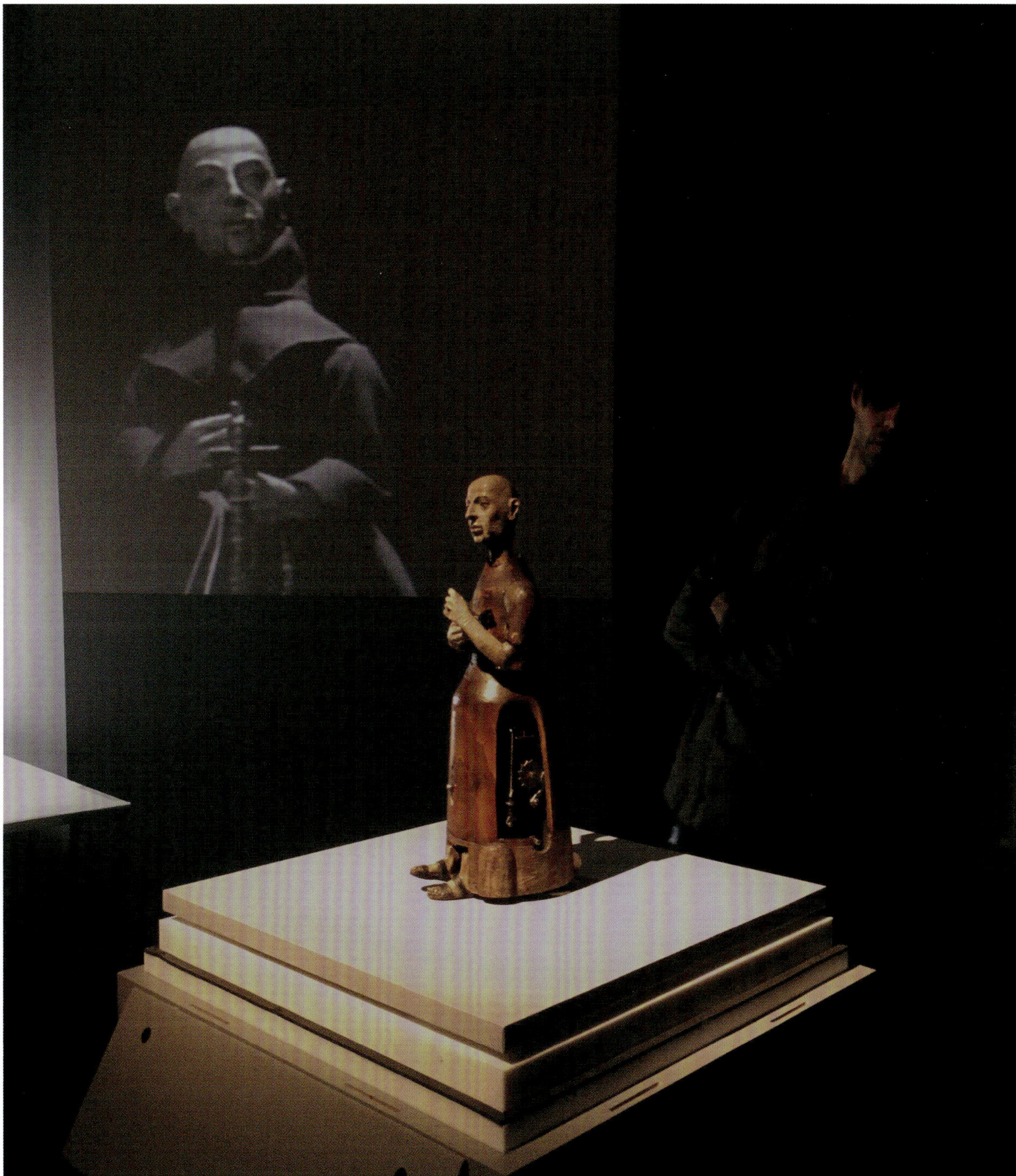

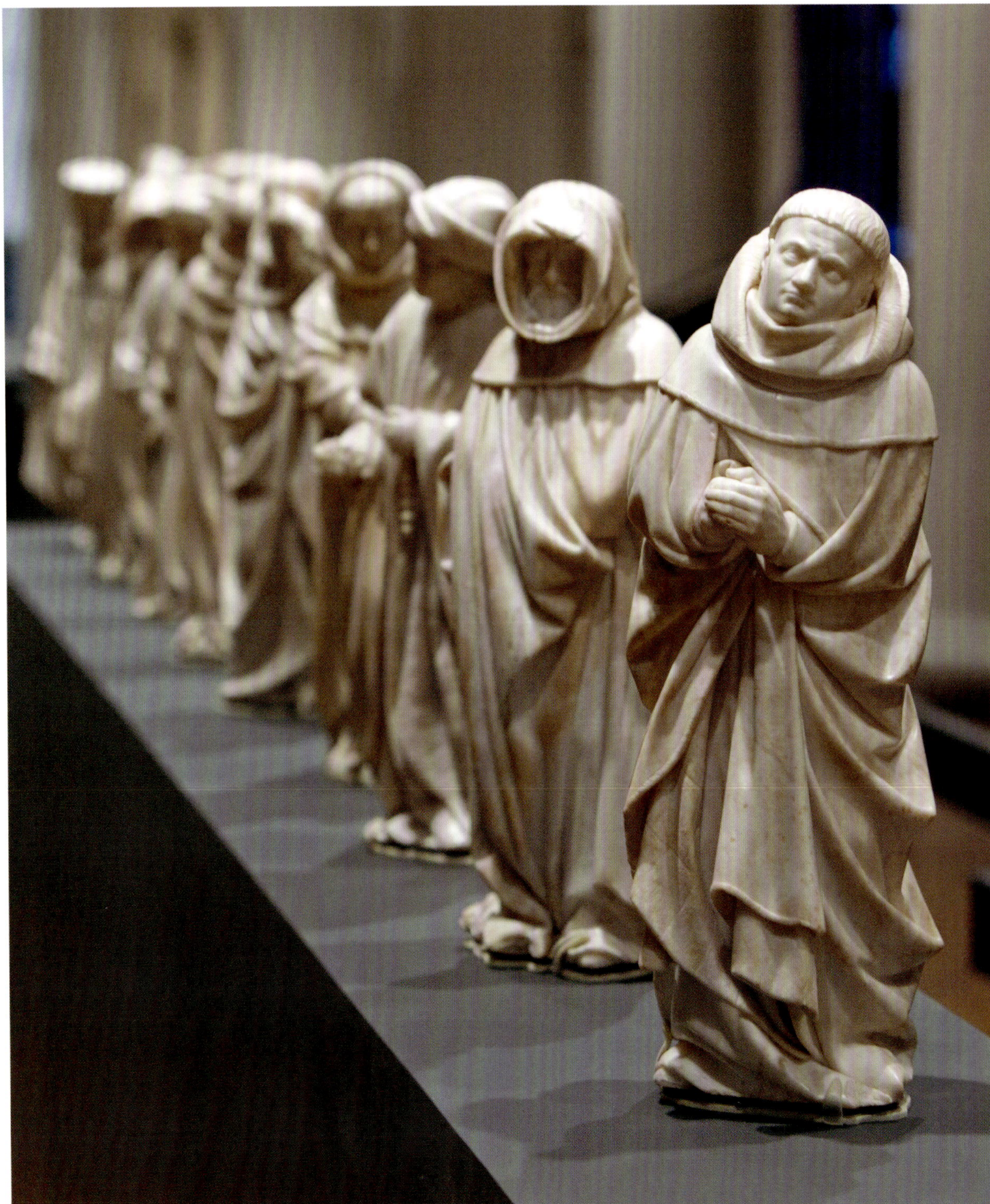

26 Perpetual Devotion

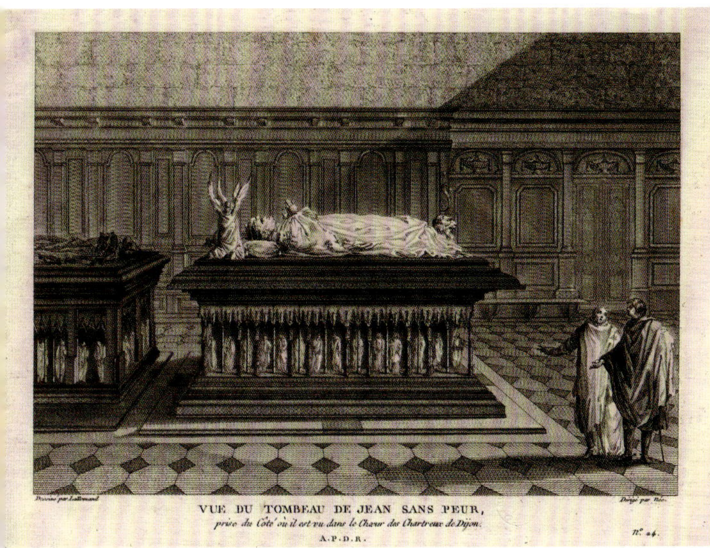

We touch finally on an unforgettable group of sculptures that portray the processional parade as an image in its own right. The Mourners of Dijon they are called: no less than eighty small alabaster figures carved for a pair of great tomb monuments produced in fifteenth-century Burgundy. Thirty-seven of these diminutive figures were recently borrowed from one of the tombs and exhibited by themselves in a show that toured the United States and France, and they captured hearts everywhere they went (**FIG. 26.1**).

The monuments, made for two powerful dukes of Burgundy, are similar in overall form (**FIG. 26.2**). A massive black marble slab supports a recumbent life-size polychrome effigy of the duke, the slab itself raised and supported on

FIGURE 26.1. (opposite)
Juan de la Huerta and Antoine Le Moiturier, Mourners, from the Tomb of John the Fearless, 1443–70. Musée des Beaux-Arts, Dijon. Thirty-seven small processional figures, carved in alabaster, each (15 to 16 in.) in height. When the tomb monument was put in storage during museum restoration in 2009, the diminutive mourners were briefly freed from their places beneath the marble catafalque and exhibited on their own in the touring show *The Mourners: Tomb Sculptures from the Court of Burgundy*. Here seen in installation at the Minneapolis Institute of Art in 2011.

FIGURE 26.2. (above)
François Denis Née, after Jean Baptiste Lallemand, *The Tomb of John the Fearless, Duke of Burgundy*, 1780s or 1790s. Line engraving. Plate size: 187 × 248 mm (7⅜ × 9¾ in.). Paper size: 206 × 286 mm (8⅛ × 11¼ in.). The site of the dukes' fifteenth-century tomb monuments in the Carthusian monastery of Champmol outside Dijon, before their partial loss and destruction during the French Revolution. The Mourners can be seen in procession within the pediment. National Portrait Gallery, London.

an impossibly delicate openwork cloister carved in white marble. Through the arcades of the cloister winds a cortege of forty humble figures. Priest, choirboys, acolyte, deacon, bishop, cantors, monks, and members of the family and court march in hooded cloaks. Each figure is between fifteen and sixteen inches in height, carved in the round, and bears traces of former color and gilding. Some clasp their hands in sorrow; others clutch rosaries, cinctures, or pouches; many hold books (one saving a place with a finger); some dab their eyes with their sleeves, raise their faces to heaven, cover their mouths, or open their mouths in prayer. Their twisted and wrung cloaks heighten the drama of the poses. Drawn hoods leave many faces in shadow or occlude heads entirely, allowing onlookers to imagine their own faces of sorrow (**FIG. 26.3A**). They form a pageant of individualized gestures of grief. The first tomb was completed in 1410. By the time the second tomb was finished in 1470, more than four generations of sculptors had worked on the two monuments.[1]

The tombs are those of Philip the Bold (1342–1404) and his son John the Fearless (1371–1419), the first of four ambitious dukes of the French Valois dynasty who expanded the duchy of Burgundy into a major center of political and economic power in Europe. By the end of the fifteenth century, much of current-day Belgium, Luxembourg, and the Netherlands came under its rule, Dijon its capital, Antwerp its port to the world. Alliance with the House of Habsburg followed, and Charles I (crowned Holy Roman Emperor Charles V in 1520), born in Ghent, the grandchild of the union of two great dynasties, would be a duke himself at the age of five. Raised in the Netherlands in the "eye of northern Europe," he knew the tombs well and intended his own tomb to be placed alongside them, in the Chartreuse de Champmol.[2]

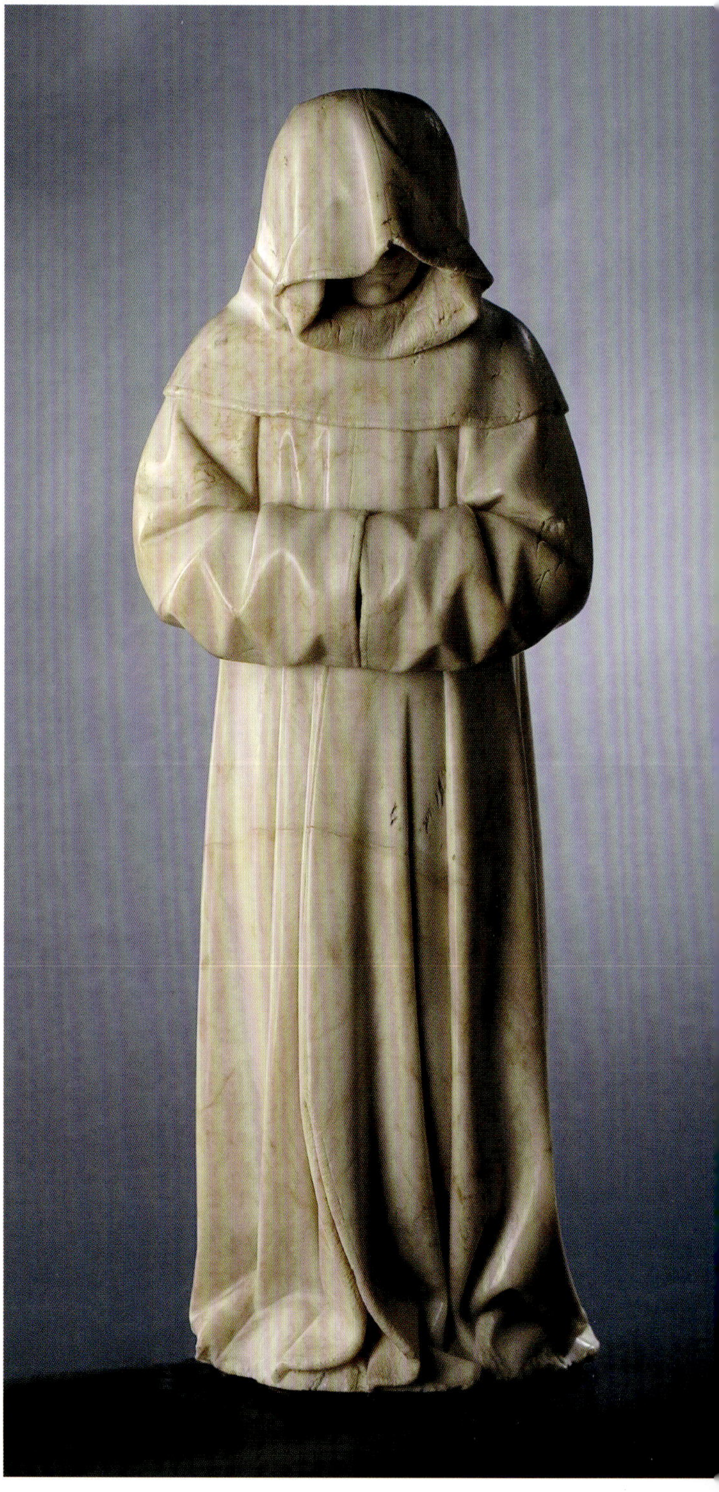

26.3A.

FIGURE 26.3A, B.
Attributed to Antoine Le Moiturier, from the Tomb of John the Fearless, 1443–70, "Mourner No. 56 with cowl pulled down, hands in his sleeves." Partially polychromed and gilded alabaster. Musée des Beaux-Arts, Dijon. Charles V would have seen these small marching tomb figures in Dijon. Peter Friess and Reinhard Steiner point to a similarly small cloaked figure in the painting by Miguel Jadraque y Sánchez Ocaña (**FIG. 23.4**).

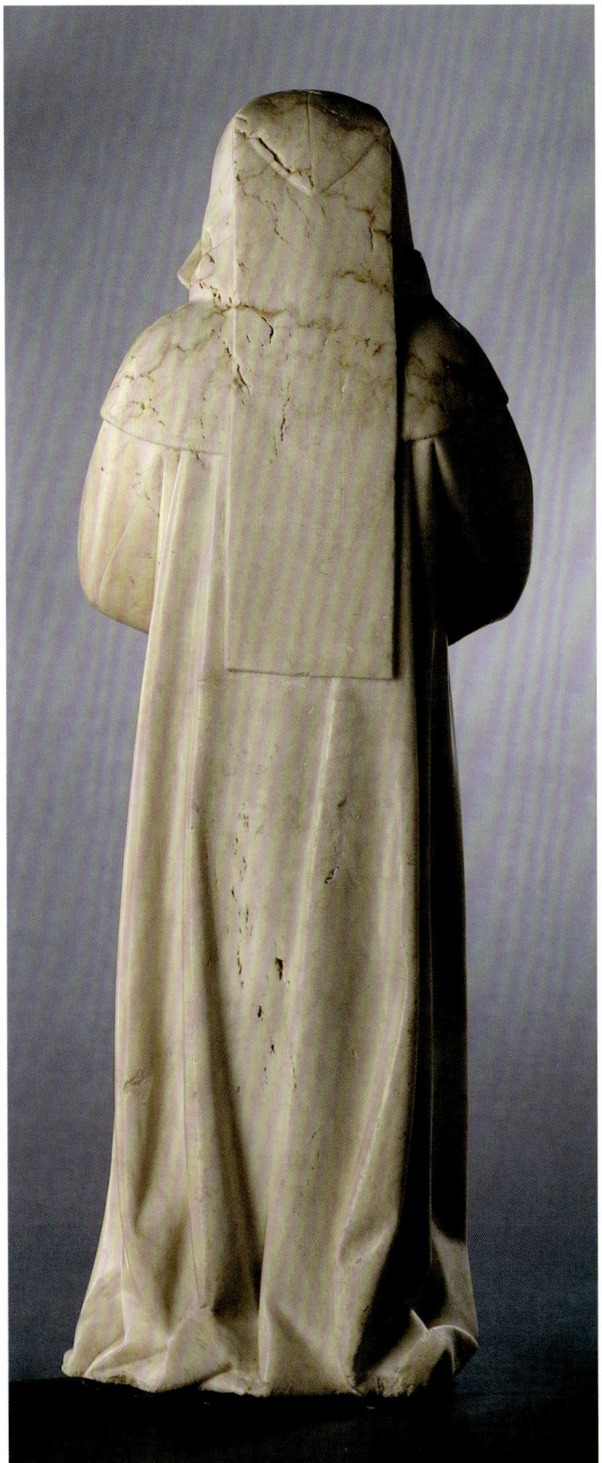

26.3B.

The monuments were originally located in the Carthusian monastery of Champmol outside Dijon but are now installed in Dijon's Musée des Beaux-Arts. They were restored in the early 2000s.[3] When the museum underwent renovation beginning in 2009, Mourners from the tomb of John the Fearless were carefully removed and assembled for an unprecedented exhibition tour of seven museums in the United States and one in Paris.[4] Released from their supporting role in the cenotaph pediment, they briefly took the spotlight as a procession of freestanding sculptures in their own right. "Tiny bravura set pieces of carving," the historian Anthony Grafton described them, reviewing the show.[5] It was an extraordinary opportunity to view them as solo objects. As Grafton pointed out, "They were carved at a time when painters finished the backs of altarpieces to perfection to please the all-seeing eye of God. To see [the Mourners] in the round is to be dazzled by the skill that their creators lavished on details that they did not expect human visitors to see." Each host museum arranged them differently, but all preserved the processional essence. The Metropolitan Museum of Art in New York built an extended runway in its Medieval Sculpture Hall and marched the figures two by two in one straight suite. The show brought their procession into the viewer's own space: "out of the church and into the street."

The exhibition curators reprised the Mourners' task: "Following the Catholic ritual, the Dijon Mourners are designed to perform for all eternity the vigil for the dead, offering up prayers for the salvation of the deceased." Indispensable for the soul of the prince on Judgment Day are the perpetual prayers of the living: the more prayers, the better. Seen without the object of their labor, the Mourners' labor itself comes into view. Unique though each figure is, the sight of them massed in cloaked procession verges on the cinematic to our modern eye, giving us a Muybridge-like sequence of walking supplication, a frame-by-frame analysis of the Catholic requiem. They teach the onlooker not only the proper performance of ritual mourning but also the power of sculpture to represent and, in its own way, perpetuate the act of prayer. Now more than 550 years old, these small medieval sculptures, among the most moving and luminous in the history of art, instruct us yet.

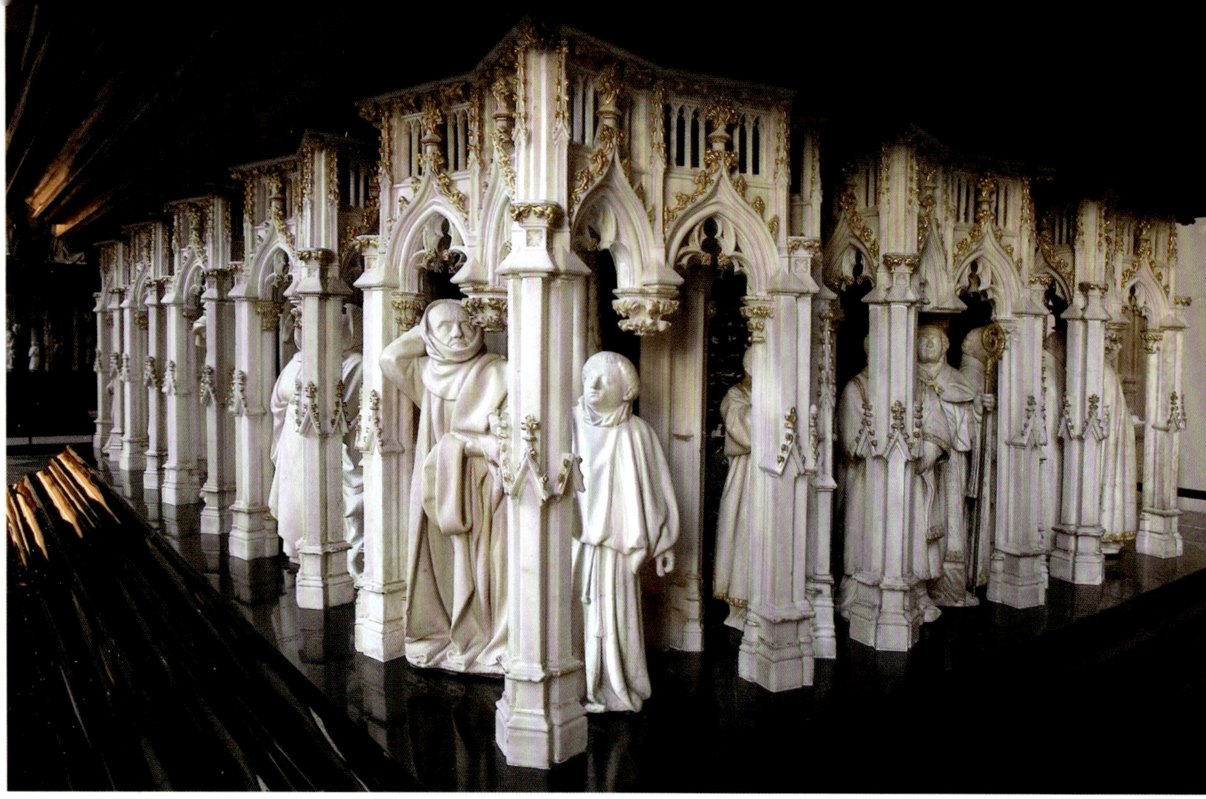

When the show finished its tour, the Mourners were returned to their assignments within the monument. By design, the size and number of the figures are properties bound to their placement within a predella-in-the-round. They inhabit the shadowy through-space beneath the bier, a dark interior broken only by the delicate miniature columns of the arcades and by the figures themselves, dependent on light available from high windows or candles. A viewer walking around the monument—one procession required to view another!—sees it with a moving eye, and glimpses the pale queue in parallax motion, as arcade columns in front of each small figure slide past faster than columns behind (**FIG. 26.4**). In the indeterminate space under the bier, a viewer might imagine even more figures hidden in the recesses, a subterranean zone full of glowing walkers. In these monuments, the living and the dead exchange places: the "corpse" on top, the mourners below. The carving of processional figures on sarcophagi was, of course, an ancient practice, but raising a colossal catafalque to place the diminutive figures *under* it transforms ornament into physical support. The practice was not new, but its iteration at Champmol was unique not only for its material grandeur but also for the opened space under the bier, for the extreme keyboard of scale, and for the full three-dimensionality and individuality of the Mourners.

Looking at "Mourner No. 56" in **FIGURE 26.3** and again at the Jadraque painting of Charles V in San Yuste, **FIGURE 23.4**, one sees why Peter Friess and Reinhard Steiner, studying the painting in 2003, noted the small icon of a hooded monk in the background among clocks and paintings—and cited the alabaster mourners. The small sculptures would have been well known to the emperor. Could Jadraque have actually intended this reference? As Friess and Steiner go on to say, the small icon is indeed the same size as the monkish automata. Elizabeth and David, for their part, dare go no further than to imagine that the processional character of the Mourners, which predate the monk by more than a century, might be something of a precedent or a script for the actions of the automaton. The monk performs in clock time what the Mourners perform in space. A single automaton is perhaps a procession of one—performing, like the Mourners, a sequence of urgent gestures: a temporal repetition to the Mourners' spatial repetition. An ensemble of mourners, on the one hand, and a solo clockwork monk, on the other: each articulates the repetition of prayer according to its nature. In both, the more prayers, the better.

FIGURE 26.4.
Tomb of Philip the Bold. Detail of the walking Mourners in procession within the shadowy arcades. Musée des Beaux-Arts, Dijon.

Walking in Procession

Procession itself, an essential part of every culture, was a fundamental ritual of public life in early modern Europe. Kings and emperors staged triumphant entries and marriage processions; bishops decreed requiems and jubilees; parishes hosted feast day processions; confraternities walked in supplication to the shrines of patron saints. William A. Christian Jr., throughout his book *Local Religion in Sixteenth-Century Spain*, recounts the enduring richness of procession in community life in New Castile, not just for Holy Week or Corpus Christi, but for many other annual imperatives such as Ascension Day, Saint Anne's Day, All Souls Day, and local saints' days.[6] There were petitionary processions, penitential processions, votive processions, and seasonal processions: to implore village saints for relief from plague, from locusts, from drought; to give thanks for such relief or observe hereditary vows to perpetuate the thanks; to honor martyrs; to visit sites where visions were reported. Pilgrimages were nothing if not forms of procession. The townspeople and clergy of Alcalá de Henares formed a procession as they bore the body of Diego from monastery to sickroom. The Spanish *imágenes de vestir* stamp and heave through the city in procession. The Mourners process through a miniature cloister in three-dimensional depth. The clockmaker designing a walking automaton is making something that is familiar after all, a figure engaged in a defining human act. Diego himself was a walker: to Rome and back from Castile in 1450, surely in the company of other pilgrims, for the great Jubilee ceremonies under Pope Nicholas V. The monk automaton shares its performative and temporal nature with the Spanish processional sculptures. But it shares its scale and repetitive nature with the Mourners. It is, we can say, many sculptures in one.

And in their compressed quarter, the Mourners enact the same choreography as do the sculptures and participants of the Spanish *pasos*, though in their own incomparable Netherlandish idiom. They too bear a platform carrying an effigy. Beyond the processional topos shared by these very different works, the subdued emotional expressiveness of the alabaster Mourners is a strong contrast to the theatrical figures of the Spanish *pasos*, though taken as a whole the tomb monuments comprise a theater of grandeur.[7] Small as the Mourners are, their manner and bearing represent the large and universal gestures of grief, and the monk has inherited their task. They perform a service: they provide a form of perpetual prayer. They succeed by virtue of their effectiveness as images, by their humanity.

In the pantheon of European Renaissance works of art are sculptures and paintings that capture and enact moments of religious mystery, ecstasy, trial, and emotion. As a figure belonging to the history of art, the monk gathers some of his outward bearing as sculpture from the expressive Mourners, from Spain's "most precise and lifelike" processional figures, from artistic depictions of the ethereally formidable apparitions of saints, and from images of the performance of prayer. These sculptures exemplify the heightened realism, skill, and observational brilliance of the best art of its kind, created at a time when such works had the power to address the terrors and splendors of the body to viewers who brought to sculpture, to painting, to art of all kinds, their deepest religious needs.

We have traced a few sculptural practices to sketch a landscape for considering the monk's character as a religious icon. The nature and varied makeup of audiences are implicit in what we have shown, but the overall question of audience—who saw the monk and where?—must remain open. The street, the church, the tomb monument, the *Kunstkammer* each call for different expectations of, and responses to, works of art, for different commissions by patrons, for different audiences and occasions.[8] What we miss, and would gladly add to the public and formal settings we have explored, is an example from the private sphere. Perhaps such privacy—outside an aristocratic or ecclesiastical setting—is a relatively modern invention. For all that he shares with the devotional sculptures illustrated in this study, the monk automaton wants a quiet room with a small audience.

27 The Machine and the Corpse

Servus Gieben's humble epiphany on the feast day of San Diego will always belong to the automaton monk. The proposal has taken on a life of its own. Just as the cardinal's description of Saint Francis standing in the crypt created a new kind of image centuries later, so has Father Servus created a new kind of story, four hundred years on, about the automaton. The difference is that the automaton—the image—came first in this case, then the story. Or so we think.

Elizabeth's father, a scientist, read an early draft of our study of the monk and called it a "dendritic" tale. One object: an automaton—and a clockmaker, an accident, a miracle, and a proposal for how they were connected—with no proofs, is this enough for a book? Yes, it is, for the ineluctable fact of the automaton itself. We never quite recover from looking at him. We look closer, and the histories of technology, medicine, art, and faith all compete for a claim to our response.

The legend that the monk is a portrait of Brother Diego serves a purpose (to us now) as useful and urgent as if it were true. Paired, the automaton and the saint permit a glimpse of competing definitions of the body and its life. Medicine and faith may be uncertain partners. But the automaton, in good running order after four and a half centuries, comes to life in animated defiance of time and decay. Like the saint's body, it still delivers after all these years.

The practice of medicine in the sixteenth century was assembled from incompatible ideas about nature and divinity. Don Carlos's suppurating head wound, the physicians' pharmacopoeia for the treatment of infection, the ascetic Franciscan and his uncanny corpse (still sweet to the nostrils), the midnight apparition, the immaculate cure—the story is a microhistory of early modern beliefs. What is the body made of, and what keeps it running? In the midst of the debate, the spring-based clock arrived. In a clockwork universe, the coiled spring is *primum mobile*, the fusee its tamer, the gear train takes the power and divides it into systems, the cams and levers are motive and memory, their multiple linkages the nerve and muscle of motion, whether to the hands of a clock, the heavenly bodies of the astrarium, or the limbs of a man. The delicate spinning fly brings judgment and governance. The clockmaker who fashioned these parts and put them together inside the wooden monk was producing a coherent equivalent of body and will. Perpetual motion would be the next frontier, the promise of eternal life.

During the twenty-two days Don Carlos lay in delirium in Alcalá de Henares, his surgeons and physicians labored to name what they saw and deduce what they couldn't see. They engaged in heated debate over whether to trepan the skull, for among the attending physicians was the Flemish anatomist Andreas Vesalius, who had conducted a postmortem in France three years before and observed how unseen damage within the cranium had proved fatal.[1] Near the end of the crisis, drilling was begun, then halted, amid fervent argument. At another point physicians noted that the patient was moving his bowels more than five times a day; taking this as a positive sign, they administered a purgative to assist nature, and within hours twenty more bowel movements ensued. On the twelfth day, still seeking signs of fracture, surgeons cut away the discolored tissues surrounding the wound and exposed the skull, but profuse bleeding prevented visibility. One day later, infection "engulfed the patient's face, glued his eyes shut, and then moved downward across his neck and upper torso."[2] In desperation, the physicians authorized the ministrations of a lauded Moorish healer from Valencia, whose ointments had the effect of burning the wound and turning the exposed bone "black as ink." The patient had been given the final sacraments, and once already the sheet had been drawn over his head. The opening of the tomb of fray Diego at the monastery's Church of Saint Francis marked the moment when medicine gave up and faith took over.

The story of Don Carlos shows us the foremost physicians of the age debating best practice in the treatment of infection. It is a harrowing document of the physical body in extremis, and the agony of witness. Those who gazed on the mortifications of the flesh—and who is spared yet?—could not but ask for the promises of faith. They found its assurance in the miraculous absence of decay in fray Diego's disinterred body. When friars of the monastery of Santa María de Jesús brought the corpse into the sickroom, fresh discussion arose, about whether to force open the patient's eyes. Instead, the prince signed for the corpse to be brought close enough to touch. The steward and friars transferred the corpse to the sickbed and folded back a part of its shroud to expose the face. The patient reached out and found the shroud and the corpse. "Having touched his hands to the face and body of the said Brother Diego, [the prince] raised them and then ran them over his own face, eyes, and mouth."[3]

Many people in the room witnessed this moment: the king on his knees, the friars, the surgeons and physicians, the royal stewards, the confessors, the bishop of Cuenca, the duke of Alba.[4] What resuscitating essence was expected to pass from the corpse to the prince? Vesalius, dissector of corpses, whose anatomical treatise *De humani corporis fabrica* would transform the practice of medicine, must have wondered.[5] Many years later, with Diego at last secured in the pantheon of saints, Philip had the tomb in Alcalá opened once more, to remove the bones of the saint's lower legs (still bearing flesh) for inclusion among his personal sacred relics.[6] He too wanted to be able to touch the life-giving body of the saint.

Such moments of ordeal and mystery are repeatedly given visual image in the paintings and sculptures of early modern Europe. The lifelike presentation of flesh and its affliction, the close observation of wounds, the body at the threshold of death, the supernatural aliveness of martyrs, saints, and savior, each recovery a resurrection, a Lazarus: Gregorio Fernández, Juan Martínez Montañés, Pedro de Mena, and masters of sixteenth- and seventeenth-century Spanish sculpture depicted, "most precise and lifelike," the ravages and miracles of the body, in full color. Thus does

Mena give substance to the apparition of Saint Francis in the crypt, caught in the visitor's torchlight. Our Iberian focus, though it may yet prove mistaken, finds in the legend of the saint the paradox that animates the automaton.

The concept of a body that comes back to life, from sickness or through holiness, is implicit in the object that is the monk too. Theologians describe miracles as events with no apparent human or natural cause. The monk models in votive intercession or sympathetic repair everything the physicians were trying to accomplish: to prevent the ebbing of life from the body, to rewind its vital spring, to put the prince back on his feet so that he might move again under his own power. The legend adheres to the monk because in both we sense the analogue forces in play between medicine, mechanics, and faith. Both are ambivalent about where the agency of craft ends and that of faith begins, about what is proper to the human realm and what proper to the divine.

A small wooden icon stands arrested on a shelf, and then one day the icon parades about and gesticulates on a table. A small miracle. Or the image of a small miracle. Or the metaphor of a large miracle. Or an artificial miracle. Our first and last definition of the storied monk: a motionless thing that comes to life.

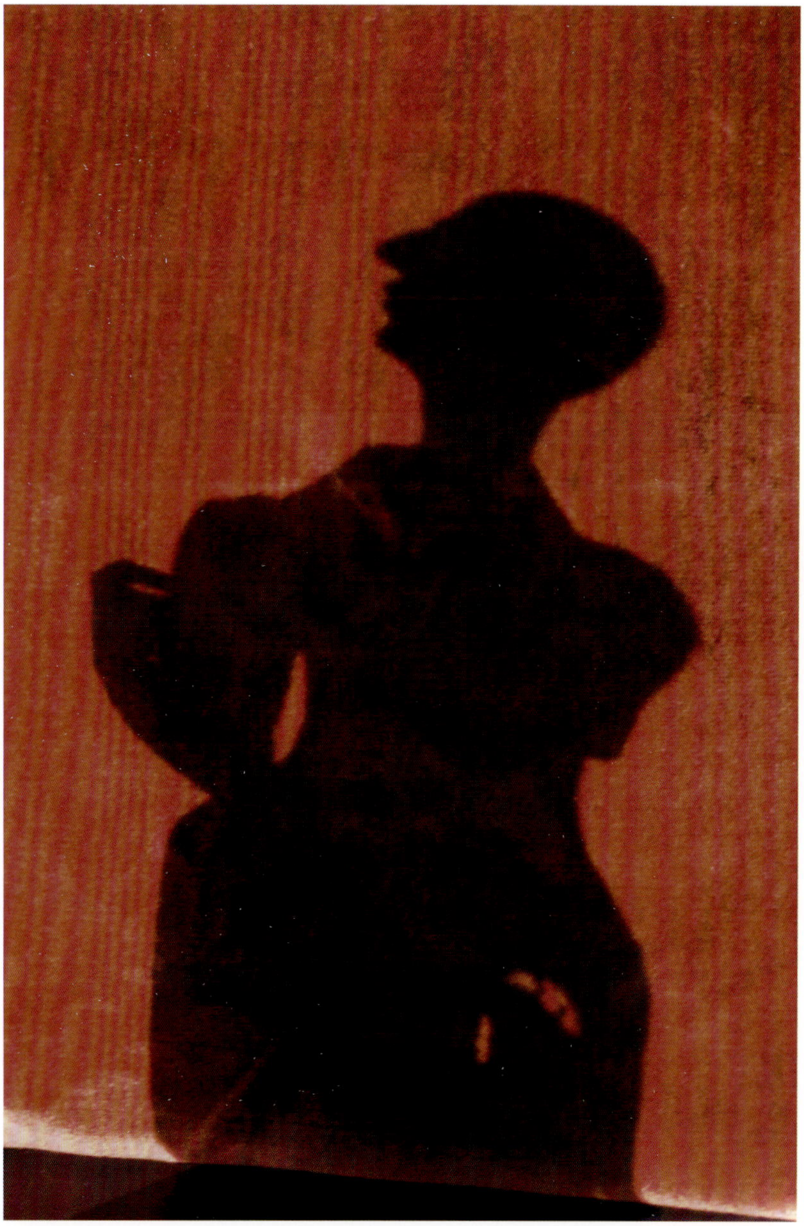

APPENDIX

Table of Comparisons: Eight Automata

The following are excerpts from an ongoing database compiled by the authors. Details are sourced from host museums and collections archives, as well as historians, curators, and conservators named in the chapters on each automaton.

AUTOMATON (As currently identified by host institution)	FIRST DOCUMENTED APPEARANCE	MATERIALS
Automaton of a friar Mid-16th century, Spain or Germany Height: 40.64 cm (16 in.) National Museum of American History, Smithsonian Institution, Washington, DC https://americanhistory.si.edu/collections/search/object/nmah_855351 ID NO. ME.336451	1975 Zurich (Uto Auktions AG, Sept. 29, 1975, auction catalog); acquired 1977, Smithsonian Institution, Washington, DC	Iron clockwork; wood fusee. Head, arms, hands, feet of carved wood; stained beech or poplar (by visual analysis), polychromed. Body carved and lathe-turned from single block of wood. Eyes made of iron, painted. Cloth habit of Franciscan friar when acquired not original. Modern Franciscan habit was made at Smithsonian after purchase.
Automaton: Preaching monk Automatenfigur: Predigender Mönch ca. 1560, Spain or Germany Height: 39 cm (15 5⁄16 in.) Deutsches Museum, Munich https://www.deutsches-museum.de/en/museum-island/exhibitions/robotics#c21914 INV NO. 84/18	1983 London, Christie's auction (Financial Times, Oct. 15, 1983); acquired 1985, Deutsches Museum, Munich	Iron clockwork; wood fusee. Signs of a former internal glockenspiel. Head, arms, feet of carved wood; poplar (by sample microanalysis), polychromed. Body carved and lathe-turned from single block of wood. Decorative pattern in red paint, now faded, on lower front of body shell, unique to this automaton. Unclothed when acquired.
Musical automaton-male saint Zenélő automata-férfi szent 16th century, Spain Height: 41 cm (16 1⁄8 in.) Iparművészeti Múzeum (Museum of Applied Arts), Budapest https://collections.imm.hu/gyujtemeny/zenelo-automata-jezsuita-ferfiszent/869 ACCESSION NO. 11254.a-b	1915 Budapest, Iparművészeti Múzeum	Iron clockwork; wood fusee. Glockenspiel bells of bronze and pewter. Head, hands, feet of carved wood with gesso primer; linden (by visual analysis), polychromed. Body carved and lathe-turned from single block of wood. Eyes made of pewter, painted. Silk damask cloak and cape in the Italian style, considered 18th-century replacements.

PERFORMANCE (if mainspring is fully wound)	**CONDITION BEFORE CONSERVATION**	**CONSERVATION**
Walks ≈19 in., turns right, repeats 6 more times, tracking irregular polygonal path. Per leg of path, head moves (1x: left, center, bow), eyes (3x: left, center), mouth (6x: open, close; 1x: kiss), right elbow (3x: to strike chest), left elbow (6x: to raise and lower cross), left elbow and shoulder (1x: to raise cross to kiss), feet (4x each: lift, fall). Full performance would last 3 to 4 minutes. Automaton is functional, but to protect the mechanism the mainspring is not wound fully. For extended description, see chap. 10.	Little corrosion present on iron clockwork surfaces. Body and clockwork in original condition, though mainspring and four chain links were replaced at some time in the past. Two gut lines (fusee line and line to left elbow) as well as cassock, cowl, belt, and rosary are modern.	Todd (1978): "Remove, clean and re-grease mainspring. Fit one new line. Close up all strained chain links. Make three chain links and one hook."
Walks ≈15¾ in., designed to turn right and repeat 10 more times. Missing turning pivot, hence geometry of walking path unknown. Per leg of path, head moves (1x: left, right, nod), eyes (8x: left, right), mouth (4x: open, close), right wrist (oscillates 40.5x), left wrist (oscillates 22.5x), feet (8x each: lift, fall). Automaton is partly functional. Mainspring is not wound fully, to protect the mechanism. For extended description, see chap. 12.	Friess (1988): "The surface of the iron mechanism was rusted but aside from the torn gut string and the residue of resinous oil, it was functional. All that is left of an original glockenspiel mechanism is the driving wheel of the cylinder on the drive-shaft.... Four fractures [in the wooden body] had been previously repaired.... One arm was broken at the elbow, the other arm at the spot where the bearing for the hand is screwed in.... One hand and both feet were missing, two fingers of the remaining hand were broken off. The head has been refinished several times."	Friess (1988): "Simple cleaning and rust prevention on mechanism. Most of the labor involved was devoted to the stabilization of cracks, gluing of fractures, and the replacement of missing fragments of the wooden casing.... The elbow of the broken arm also received a replacement fashioned out of poplar wood." Missing fingers, hand, and feet were remade and painted, but no details about this work appear in Friess's report.
Stationary figure containing an internal glockenspiel (unique to this automaton). At full wind, a single sequence: head moves (1x: right, left), eyes (16x: side to side), mouth (26x: open, close), right hand (wields bell, oscillates at wrist to ring 36x). Movement corresponds with 45-second glockenspiel melody delivered twice; head bows at end. Total performance time is ≈ 1 minute 48 seconds. Automaton is functional. For extended description, see chap. 13.	Not known. Soós and Rácz (1990) reported nonfunctioning clockwork, signs of former repair in the right arm movement, and body shell split in many places. Missing fingers on both hands. Missing bell handle. Garments damaged. Report does not comment on age or condition of paint. Feet presumed original. Glockenspiel wooden barrel had been repegged, hence original melody considered lost.	Soós and Rácz (1990): clockwork entirely dismantled, inventoried, corrosion removed. Mainspring replaced. Halo repaired with brass and re-gilded. Glockenspiel bells cleaned. Missing wood fingers of the bell hand and missing bell handle remade. Splits in the body shell closed and glued. Garments, though not considered original, were conserved, with new material clearly identified.

AUTOMATON (As currently identified by host institution)	**FIRST DOCUMENTED APPEARANCE**	**MATERIALS**
Automaton, Lady lute player 1640–60, Germany Height: 31 cm (12 3/16 in.) National Museum of American History, Smithsonian Institution, Washington, DC https://americanhistory.si.edu/collections/search/object/nmah_856653 ID NO. 1984.0865.01	1978 Geneva (Galerie d'Horlogerie Ancienne, Apr. 9, 1978, auction catalog); acquired 1984, Smithsonian Institution, Washington, DC	Brass clockwork; wood fusee. Head, arms, hands, feet of carved wood, polychromed. Body carved and lathe-turned from single block of wood. Cloth dress considered 18th-century replacement. Intact but tattered at hem and sleeves. Hair wig.
Lute player Lautenspielerin ca. 1600, South Germany Height: 39.7 cm (15 5/8 in.) Bayerisches Nationalmuseum (Bavarian National Museum), Munich INV NO. 27/30	1927 Munich, Bayerisches Nationalmuseum	Brass clockwork; wood fusee. Wax head, not original. Arms, hands of carved wood, polychromed. Wood torso. Separate lower body shell of lathe-turned wood, identified as mountain ash (by sample microanalysis), sections missing. Dress considered 19th-century replacement. Traces of original cloth remain on upper body: green cotton or silk with pale gold or silver metallic thread.
Automaton: Mandolin player Automate: Joueuse de mandoline ca. 1775–99, France Height: 35 cm (13 3/4 in.) Conservatoire national des arts et métiers (CNAM), Paris https://collections.arts-et-metiers.net/?queries=query=search=N°%20d'inventaire=[06152-0001-]&showtype=record INV NO. 06152	1853 Paris, Conservatoire national des arts et métiers	Brass clockwork; brass fusee. Head, arms, hands, feet of carved wood, polychromed. Lower body shell of wood, only remnants remain. 18th-century peasant costume.
Automaton, Cittern player Automat, Cisterspielerin 2nd half of 16th century, Spain (?) Height: 44 cm (17 5/16 in.) Kunsthistorisches Museum, Vienna https://www.khm.at/objektdb/detail/96017/ INV NO. Kunstkammer, 10000	1920s The Netherlands; acquired 1934, Kunsthistorisches Museum, Vienna	Iron clockwork; metal fusee, frame, and gear train layout unique to this automaton. Head, arms, hands, upper torso, feet of carved wood, polychromed. Lower body shell of wood, repaired with patches of leather and brass. Linen and silk brocade garments in 17th-century Spanish court style.
Automaton: Noble matron playing a drum Automa: Nobil matrona in atto di suonare un timpano ca. 1560, Switzerland or Germany Height: 38 cm (15 in.) Private collection, Milan https://www.youtube.com/watch?v=jAtkSqRH6-4	2005 Trent (Italy) (Exhibition: *La misura del tempo: L'antico splendore dell'orologeria italiana dal XV al XVIII secolo*)	Iron clockwork; wood fusee. Head, arms, hands, feet of carved wood; poplar (by visual analysis), polychromed. Body carved and lathe-turned from single block of wood. Eyes made of carved wood, painted. Silk dress, cap, and fan, probably early 18th century replacements. Intact but threadbare at sleeves and bodice. Wig of real hair

PERFORMANCE (if mainspring is fully wound)	CONDITION BEFORE CONSERVATION	CONSERVATION
Designed to walk ≈24 in., turn right and repeat 10 more times, tracking a roughly triangular course. Per leg of path, head moves (1x: left, right, nod), eyes (5x: right, center), right wrist (oscillates to strum lute 30x), left elbow (to carry forearm and hand up and down fretboard 5x), feet (5x each: lift, fall). Automaton considered functional. For extended description, see chap. 14.	Wood surfaces original and apparently untouched. SI registrar's notes (1979) reported, "Figure is in working order. Figure's costume brittle, torn and badly faded. Wooden frame encasing clockwork was damaged and repaired at some point in the distant past." Todd (1979): clockwork "has been repaired on a number of occasions with varying degrees of competency. [Clockwork] has also suffered recently through being 'cleaned' with a wire brush and consequently the original finish and patina has been lost."	Todd (1979): "No work done. Observation, comparison, photography, drawing and documentation only."
Designed to walk and turn, though feet, stepping levers, and turning pivot are now missing, hence geometry of walking path unknown. Head now replaced, but a missing former wooden head once moved (side to side, nod), and eyes, mouth, right hand (to strum lute), and left forearm (to carry hand up and down fretboard) would have been movable. Signs of a former internal bell, sounding once at each turn. Automaton no longer functional. For extended description, see chap. 14.	Not known. Todd (1980) indicated the original head (now wax) would have been wood; missing components in the clockwork show evidence of former movable mouth and eyes as well as stepping levers and feet. Evidence of a former single bell with strike hammer (both missing) within the body. Mechanism for head turning no longer functional. Extensive damage to right arm from insertion of replacement mechanism. Overpainting on hands.	No museum records other than Todd's 1980 documentation of the automaton and its clockwork.
Designed to walk and turn. "A walking automaton making the gestures of playing the mandolin, while turning its head and eyes" (CNAM 1960 catalog). Right hand strums, left hand moves at mandolin fretboard, feet move back and forth. Automaton no longer functional. For extended description, see chap. 14.	Not known.	Dress was restored in 1997 and 2016. Former CNAM database listed three previous restorations of the automaton: Auricoste, ingénieur mécanicien, Paris [no date]; Hamm, 1970, Paris; Robert-Houdin, 1859, France.
Walking path designed to vary between straight line and circular path, depending on manually-set trailing wheel, unique to this automaton. On a full wind of the mainspring the figure could once walk 20 meters (KHM 1935 catalog). "[She] moved with small tripping steps, strumming the instrument with her right hand, and turning her head from right to left" (KHM 1966 catalog). Rack-and-pinion design for the head motion gives it an elegant tilting side-to-side turn. No eye or mouth motion. Automaton functioned until 1964, now too fragile to operate. For extended description, see chap. 15.	Bassermann-Jordan (1926): "The automaton was treated carefully and kept with respect: when its original dress was worn out and no longer in fashion, a new one was made about 40 years later, in the style of international court costumes of the nobility coming out of Spain at the time.... The face, finished in enamel-like colors, has been modified in later times; the lute is still the original."	Textile costume conservation in 1963.
Walks and turns, tracking the path of a five-pointed star. Per leg of path, head moves (10x: side to side; 1x: nod), eyes (≈30x: side to side), mouth (≈30x: open, close), right hand (oscillates at wrist 4x per second), feet (10x: lift, fall). The most rapid head movements of all the automata. Automaton is functional. For extended description, see chap. 16.	Significant damage: missing feet, left forearm and hand, and sections of wooden body shell. Original paint on face appears intact. Clockwork very corroded and nonfunctioning. No signs of past repair efforts.	Complete clockwork cleaning and overhaul in 1989 by Giorgio Gregato. Reconstruction of missing sections of wooden body shell. Left forearm and hand remade along with drumstick and drum. Drum secured via brass bracket to automaton's waist. Both feet remade.

Notes

Introduction

1. A note on the word *monk*. In Catholic nomenclature, monks are members of monastic orders, defined by seclusion. Friars belong to the mendicant orders, including Franciscans, who live a communal life defined by works in the secular world. Though the cassock and sandals, and the legend itself, urge a Franciscan character, the automaton received the generic label "monk" in auction and acquisition records and in much of the subsequent literature. The colloquial name has adhered, and we use it too. The museum has recently rechristened it a *friar*.

2. Otto Mayr, "Purchase of 16th Century Automaton," April 1, 1977, accession 1977.1191, Registration Services, National Museum of American History, Smithsonian Institution, Washington, DC. On the French inventor Jacques de Vaucanson (1709–1782) and the Swiss clockmakers Pierre Jaquet-Droz (1721–1790) and his son Henri-Louis Jaquet-Droz (1752–1791), see chap. 17, below. Vaucanson went on to invent a prototype automated loom, using interchangeable perforated punch cards to program the lift and fall of hooks attached to warp threads. The loom can be seen today in the Musée des arts et métiers in Paris; Vaucanson's inventions and others he collected were key to the founding of the Conservatoire national des arts et métiers in 1794. The French inventor and textile merchant Joseph-Marie Jacquard (1752–1834) patented a loom in 1804 based on Vaucanson's machine; the Jacquard loom would play a prime role in the Industrial Revolution. The lowly punch card, the first digital command and data storage medium, interacting with a compounded matrix of levers, gear wheels, and pinned barrels, in turn played a central role in early computing, famously in the work of the English mathematician and engineer Charles Babbage (1791–1871), inventor of the first mechanical computer. See Douglas R. Hofstadter, *Gödel, Escher, Bach: An Eternal Golden Braid* (New York: Basic Books, 1979), 24–25.

3. Otto Mayr, "Automatenlegenden in der Spätrenaissance," *Technikgeschichte: Beiträge zur Geschichte der Technik und Industrie* 41, no. 1 (1974): 20. Thanks to Henning Bauer for his translation from the German, which Otto Mayr read and approved. See also Otto Mayr, *Authority, Liberty, and Automatic Machinery in Early Modern Europe* (Baltimore: Johns Hopkins University Press, 1986); and Otto Mayr, "A Mechanical Symbol for an Authoritarian World," in *The Clockwork Universe: German Clocks and Automata, 1550–1650*, exh. cat., ed. Klaus Maurice and Otto Mayr (New York: Neale Watson Academic Publications, 1980), 1–8.

CHAPTER 1

To Set the Stage

1. Hugh Trevor-Roper, *Princes and Artists: Patronage and Ideology at Four Habsburg Courts 1517–1633* (London: Thames & Hudson, 1976), 81. Trevor-Roper cites José de Sigüenza, *Historia de la Orden de San Gerónimo* (1605), reprinted in *Nueva Biblioteca de Autores Españoles* (Madrid, 1909), 2:635–36.

CHAPTER 2

A Miracle and a Vow

1. To extricate truth from myth concerning the death of Don Carlos, see Henry Kamen, *Philip of Spain* (New Haven, CT: Yale University Press, 1997), 120 ff. A more recent and granular narrative is Geoffrey Parker's *Imprudent King: A New Life of Philip II* (New Haven, CT: Yale University Press, 2014), chap. 10, "The Enigma of Don Carlos." Just published is a new study by Maria-Cristina Necula, *The Don Carlos Enigma: Variations of Historical Fictions* (Washington, DC: Academica Press, 2020). Among earlier biographies, *Don Carlos et Philippe II*, 2 vols. (Brussels: E. Devroye,imprimeur du roi, 1863), by the nineteenth-century Belgian historian Louis-Prosper Gachard (1800–1885), is essential. Verdi's conceit and Friedrich Schiller's before him were based on the novel *Don Carlos* (1672) by the seventeenth-century French writer César Vichard de Saint-Réal, who in turn had incorporated allegations from a polemical tract written by William of Orange (1533–1584).

2. Geoffrey Parker attributes the fall to the prince's physical deformity (*Imprudent King*, 182). The version of Charles Donald O'Malley, *Don Carlos of Spain: A Medical Portrait* (Los Angeles: University of California, 1970), follows earlier accounts of the prince in pursuit of the daughter of the house concierge:

> It happened that the caretaker of the royal quarters in Alcalá had several daughters, one of whom appears to have taken the fancy of Don Carlos, although for what reason we are not sure.... [O]n 19 April, a Sunday, Don Carlos caught sight of the girl walking in the garden outside the palace, and in his haste to pursue her ran down a flight of steps, missed his footing, turned a complete somersault and struck his head against the door at the bottom of the staircase. In his report to Queen Elizabeth, [Thomas] Challoner, the somewhat unsympathetic English ambassador[,] ... told the story succinctly: "The prince in hasty following of a wench, daughter to the keeper of the house, fell down a pair of stairs [and] broke his head." ... Don Carlos' haste was all in vain since, as it turned out, the door at the foot of the stairs was locked; and whatever the purpose of the prince— ... as likely brutality as seduction—the girl, who would otherwise have been lost in anonymity, is known to us as Mariana de Garcetas. She was promptly clapped into a convent with a consolation of 1,000 crowns. (7)

3. William H. Prescott, *History of the Reign of Philip the Second, King of Spain*, 3 vols., ed. John Foster Kirk (Philadelphia: J. B. Lippincott, 1874), 2:467–68. Volumes 1 and 2 were first published in 1855; volume 3, in 1858. Prescott mentions Schiller's play (2:458). Verdi's opera would premiere in Paris in 1867. *Erysipelas*, also called St. Anthony's fire, was a spreading inflammation of the skin and subcutaneous tissues, now understood as infection caused by *Streptococcus* bacteria.

4. See, e.g., Kamen, *Philip of Spain*, 91–92; Charles Donald O'Malley, *Andreas Vesalius of Brussels, 1514–1564* (Berkeley: University of California Press, 1965), 296–302; and Bjørn Okholm Skaarup, *Anatomy and Anatomists in Early Modern Spain* (Surrey: Ashgate, 2015), 128–38. Vesalius (1514–1564) was a royal physician to Charles V from 1543 to 1556 and then physician in the court of Philip II in Madrid from 1559 to 1564 (O'Malley, *Andreas Vesalius*, 189, 267, 288; Skaarup, *Anatomy and Anatomists*, 3).

5. L. J. Andrew Villalon, "Putting Don Carlos Together Again: Treatment of a Head Injury in Sixteenth-Century Spain," *Sixteenth Century Journal* 26, no. 2 (1995): 347–65. Chief among the primary texts Villalon consults is the document written by the prince's surgeon, Dionisio Daza Chacón: "Relación verdadera de la herida de cabeza del Serenísimo Príncipe D. Carlos nuestro Señor, de gloriosa memoria, la cual se acabó en fin de julio del año de 1562," in *Colección de documentos inéditos para la historia de España*, 112 vols. (Madrid: Academia de la Historia, 1842–95), 18:537–63. O'Malley, *Andreas vesalius*, 298, calls Daza Chacón's chronicle "one of the fullest pictures we have of the medical procedure of that age" and prints an English translation of the entire report in his appendix, 407–19: "True Account of the Head Injury of the Most Serene Prince Don Carlos, Our Sire of Glorious Memory, Which Occurred the End of July [*sic*] 1562."

6. Villalon, "Putting Don Carlos Together Again," 356.

7. Prescott, *Philip the Second*, 469.

8. Villalon, "Putting Don Carlos Together Again," 354. See Daza Chacón, "Relación verdadera," 543; and O'Malley, *Andreas Vesalius*, 409. Betony or bishopwort: a medicinal herb "once the sovereign remedy for all maladies of the head" (botanical.com).

9. For eyewitness accounts of Don Carlos's report, we relied on Fr. Lucio María Núñez, OFM, ed., "Documentos sobre la curación del príncipe D. Carlos y la canonización de San Diego de Alcalá," *Archivo Ibero-Americano* 1, no. 4 (July–August 1914): 424–46. This four-part series continues under the same title to 2, no. 9 (July–August 1915): 374–87; 3, no. 13 (January–February 1916): 107–26; and 4, no. 19 (January–February 1917): 421–31. All are dedicated to the detailed account of Don Carlos's illness and cure and the subsequent canonization of Diego, compiling sources and quoting contemporary witnesses, ambassadors' letters, physicians' testimonies, and original *proçeso* documents (documents in support of canonization) then kept in the Archivo Central de Alcalá de Henares. Núñez gained significant portions of his narrative from Gachard's *Don Carlos et Philippe II* (see esp. vol. 1, chaps. 4 and 6); and Daza Chacón, "Relación verdadera."

We hesitate to interpret Carlos's midnight cry, "Basta, basta, padre, aparta el cordon," for the word *cordon* has many meanings, even within Franciscan symbolism. The witness who heard and reported it was Hernán López "el Portugués," the king's surgeon, among the physicians who attended the prince throughout the ordeal (Núñez, "Documentos sobre la curación del príncipe D. Carlos," 1914, 433). The prince related his waking description of the vision to his chamber physician Cristóbal de Vega: "Esta noche se me ha aparecido en la cama, de pies, el santo fray Diego, con una cruz de caña en la mano. Y le pregunté si era Sant Francisco, porque me pareció que lo era; y respondió que no era Sant Francisco, mas que tuviese confiança en Dios que sanaria de aquesta enfermedad" (Núñez, "Documentos sobre la curación del príncipe D. Carlos," 1914, 434; and noted in Daza Chacón, "Relación verdadera," 558–59).

CHAPTER 3

Diego de Alcalá

1. L. J. Andrew Villalon, "The Miracle Book of San Diego de Alcalá, or, the Fifteenth Century Failure to Canonize the First Counter-Reformation Saint," *Mediterranean Studies* 10 (2001): 9–35. Thanks to Professor Villalon for permitting Elizabeth King to see his essay before publication and to quote from it in her early publications on the automaton monk. Villalon located the miracle book in the Real Biblioteca de El Escorial, he writes, designated only by reference number (&.II.14) in the

three-volume *Catálogo de los manuscritos castellanos de la Real Biblioteca de El Escorial* (San Lorenzo de El Escorial, 1929), v. 1, 271–72. He identifies it as a copy of the now lost original fifteenth-century manuscript, describing it as a leather-bound volume of 109 folios, transcribed in 1562 by a scribe named Andrés Muñoz Corchado, at the express command of Prince Carlos. See also Villalon's "San Diego de Alcalá and the Politics of Saint-Making in Counter-Reformation Europe," *Catholic Historical Review* 83, no. 4 (October 1997): 691–715. Villalon revisited and extended his work in 2014: "The 1562 Head Injury of Don Carlos: A Conflict of Medicine and Religion in Sixteenth-Century Spain," *Mediterranean Studies* 22, no. 2 (2014): 95–134.

2. Villalon, "The Miracle Book of San Diego de Alcalá," 22–24.

3. Villalon, "The Miracle Book of San Diego de Alcalá," 26.

4. Peter Burke, *The Historical Anthropology of Early Modern Italy: Essays on Perception and Communication* (Cambridge: Cambridge University Press, 1987), chap. 5, "How to Be a Counter-Reformation Saint," 49–50.

5. The circumstances of Don Carlos's confinement and death have been the subject not only of literature and opera but of research by generations of historians. Every chronicle of the sixteenth-century House of Habsburg includes reports of the prince's infirmity and strange behavior, with some modern-day scholars venturing clinical diagnoses. Don Carlos did in fact plot a conspiracy against the crown and threaten the lives of ministers who tried to dissuade him and was poised to leave Madrid on the night the king confronted him. The king issued immediate notice to foreign courts of his confinement of Don Carlos and, six months later, of the prince's death. But he remained steadfastly silent about the prince's actions and affliction. William H. Prescott, writing in the mid-nineteenth century, heralds the opening of the Royal Archives in Simancas ("which have held the secrets of the Spanish monarchy hermetically sealed for ages") as a great service to scholars of his day (*Philip the Second*, 1:vi). Prescott's chap. 7, "Death of Don Carlos," 2:493–545, opened the court for the English-speaking reader.

CHAPTER 4

An Automaton by Juanelo Turriano

1. Fr. Lucio María Núñez's 1914–17 articles have become even more valuable as the *proçeso* documents he indexed and quoted were destroyed in the Spanish Civil War when the Archivo Central de Alcalá de Henares burned to the ground. Villalon notes the loss, but was able to locate copies of sections of the *proçeso* in libraries in the Escorial, the Vatican, and the Archivo General de Simancas, in 2001 and 2014. The full *proçeso* was produced in copies, both in Spain and in Rome, across the reigns of four popes, but a surviving complete copy was apparently found only in 1991 by Alejandro Recio Veganzones, OFM. He discovered it not in the Archivo Secreto Vaticano where processes of beatification and canonization are kept but among manuscripts filed in the Biblioteca Apostolica Vaticana under the name Francisco de la Peña, whose papers had entered the library in 1612, many years after Diego was canonized. Peña, canon lawyer appointed by Philip II as Spain's auditor of the Rota in Rome, had played a key role at the papal court, representing the king and advancing the prospects for Diego's case. Recio Veganzones marks its location in the library, and lists its full contents, in his article, "Proceso de beatificación y canonización de S. Diego de Alcalá, conservado en el Ms. Vat. Lat. 7008 e instruido en 1567 por Ambrosio de Morales, su procurador en Alcalá de Henares," *Archivo Ibero-Americano* 51, no. 203–4 (1991): 767–97. In a second article, he adds an extensive bibliography for San Diego: "Ensayo biobibliográfico sobre San Diego de Alcalá," *Archivo Ibero-Americano* 60, no. 236 (2000): 259–305.

2. Ambrosio de Morales, *Vita B. Didaci Complutensis*, in volume 3 of Morales's collected works, *Opuscula historica, quorum exemplaria in R. D. Laurentii Bibliotheca, vulgo del Escorial custodiuntur*, ed. Francisco Valerio Cifuentes (Madrid, 1793), 206–11. The original vita was written in 1567 (see n. 1, above). Subsequent accounts of the saint's life include Melchor de Cetina, *Discursos sobre la vida y milagros del glorioso padre San Diego, de la orden del serafico padre S. Francisco* (Madrid, 1609); and Antonio Rojo, *Historia de San Diego de Alcalá: Fundación y frutos de santidad que ha produzido su convento de Santa María de Iesús [...]* (Madrid: Imprenta Real, 1663). Both Cetina and Rojo were later guardians of the monastery of Santa María de Jesús, Diego's home when he died in 1463. In the twentieth century, the saint's life is told by Thomas E. Case, *La historia de San Diego de Alcalá: Su vida, su canonización y su legado* [bilingual edition in Spanish and English] (Alcalá de Henares: Universidad de Alcalá, 1998). A detailed biography of San Diego narrated through depictions of him in works of art is José Luis Romero Torres, "Iconografía y devoción regia de un santo franciscano andaluz: San Diego de Alcalá," in *Actas del Simposio Cuatro siglos de presencia de los franciscanos en Estepa, celebrado el 11–13 de septiembre de 2003* (Estepa: Ayuntamiento, 2007), 94–113. See also Thomas James Dandelet, *Spanish Rome 1500–1700* (New Haven, CT: Yale University Press, 2001), 170–78.

3. Ambrosio de Morales, *Las antigüedades de las ciudades de España* (Alcalá de Henares, 1575). A facsimile reprint of a 1792 edition is available in two volumes (Valencia: Librerias "Paris-Valencia S.L.," 2001). Our thanks to the historian Michael Schreffler, who interpreted for us the scope and significance of this work by Ambrosio de Morales. See Guy Lazure, "Possessing the Sacred: Monarchy and Identity in Philip II's Relic Collection at the Escorial," *Renaissance Quarterly* 60, no. 1 (Spring 2007): 76–77.

4. "Tambien ha querido Janelo por regocijo renovar las estatuas antiguas, que se movian, y por eso las llamaban los Griegos Automatas. Hizo una dama de mas de una tercia en alto, que puesta sobre una mesa danza por toda ella al son de un atambor, que ella misma va tocando, y da sus vueltas, tornando adonde partió. Y aunque es juguete y cosa de risa, todavía tiene mucho de aquel alto ingenio." Morales, *Las antigüedades*, 93v, as transcribed in the 1792 edition reprint, p. 341. The English translation is from José A. García-Diego, *Juanelo Turriano, Charles V's Clockmaker: The Man and His Legend*, trans. Charles David Ley (Madrid and Sussex: Editorial Castalia and Antiquarian Horological Society, 1986), 101. A *tercia* is one-third of a *vara*, or rod. The *vara* is a Spanish linear measure of .84 meter or 33 inches.

5. But see the recent biography of Juanelo by Cristiano Zanetti: *Janello Torriani and the Spanish Empire: A Vitruvian Artisan at the Dawn of the Scientific Revolution* (Leiden: Brill, 2017). Chap. 1, "Janello Torriani's First Education," 23–70, offers a detailed description of the milieu and ethos of Cremona at the time Juanelo came of age.

6. Bernado Sacco, *De Italicarum rerum varietate et elegantia [...]* (Pavia, 1565), 150–51. Quoted and translated from the Latin by Silvio A. Bedini and Francis R. Maddison, "Mechanical Universe: The Astrarium of Giovanni de' Dondi," *Transactions of the American Philosophical Society*, n.s., 56, no. 5 (1966): 38–39 (brackets theirs). Bedini and Maddison declare that the clock, corroded and forgotten in the castello in Pavia, was "almost certainly" the once-celebrated fourteenth-century astrarium of Giovanni de' Dondi. In the ensuing decades, Juanelo indeed made a new clock for the emperor, improving on de' Dondi's machine, though this clock too is now lost to the ages.

7. Girolamo Cardano, *De subtilitate [...]* (Lyons, 1554), 611–12. Vida's memorable portrait of Juanelo is presented in chap. 20, below. An earlier text clearly describing tabletop automata, though not specific to Juanelo Turriano, is Cristóbal de Villalón, *Ingeniosa comparación entre lo antiguo y lo presente* (1539) (repr. Madrid: Manuel Serrano y Sanz, La Sociedad de Bibliófilos Españoles, 1898), 173–74.

8. Antonio Campi, *Cremona: Fedelissima città et nobilissima colonia de Romani rappresentata in disegno col suo contado et illustrata d'una breve historia delle cose più notabili appartenenti ad essa et dei ritratti naturali de duchi et duchesse di Milano e compendio delle lor vite* (Milan: G. B. Bidelli, 1645), Libro Terzo, 198–99. Originally published in 1585 and dedicated to Philip II. Our thanks to Sally Hill for translation from the Italian.

9. Famianus Strada, *De bello Belgico: The history of the Low-Countrey warres*, trans. Sir Robert Stapylton (London: Humphrey Moseley, 1650), bk. 1, p. 7. Stapylton translated the first of two volumes, or "decades" of Strada's title. The Latin original is *De bello Belgico decas prima ab excessu Caroli V. Imp. usque ad initia Præfecturæ Alexandri Farnesii [...]* [Concerning the Belgian war, first decade, from the abdication of the Emperor Charles V to the beginnings of the Prefecture of Alexander Farnesius...] (Leyden, 1643).

10. Strada, *De bello Belgico*, in Stapylton trans., 6.

11. William Stirling-Maxwell, *The Cloister Life of the Emperor Charles V*, 4th ed. (London: John C. Nimmo, 1891), 179–80. The first edition of this book was published in 1852.

12. Bedini and Maddison, "Mechanical Universe," 57.

CHAPTER 5

The Attribution of the Monk

1. Ernst von Bassermann-Jordan, *Alte Uhren und ihre Meister* (Leipzig: Wilhelm Diebener, 1926), 58, 66–69. This volume collects twenty years of Bassermann-Jordan's essays in chapters on the history of timekeeping, clocks, and master clockmakers from the early medieval period forward, with emphasis on masterpieces from the high Renaissance. His essay "Eine Automatenfigur aus der Zeit Kaiser Karls V" [An Automaton from the Era of Emperor Charles V] is condensed from his earlier article, "Eine Automatenfigur aus der Zeit Kaiser Karls des Fünften" published in *Der Kunstwanderer: Zeitschrift für alte und neue Kunst, für Kunstmarkt und Sammelwesen*, Adolph Donath, ed., (July 1922), 489–91.

2. Alfred Chapuis and Édouard Gélis, *Le monde des automates: Étude historique et technique*, 2 vols. (Paris, 1928; repr. Geneva: Editions Slatkine, 1984), 2:215.

3. José A. García-Diego, *Los relojes y autómatas de Juanelo Turriano* (Madrid: Albatros Ediciones, 1982), vol. 6 in the series Tempvs Fvgit, Monografías Españolas de Relojería. Elizabeth encountered this pioneering book in 1996 when its unexpected title popped up in the catalog holdings at Harvard's Widener Library from a then none too hopeful key word search for "Juanelo Turriano." Reading it launched her pursuit of the legend and precipitated a looping sequence of discoveries that she published in 2002 as "Clockwork Prayer: A Sixteenth-Century Mechanical Monk," *Blackbird: An Online Journal of Literature and the Arts* 1, no. 1 (Spring 2002). High points of that sequence are preserved here, for the way that the legend emerged from an open search before our later discoveries. Thanks to Adam Meuse and Carlena Kirkpatrick for translation from the Spanish of material we quote from García-Diego's book.

4. García-Diego, *Los relojes y autómatas de Juanelo Turriano*, 104 and LV–LX.

5. *Uto Auktionen* [auction catalog], Uto Auktions AG, Zurich, September 29, 1975, item no. 351.

6. García-Diego, *Los relojes y autómatas de Juanelo Turriano*, 105–6.

CHAPTER 6

A Letter

1. *Très importante collection de tableaux Espagnols du XIIIe au XVIIIe siècle...* [auction catalog] (Geneva, 1976). (Maarten van Heemskerck's surname is correct elsewhere in the catalog.) The auctioneer, Pierre Cornette de Saint-Cyr, wrote the introduction to the catalog. He thanks a number of scholars for their contributions to the text entries, which otherwise are not identified by author. For the entry on the monk, see pp. 83–86.

2. *Très importante collection*, 84. Elizabeth King's translation from the French.

3. *Très importante collection*, 84–86. The catalog footnotes two sources for the story of Don Carlos: "Cf. sur la guérison du prince, Fr. Lucio Mª Nuñez, 'Documentos sobre la curación del príncipe Don Carlos y la canonización de San Diego de Alcalá,' *Archivo Ibero-Americano* 2 [1914]: 424–46, et, sur sa mort, Elías Tormo, 'La tragedia del príncipe don Carlos y la trágica grandeza de Felipe II,' *Bol. de la Real Academia de la Historia* 112 [1943]: 161–209." The article by Elías Tormo y Monzó (1869–1957), Spanish art historian, Prado Museum board member, and minister of education and fine arts in the Spanish National Assembly in the late 1920s, was delivered as a lecture at the Prado in spring 1943; it is a politically nuanced review of the many versions of the story of Don Carlos.

4. Thanks to Nancy Card of the Registrar's Office of the Smithsonian's National Museum of American History for arranging for us to view the photocopy of this letter and the earlier letter from Georges Sedlmajer. Elizabeth's translation from the French, with help from Lory Frankel. "Lay brother" is given as *frère convers*—which should not be confused with "converso." Servus Gieben's suggestion to consult the "Twenty-One Books" was a reasonable one in 1975, but since then scholars have found enough to doubt an attribution of this work to Juanelo Turriano that the anonymous author is now sometimes identified as Pseudo-Turriano. The letter bore no return address. Gieben refers to the article by Fr. Lucio María Núñez in *Archivo Ibero-Americano* 2 (1914). A small error? The article is cited in turn in the Geneva auction catalog. We find it in 1 (2014), as in chap. 2,n.9, above.

CHAPTER 7

A Monk's Opinion

1. García-Diego, *Juanelo Turriano*; see chap. 4, n. 4, above.

2. García-Diego, *Juanelo Turriano*, 112–13, 115–16.

3. *The Clockwork Universe: German Clocks and Automata, 1550–1650*, exh. cat., ed. Klaus Maurice and Otto Mayr (New York: Neale Watson Academic Publications, 1980), 170. Published simultaneously in German: *Die Welt als Uhr: Deutsche Uhren und Automaten, 1550–1650* (Munich: Deutscher Kunstverlag and Bayerisches Nationalmuseum, 1980). The entry on the monk, written by Maurice and Mayr, appends references as follows: for the Spanish ascription, "Ambrosio de Morales, *Las antigüedades de las ciudades de España*, Alcalá, 1575, pp. 91 ff." For Jakob Bulmann in Nuremberg, "Neudorfer, 1547; cf. Maurice I, p. 58" (referring to Maurice's book *Die deutsche Räderuhr* [Munich, 1976], vol. 1). For Emperor Rudolph II, "cf. Nos. 2193 and 2194 in the 'Kunstkammerinventar Kaiser Rudolf II, 1607–1711,' by Rotraud Bauer and Herbert Haupt, in *Jahrbuch der Kunsthistorischen Sammlungen in Wien*, 1976, 72:113." (The dates of the inventory are a misprint; they should be 1607–1611.)

4. Servus Gieben, email to Elizabeth King, July 29, 1999.

5. A review of the works of Juanelo Turriano is Giuseppe Brusa, "Giovanni Torriano e i suoi favolosi orologi microcosmici," *La voce di hora: Pubblicazione dell'Associazione Italiana Cultori di Orologeria Antica*, no. 15 (December 2003): 3–22. Brusa argues that there is one identifiable extant piece, an armillary sphere in Milan, in the Pinacoteca Ambrosiana, signed and dated *Janellus, 1540, Mediolani* (7–8).

6. Morales, *Las antigüedades*, 93v.

CHAPTER 8

An Immaculate Object

1. This is not atypical; there are many outstanding sixteenth-century clocks that do not carry a signature. More on the presence or absence of signatures or makers' stamps in chap. 19.

2. A tentative analysis of the paint was made in 2013; see chap. 11 n. 3, below.

3. How in fact did the clockmaker calculate the amount of spring power needed to drive a given clock or automaton? Even today, clockmakers may make these judgments intuitively. There are tables and formulas to calculate force-to-weight ratios, but David prefers to trust his instincts. He has said, "I truly do not know if the maker of the monk calculated or 'guesstimated' the strength of his spring, but I certainly 'guesstimated' the model's spring and was lucky. Imagine holding the ends of a blade spring in your two hands and flexing it. Your hands can distinguish very subtle differences in tension from one spring to the next. So an instinctive judgment of the power of a coiled mainspring is just an extension of this kind of knowledge. However, if I had arbitrarily made the model 1½ times bigger than the monk, rather than 1¼, I would not have been able to find a spring strong enough to drive it." Excerpt from a conversation between Elizabeth King and David Todd, recorded by Elizabeth King on January 23, 2004.

4. From a conversation between Elizabeth King and David Todd, recorded by Elizabeth King on January 23, 2004.

CHAPTER 9

The Monk Performs

(no notes)

CHAPTER 10

A Mechanical Anatomy

1. Our profound gratitude to Carlene Stephens, curator, Division of Work and Industry; Scott Nolley, conservator, Exhibits and Loans; Dawn P. Wallace, objects conservator; and the Conservation Department of the National Museum of American History at the Smithsonian for making this essential video for us in late December 2021. Carlton Newton transcribed the bird's-eye-view video into a precise map of the monk's walking path for this specific performance. Each turn angle will vary by several degrees, from one performance to the next.

2. Leonardo da Vinci knew about it and sketched it, and the makers of crossbows and catapults have used it from antiquity. See Peter Honig, "History and Mathematical Analysis of the Fusee," in Maurice and Mayr, *The Clockwork Universe*, 114–20; and Eric Bruton, *The History of Clocks and Watches* (London: Orbis, 1979; New York: Crescent Books, 1989), 53.

3. Such a line or cord would have been made of sheep intestine cut lengthwise into thin strips and twisted before being cured. Traditionally known as "catgut," it is made in many diameters and is still used, principally on stringed musical instruments but also by clockmakers. It is very strong and if kept well oiled will last a century or more. The wood fusee absorbed and held oil, and this kept the gut supple.

4. The fusee with its trumpet-shaped profile must be individually cut for its mainspring, for not only does a given spring deliver a diminishing pull as it unwinds, but it does so at a unique, or characteristic, rate. Clockmakers cut the fusee's graduated diameter and groove with an elegant jig called a fusee-adjusting tool. With the fully wound spring and uncut fusee "blank" in situ in the clockframe—gut line in place but gear train not yet connected—the tool was attached to the fusee's winding square. A sliding weight on the long tool arm would indicate, with each quarter-turn release of the spring, the necessary groove depth for the gut line on the fusee, to balance the pull

of the spring at that point. The groove would then be cut and tested for balance, and, when perfect, the spring would be unwound another quarter turn for the next cut. And so on, until the spring reached its not quite spent release, and the fusee's spiral groove was mapped out and completed. The clockmaker plied his chisel with caution, bit by bit, for if the groove was cut too deeply, destroying the balance, he had to start over with a new blank. Every spring thus had its dedicated fusee—and a no less dedicated clockmaker to make it.

5. The clock- and watchmaking trades are replete with onomatopoeic words, which helped craftspeople identify mechanical assemblies with names that imitate their functions.

6. We use the term "gear train" here, rather than the clockmaker's preferred term "wheel train," to avoid any confusion with the monk's perambulation wheels.

7. We have noted that the cross that arrived with the monk at the time of its acquisition is not original. This small cross, which we see in the monk's hand in FIGURE I.1, does not rise the full distance to the monk's lips for the kiss. The original cross would have been taller, for there is no sign that the clockwork mechanism is at fault. Servus Gieben, in his letter proposing that the automaton represents Diego de Alcalá, suggested that the monk carried a cross *and* a rosary, rather than a cross *of* the rosary, for a closer match with Diego's attributes (see chap. 6, above). This is a worthwhile distinction, as it would predict a larger and taller original cross, calculated for the arc of the kiss.

8. A case in point: When the monk arrived at the Smithsonian, the perambulation wheels were reversed. The serrated wheel was thus lifted, rather than the smooth wheel, for the walking turn, defeating the purpose of the serrated wheel in traction on the tabletop. David reassembled them correctly. But photographs taken before and after his work reveal the error, as does the original film of the monk in action. FIGURE 10.8B shows the error. David points out that each wheel mounts to the perambulation arbor with differently sized square mounts; hence the entire arbor would have been improperly assembled sometime in the monk's past.

CHAPTER 11

Monks, Saints, and Musical Ladies

1. See the exhibition and catalog text in chapter 7, above. See also Otto Mayr, "An Offering of Renaissance Clockwork Masterpieces," *Smithsonian* 11, no. 9 (December 1980): 44–53. Otto Mayr was then curator, Division of Mechanisms, in the Smithsonian's National Museum of History and Technology. Appointed in 1970 curator of Mechanical and Civil Engineering in the Division of Engineering and Industry at the museum, he went on to be chair of the Department of Science and Technology (1976–77) and then acting director of the museum (1978–79). *The Clockwork Universe* was among the last large exhibitions planned for the Smithsonian's National Museum of History and Technology before the name change that would alter its mission, in 1980, to the National Museum of American History (see chap. 17). In 1982 Mayr took the reins as general director of the Deutsches Museum in Munich and served until his retirement in 1992. His books, including his 1986 volume, *Authority, Liberty, and Automatic Machinery in Early Modern Europe*, are widely cited as groundbreaking studies connecting the history of technology to the history of philosophy, culture, and political science. Klaus Maurice at the time of the exhibition was curator of horology and metal crafts at the Bavarian National Museum. He would later become director of that museum and ultimately secretary general of the Cultural Foundation of the German States, Berlin. He trained as a clockmaker, then a historian of clocks; his two-volume opus, *Die deutsche Räderuhr: Zur Kunst und Technik des mechanischen Zeitmessers im deutschen Sprachraum* [The German wheel clock: On the art and technology of the mechanical timepiece in the German-speaking world] (Munich: C. H. Beck, 1976), is one of the most respected works in the field of horological scholarship.

2. Adelheid von Herz, "Androiden des 16. Jahrhunderts" (master's thesis, University of Hamburg, 1990).

3. In 2013, to examine the monk's paint, Elizabeth organized an informal meeting of curators and conservators from the National Gallery of Art (NGA) in Washington, DC, and the Smithsonian's National Museum of American History (NMAH), where the monk is kept. Carlene Stephens, curator, Division of Work and Industry, NMAH, and Beth Richwine, senior objects conservator, gave kind permission for the meeting and welcomed us to the NMAH conservation lab on August 20. From the NGA, Daphne Barbour and Judy Ozone, senior objects curators; Alison Luchs, curator of Early European Sculpture; and Mary Levkoff, head, Sculpture and Decorative Arts, joined us. Fredrika Jacobs, Renaissance cultural historian and professor, Virginia Commonwealth University, traveled to Washington with Elizabeth and David, who brought with them a portfolio of photographs of the seven kindred automata for comparison. In preparation, Beth Richwine performed an X-ray fluorescence (XRF) analysis on the paint surfaces of the monk's head, hands, and feet. Her readings showed the presence of lead, arsenic, mercury, and iron, but she would like to see the paint analyzed with more sophisticated equipment, including a scanning electron microscope, for more definitive data on medium, ingredients, and date, by paint layer. She detected the presence of titanium and zinc and suggested these could indicate recent paint touch-up. Daphne Barbour and Judy Ozone identified at least three layers of paint on the monk's head and explained the horizontal paint cracking as characteristic of the paint medium itself, not the result of wood grain movement. They did not say if the layers were applied separately over time or were applied in sequence when the figure was first painted. Richwine said her paint analysis showed ultraviolet light markers on the paint of the feet that were different from those on the paint of the rest of the figure but said that by itself this does not offer firm information on ingredients or date. We noted that the paint breaks on the feet are different from those on the head and arms. The feet also show chips and remnants of a prior layer of some kind under the paint. All members of the NGA group commented on the small size of the sandaled feet for the overall proportion of the monk. Photographs of the other automata show similarly small feet. Elizabeth pointed out that if the feet are not original, their role in helping to identify the monk as a Franciscan icon—i.e., the sandals—is less certain. However, all agreed that the

carved precision and anatomical subtlety of the feet matches that of the hands and head. Barbour and Ozone pointed to the transverse seam between front and back of the head, visible beneath the paint and said that when the head was made the wood was sawed in half for hollowing out the inside of the head rather than split with a chisel (the split with a chisel would have been easier to hide under paint). Mary Levkoff commented on the similar stylistic character of the Smithsonian monk and the Milan "matron" automaton. She also brought up the work of Johan Gregor van der Schardt in Nuremberg as an example of a sculptor who worked in polychromy and often in small scale (see FIG. 19.1 and chap. 19, note 5, below). Discussion moved to the subject of wood itself. Without microanalysis, the identity of the species must remain unknown for now. Levkoff's strongest message for our meeting was to urge a wood analysis of head, hand, foot, and bell-shaped body case. Barbour, Ozone, and Richwine asserted that contemporary analysis can now be done on extremely tiny specks of wood. (In 2019 Smithsonian NMAH staff again examined the monk's wood, stating it is likely poplar but possibly beech. Only microanalysis will confirm the species.)

CHAPTER 12

A Second Monk?

1. Peter Friess, "Restaurierung einer Automatenfigur," *Alte Uhren und moderne Zeitmessung* 4 (August 1988): 40–50. The automaton made its first appearance in London in 1983 at auction at Christies. The *London Financial Times* published an article on it on October 15, 1983. The Deutsches Museum purchased it in 1985. David had met Peter Friess in 1980 when the two of them helped install the exhibition *The Clockwork Universe* at the Bavarian National Museum in Munich, where Friess then worked as a young conservator before taking a position at the Deutsches Museum. They continued a professional friendship for many years.

2. Von Herz, "Androiden des 16. Jahrhunderts," 69, recorded the wrist motions as even more rapid at 61 (right hand) and about 33 (left).

3. Von Herz, "Androiden des 16. Jahrhunderts," 65.

4. Friess, "Restaurierung einer Automatenfigur," 45.

5. The chemistry and technical methodology of virtuoso polychromy is detailed by Michael Baxandall, *The Limewood Sculptors of Renaissance Germany* (New Haven, CT: Yale University Press, 1980), 41–48. For a close description of Spanish polychromy, see Daphne Barbour and Judy Ozone, "The Making of a Seventeenth-Century Spanish Polychrome Sculpture," in *The Sacred Made Real: Spanish Painting and Sculpture 1600–1700*, ed. Xavier Bray (London: National Gallery, 2009), 58–71, the catalog for an exhibition that opened in London in 2009 and traveled the next year to the National Gallery of Art in Washington, DC. Barbour and Ozone are senior objects conservators at the National Gallery of Art. In 2013 we were extremely fortunate that both responded to our invitation to step across town to view the monk, together with curators and conservators from the Smithsonian and the National Gallery. See chap. 11, n.3, above.

6. Wolfram Koeppe, Marina Kellen French Curator in the Department of European Sculpture and Decorative Arts at the Metropolitan Museum of Art in New York, has commented that the painted pattern is similar in its floral design to the Italian velvet brocades widely produced and exported throughout Europe in this period. He provided Elizabeth with this reference, describing such brocades: Wolfram Koeppe, *Die Lemmers-Danforth-Sammlung Wetzlar: Europäische Wohnkultur aus Renaissance und Barock* (Heidelberg: Edition Braus, 1992), 283.

7. Raphael Beuing's full title at the Bavarian National Museum in Munich is curator for Arms and Armour, Timepieces, Scientific Instruments and Base Metal.

8. When Elizabeth could finally view the automaton in person, she saw that the new fingers on the right hand were given a different shade of paint, to mark them as new. But how were decisions made for replacing the left hand, its pose, and the sandaled feet? What kind of wood was used, who did the carving, and who did the painting?

CHAPTER 13

A Saint in Budapest

1. Katalin Soós and Jenö Rácz, "Eine Automatenfigur in Budapest," *Zeitschrift für Kunsttechnologie und Konservierung* 4 (1990): 207–14; for the melody, 209–10. Soós and Rácz consulted the music historian Ilona Ferenczy, M.T.A. Zenetudományi Intézete (Institute of Music Sciences).

2. Unable to see this automaton in the flesh, the authors are grateful to Anna Ecsedy, historian and curator of the Metalwork Collection at the Iparművészeti Múzeum, for detailed correspondence in 2021, clarifying the saint's precise performance sequence. Equal thanks to Ágnes Prékopa, former curator at the Iparművészeti Múzeum, for her generous responses in 1999 and 2000 to Elizabeth's many questions. Hanoch Ben-Yami, philosophy professor at Central European University in Budapest (and Vienna), has generously shared with us over time his observations of the saint in relation to his own interest in the history of artificial life.

3. Soós and Rácz, "Eine Automatenfigur in Budapest," 209 and n. 5. Also cited is the cittern player in the Kunsthistorisches Museum, together with references to Juanelo Turriano in Bassermann-Jordan (*Alte Uhren und ihre Meister*) as well as Chapuis and Gélis (*Le monde des automates*), all of which, tantalizingly, would have been available to curators in the 1920s, soon after the automaton itself arrived in the museum's collection. García-Diego (*Juanelo Turriano*, 109–10) saw photos of the saint around 1981, and Soós and Rácz cite his book to support a probable attribution to Juanelo, apparently unaware that he changed his opinion in his second book. Until the museum closed in 2017 for major renovation, the saint was on display together with photographs of its restoration and a video showing it in motion.

4. Soós and Rácz, "Eine Automatenfigur in Budapest," 213.

5. Soós and Rácz, "Eine Automatenfigur in Budapest," 214.

6. Von Herz, "Androiden des 16. Jahrhunderts," 120. Von Herz completed her thesis in the same year that Soós and Rácz published their report, hence it is not certain if she saw the saint before or after the full restoration was completed. Her bibliography lists the restoration report but as "unpublished." The quotations from von Herz throughout this book were translated for us by Henning Bauer.

7. Von Herz, "Androiden des 16. Jahrhunderts," 120.

8. See appendix, "Table of Comparisons," for sources.

9. For the saint with glockenspiel, see http://www.kfs.oeaw.ac.at/publications/1999_deutsch_remote_access_and_transfer_of_audio_recordings/dli/mech/turriano.htm.

10. Correspondence from Helmut Kowar to Elizabeth King, March 19, 2021.

11. Soós and Rácz, "Eine Automatenfigur in Budapest," 211, for pinion leaf count. This strengthens the clockwork connection between the two automata if in a much hidden way. The Deutsches Museum automaton has a reasonable six-leaf fly pinion.

CHAPTER 14

Three Sister Musicians

1. The seller was the Zurich antiquities dealer Edgar Mannheimer, from whom the monk too was purchased. For more about Edgar Mannheimer, see the obituary by Sebastian Whitestone in *The Independent*, London, Wednesday, January 12, 1994, https://www.independent.co.uk/news/people/obituary-edgar-mannheimer-1399413.html.

2. The automaton was labeled eighteenth-century Italian in the exhibition catalog, *Mechanisches Spiel und Theater*, Puppentheatersammlung im Münchner Stadtmuseum, Munich, July 4–October 15, 1972, fig. 69, cat. no. 242. The catalog cites Karl Gröber, *Kinderspielzeug aus alter Zeit: Eine Geschichte des Spielzeugs* (Berlin: Deutscher Kunstverlag, 1928). The automaton's date was moved to 1550 by Peter Friess in 1988 and changed shortly thereafter to ca. 1600.

3. *Mechanisches Spiel und Theater*, fig. 69.

4. Roswitha Schwarz, head of furniture conservation at the BNM, performed the analysis. Perhaps this wood species is the widespread European *Sorbus aucuparia*, once believed to have magical properties. The upper body and arms of the figure appear to be made of a different species of wood, and all present agreed that most of this wood seems also to be original. The wooden hands, under paint, are of course more mysterious. Each part would want its own microanalysis, something to be desired for all the automata.

5. *Conservatoire national des arts et métiers, catalogue des collections, septième édition* (Paris: Dunod, 1882), 99: "Joueuse de mandoline. Automate attribué à Vaucanson, réparé par Robert-Houdin en 1859." (On Vaucanson and the conservatoire, see our introd. n. 2.) Robert-Houdin (1805–1871) was a clockmaker who later became famous as a stage magician and illusionist. (The American magician Eric Weiss so admired Robert-Houdin that he took the stage name "Houdini.") In fact, automata often played roles in magic performances in this period. The late magician Ricky Jay assembled a collection of such automata and performed with them. A biographical profile on Jay by Mark Singer mentions an automaton by Robert-Houdin: "Two automatons stood on the table. One, called 'The Singing Lesson,' was the creation of Jean-Eugène Robert-Houdin, the nineteenth-century watchmaker-turned-conjurer, who is considered the father of modern magic. The other was a Chinese cups-and-balls conjurer built by Robert-Houdin's father-in-law, Jacques Houdin." ("Secrets of the Magus," *New Yorker*, April 5, 1993, 54–73).

6. "Cependant la facture très simple et la technique nettement XIXe siècle du mécanisme prouvent qu'il n'en est rien" (However, the very simple workmanship and the clearly nineteenth-century mechanism proves she could not be [Vaucanson's] work). Chapuis and Gélis, *Le monde des automates*, 2:218–19. Elizabeth's translation from the French for all quotes from Chapuis and Gélis.

7. *Conservatoire national des arts et métiers, catalogue du musée, section Z, automates et mécanismes à musique* (Paris, 1960), 53–56. This edition of the catalog gives an expanded and detailed description of the figure and an excellent schematic diagram of the mechanism (see FIG. 14.3C here).

8. F. J. Britten, *Britten's Old Clocks and Watches and Their Makers*, 9th ed., ed. G. H. Baillie, Courtenay Ilbert, and Cecil Clutton (London: Bloomsbury Books, 1989), 454.

9. Alfred Chapuis and Edmond Droz, *Les automates, figures artificielles d'hommes et d'animaux: Histoire et technique* (Neuchâtel: Éditions du Griffon, 1949), 325–26. This volume was translated into English as *Automata: A Historical and Technological Study*, trans. Alec Reid (Neuchâtel: Éditions du Griffon, 1958), where the Gautier figure appears on 317–18. In both editions, the caption under the photo is "Poupée marchant, 1764 (Coll. Charliat, Paris)" but the text dates the figure to 1751. The authors note that the figure was signed by P. Gautier. Georges Charliat was a Parisian antiquarian and restorer of automata, whose workshop on Rue Miromesnil opened in 1937. See John Kobler, "The Strange World of M. Charliat," *Saturday Evening Post*, March 26, 1955. We have been unable to find any further reference to this figure.

10. As of 2022, the database for the *Joueuse de mandolin* can be found at https://collections.arts-et-metiers.net/?queries=query=search=N°%20d'inventaire=[06152-0001-]&showtype=record

11. See Adelheid Voskuhl, *Androids of the Enlightenment: Mechanics, Artisans, and Cultures of the Self* (Chicago: University of Chicago Press, 2013), for a groundbreaking study of two of the greatest automata of the Enlightenment, one of which is "La joueuse de tympanon."

12. With permission from the museum, we sought another opinion and showed the microphotographs to Jeremie Ryder, Guinness Collection Conservator of the Morris Museum in New Jersey. He found signs that could confirm either method. Final hand-filing of engine-cut teeth can introduce small irregularities of its own.

CHAPTER 15

The Vienna Cittern Player

1. Morales's description is quoted in chapter 4, above.

2. Elizabeth learned only recently that the fusee in the cittern player is made, not of wood, but of metal, of uncertain type (but apparently not brass, as in the mandolin player in Paris). This was confirmed in an exchange with Paulus Rainer, curator of the *Kunstkammer* at the Kunsthistorisches Museum, who kindly sent us snapshots of the mechanism in late 2021.

3. Ernst von Bassermann-Jordan, "Eine Automatenfigur aus der Zeit Kaiser Karls V [An automaton from the era of Emperor Charles V]," in his *Alte Uhren und ihre Meister*, 67. Thanks to Henning Bauer for translation from the German. Bassermann-Jordan and later authors refer to the musical instrument as a lute (*Laute*) or a mandolin. Given the variety of terms across language and time for a stringed instrument held in the hands and plucked or strummed, we ourselves had assumed some interchangeability among the words *cittern*, *cister*, *zither*, *mandolin*, and *lute* in the period descriptions of instruments played by automata (the zither is an instrument that later gave its name to the guitar). In response to questions from Elizabeth, Paulus Rainer at the Kunsthistorisches Museum, wrote: "Our lady plays a cittern, not a lute. Lutes have a round back, while citterns have a flat back" (email to Elizabeth King, November 9, 2021).

4. Chapuis and Gélis, *Le monde des automates*, 2:215–17.

5. Bassermann-Jordan's work was revisited in 1966 by the historian Hans von Bertele (1903–1984), who furthered the argument in favor of Juanelo as maker of the automaton: "Zur Provenienz des Mandoline spielenden Automaten im Kunsthistorischen Museum in Wien," *Schriften des Historisch-Wissenschaftlichen Fachkreises "Freunde alter Uhren" in der Deutschen Gesellschaft für Chronometrie* 6 (1966): 6–9. The historian Silvio Bedini mentioned the Vienna automaton in 1966 but stayed within the realm of surmise: "An automaton of a lady which dances and plays a lute, in the collection of the Kunsthistorisches Museum in Vienna, is believed to have been the creation of Torriano. If so, it is the only example of his automata which has survived." Bedini and Maddison, *Mechanical Universe*, 56–58. See also Bedini's "The Role of Automata in the History of Technology," *Technology and Culture* 5, no. 1 (1964): 32. A decade later, the monk arrived at the Smithsonian's Museum of History and Technology, where Bedini was then acting director. Though he never wrote about the monk himself, Bedini took a lively interest in our research and gave us priceless help and advice.

6. L. Planiscig and E. Kris, eds., *Katalog der Sammlungen für Plastik und Kunstgewerbe* (Vienna: Kunsthistorisches Museum, 1935), 1:45. The entry in full:

> Giovanni della Torre (Juanelo Torreano, gest. 1580 Toledo): Automatenpuppe aus Holz, in alter Fassung und Kleidung; die Figur legt nach einmaligem Aufziehen eine Strecke von etwa 20 m zurück, bewegt Kopf und Füsse und schlägt mit den Händen die Laute; angefertigt angeblich für Karl V. (Legat K. v. Wischnitz-Naszod, 1934.)

> [Automaton made of wood, clothed in the older style; travels a distance of approximately 20 meters in a single winding, moving head and feet, hands playing the lute; supposedly made for Charles V. (Gift of K. v. Wischnitz-Naszod, 1934.)]

It is intriguing to note, according to this entry, that the lady could muster a 20-meter straight run at only one full wind of the mainspring. The Smithsonian monk, were he to walk in a straight line, would only cover a distance of just under 14 feet—a little over 4 meters—on a single wind.

7. Fritz Saxl, "Costumes and Festivals of Milanese Society under Spanish Rule," in *Proceedings of the British Academy* 23 (London, 1937), 418.

8. Saxl, "Costumes and Festivals," 418–19.

9. The Kunsthistorisches Museum now keeps a digital copy of the film. The original film remains in the Warburg archives.

10. *Katalog der Sammlung für Plastik und Kunstgewerbe*, vol. 2, *Renaissance* (Vienna: Kunsthistorisches Museum, 1966), 118–19:

> Cister-Spielerin . . . der Nürnberger Uhrmacher Jakob Bullmann (vor 1535) hat anscheinend solche angefertigt; eine andere derartige Puppe, das Werk des kaiserlichen Hofuhrmachers Christoph Margraf, befand sich in der Kunstkammer Kaiser Rudolfs II. in Prag (entstanden um 1600).

> [Cittern Player . . . the Nuremberg clockmaker Jakob Bullmann (before 1535) evidently made them; another such doll, the work of the imperial court clockmaker Christof Margraf, was located in the Kunstkammer of Emperor Rudolf II in Prague (made around 1600).]

Erwin Neumann, director of the KHM Collection of Sculpture and Decorative Arts from 1966 to 1975, is credited with writing the catalog entries that include no. 391, the *Cister-Spielerin*. The cittern player was shown in a gala exhibition in 1964 in Versailles and is pictured and described in the exhibition catalog, *Vienne à Versailles: Les grandes collections autrichiennes au Château de Versailles* (Paris: Ministère d'État Affaires culturelles, 1964), 45 and cat. no. 55. Here too Erwin Neumann is named among catalog authors, and it is reasonable to assume he was again responsible for the description of the automaton. This catalog marks the first time Rudolf's cabinet is mentioned in relation to the cittern player, and Neumann's role in bringing it forward is key.

CHAPTER 16

A Late Arrival from Milan

1. Giuseppe Brusa, ed., *La misura del tempo: L'antico splendore dell'orologeria italiana dal XV al XVIII secolo*, exh. cat. (Trento: Castello del Buonconsiglio, monumenti e collezione provinciali, 2005). Bedini's essay, "L'orologio notturno: Un'invenzione italiana del XVII secolo," addresses a number of celebrated clocks David Todd conserved for Bedini, e.g., FIG. 3, p. 205, a 1657–59 night clock attributed to Pier Tommaso and Giuseppe Campani. We are indebted to Bedini for bringing this show and this automaton to our attention. The entry on the automaton is on pp. 451–53.

2. Cristiano Zanetti, ed., *Janello Torriani: Genio del Rinascimento*, exh. cat. (Cremona: Museo del Violino, 2016). For the automaton, see pp. 137–40, 151, 268.

3. Gregato, in Zanetti, *Janello Torriani: Genio del Rinascimento*, 268: "Si tratta dell'unico esemplare di automa tardo-rinascimentale che corrisponda nei movimenti e nel sembiante all'unico attribuito a Janello Torriani dalle fonti a lui contemporanee (Ambrosio de Morales, 1575)."

4. Gregato, in Zanetti, *Janello Torriani: Genio del Rinascimento*, 268:

> Più volte citato nell'inventario del Museo Settaliano (Terzago, p. 33) come nobil matrona in atto di suonare il timpano, l'automa vi é descritto come facente parte di una scena allargata, posata su un vassoio assieme ad un altro automa che plaude all'esibizione musicale del nostro.
>
> [Mentioned several times in the inventory of the Museo Settaliano (Terzago, p. 33) as noble matron in the act of playing a tambourine, the automaton is described as being part of a larger scene, placed on a tray together with another automaton that applauds the musical performance.]

Similar text also in Brusa, *La misura del tempo*, 451.

5. Paolo Maria Terzago, *Museo, ò Galeria, adunata dal sapere e dallo studio del signor canonico Manfredo Settala nobile milanese* [Museum, or Gallery, gathered from the wisdom and study collection of Canon Manfredo Settala, Milanese nobleman], translated from the Latin by Pietro Francesco Scarabelli (Tortona, 1666), 33. The citation in full:

> Istromento Musicale ch'in tre ombreggiamenti à foglie distinto, suona: danzano in un di effi cinque figure: in un altro sedendo à mensa alcuni convivanti mostrano di bere con atteggiamento sì proprio, che se la finta loro sete non estinguono, quella almeno de'riguardanti curiosamente eccitano: nel terzo pompeggia nobil Matrona in atto di suonare un Timpano; dirimpetto à cui vago garzone s'osserua con ceteretta in mano per applaudere anch'esso à si imaginaria melodia.
>
> [Musical instrument that plays in three distinct parts: in one, five figures dance; in another, several couples sitting at a dining table display the act of drinking, each with their own expression, who, if unable to satisfy their thirst, at least curiously excite it in the viewers; in the third, a noble matron thumps her hand in the act of playing a tambourine; facing her, a charming boy observes her with a ceteretta (a musical instrument, possibly a cetra or a cittern) in hand to cheer on the imaginary melody.] (Translation from the Italian for us by Evelyn Lincoln.)

For Manfredo Settala and his collection, see Paula Findlen, *Possessing Nature: Museums, Collecting, and Scientific Culture in Early Modern Italy* (Berkeley: University of California Press, 1994); Terzago, 42, 220; Settala, 33–34, 325–34 passim. See also Antonio Aimi, Vincenzo de Michele, and Alessandro Morandotti, "Towards a History of Collecting in Milan in the Late Renaissance and Baroque Periods," in *The Origins of Museums: The Cabinet of Curiosities in Sixteenth- and Seventeenth-Century Europe*, ed. Oliver Impey and Arthur MacGregor (Oxford: Clarendon Press, 1985), 24–28.

6. While working in Pippa's shop, Gregato helped Pippa build a fully working reconstruction of the famous fourteenth-century astrarium of Giovanni de' Dondi of Padua. See https://www.youtube.com/watch?v=gNwjWufXthA. For the historical connection between this astrarium and Juanelo Turriano, see our chap. 4 and its n. 6 and FIG. 20.1; and Bedini and Maddison, "Mechanical Universe." See also Zanetti, *Janello Torriani and the Spanish Empire*, "The Microcosm," 270–96.

7. See chapter 4, above.

8. For a video of the noble matron, see https://www.youtube.com/watch?v=jAtkSqRH6-4.

CHAPTER 17

The Category Challenge

1. See Carlene E. Stephens, "From Little Machines to Big Themes: Thinking about Clocks, Watches and Time at the National Museum of American History," *Material Culture Review* 52 (Fall 2000), https://journals.lib.unb.ca/index.php/MCR/article/view/17867. And see Carlene Stephens, *On Time: How America Has Learned to Live by the Clock* (Washington, DC: Smithsonian Institution, 2002).

2. At purchase the automaton was added to the Deutsches Museum's Department of Mathematical Instruments, Calculators, and Microcomputers, where it was considered the earliest surviving example of a programmable control mechanism (Friess, "Restaurierung einer Automatenfigur," 41).

3. Chapuis and Gélis, *Le monde des automates*, chap. 21, "Les constructeurs d'androïdes," 2:181. For a history of the word itself, and its widening use in the language of automata, see Jessica Riskin's masterful book, *The Restless Clock: A History of the Centuries-Long Argument over What Makes Living Things Tick* (Chicago: University of Chicago Press, 2016), "The First Androids," 115 ff.

4. Chapuis and Gélis, *Le monde des automates*, 2:218.

5. Chapuis and Gélis illustrate these in detail in chapters on writing, drawing, and music-playing androids: *Le monde des automates*, 2:225–89. Two of the great Enlightenment android musicians—the harpsichord player by the Jaquet-Droz family and the dulcimer player by Abraham and David Roentgen and Peter Kinzing—receive close treatment in Adelheid Voskuhl's *Androids in the Enlightenment*. The dulcimer player is kept in the Paris Conservatoire national des arts et métiers, as we mention in chap. 14 n. 11, above. See also Simon Schaffer, "Enlightened Automata," in *The Sciences in Enlightened Europe*, ed. William Clark, Jan Golinski, and Simon Schaffer (Chicago: University of Chicago Press, 1999), 126–65.

6. Jessica Riskin's now-classic essay, "The Defecating Duck, or, the Ambiguous Origins of Artificial Life," *Critical Inquiry* 29, no. 4 (Summer 2003): 599–633. On the photograph of the duck in ruins, Riskin writes, "One of a mysterious set of photographs discovered around 1950 by the curator of the Musée des Arts et Métiers in Paris. The photographs were in a folder left by his predecessor, labeled 'Pictures of Vaucanson's Duck received from Dresden'" (607). On the life and demise of the duck, see Gaby Wood, *Edison's Eve: A Magical History of the Quest for Mechanical Life* (New York: Knopf, 2002), 26–40. The definitive history of the duck is found in Chapuis and Droz, *Automata: A Historical and Technological Study*, 233–42.

7. Maurice and Mayr, *Clockwork Universe*, 286–87. "Trumpeter Automaton," 1582, Hans Schlottheim, KHM inv. no. KK 855: =https://www.khm.at/de/object/b0812f10bd/.

8. Maurice and Mayr, *Clockwork Universe*, 278–79, 282–83. Both automata are found in the Kunsthistorisches Museum *Kunstkammer* and can be seen in action in videos on the museum website. "Clock: Diana on a Centaur," ca. 1602/06, by Augsburg goldsmith Hans Jakob Bachmann, inv. no. KK 1166: https://www.khm.at/de/object/2add3fa4ad/; "Automaton, Triumphal Chariot with Minerva," c. 1620, attributed to Achilles Langenbucher, inv. no. KK 849: https://www.khm.at/de/object/a0eef8e424/.

9. Lorraine Daston and Katharine Park, in *Wonders and the Order of Nature 1150–1750* (New York: Zone, 1998), 90, cite medieval texts describing artificial wonders and write, "What made a work of art marvelous, as characterized in these texts? In the first place, many artificial wonders exploited and depended for their operation on the magical properties of exotic substances that were already part of the canon of natural wonders: magnetic lodestones or luminescent carbuncles, the prophetic caladrius or the mysterious dindialos, precious spices or medicinal gums. To fabricate artificial wonders required access to such products, as well as deep knowledge of their natural powers. Thus the writers of romance invariably described their makers as scholars and magicians rather than as artisans."

10. "Automaton in the Form of a Ship," 1585, Hans Schlottheim, KHM inv. no. KK 874: https://www.khm.at/de/object/1b2cf3f904/.

11. A bellicose inscription in German runs around the hull's perimeter: "Year 1585. I am called the double-headed eagle, well known of all Princes and Lords. I sail with power; no wind is strong enough to harm me. I tell you, navigators, remember this with care that you do not bar my way.... I will not spare my cannons." See Nadja Zweigler, "La nef de Vienne," in *Ces curieux navires: Trois automates de la Renaissance*, ed. Julia Fritsch (Paris: Éditions de la Réunion des musées nationaux, 1999); and for the inscription on Schlottheim's ship, 43 n. 2. (An English edition of this book, which we could not find, is *Ships of Curiosity: Three Renaissance Automata*, 2001. Our thanks to Ian Greaves who has a copy and gave us the translation of the inscription.) In the same volume, see J. H. Leopold, "La construction des Nefs de Schlottheim," 61–73. Invaluable also is Leopold's "Automata and the State of Horology until c. 1650," in *Royal Music Machines*, ed. J. J. L. Haspels, exh. cat. (Zutphen, the Netherlands: Walburg Pers, 2006), 35–43 (catalog for the 2006 show at the Nationaal Museum van Speelklok tot Pierement in Utrecht). For a rare tour of the clockwork and music of both the ship and the Trompeterautomat, see Helmut Kowar, "'Wir machen Musik!': Trompeterautomat, Schiff und Bacchuswagen des Hans Schlotheim," *Jahrbuch des Kunsthistorischen Museums Wien*, 13–14, 118–59. (Thanks to Paulus Rainer for bringing this study to our attention; we were thrilled to read it, for we had only recently learned that none other than Helmut Kowar recorded the automaton saint in the Iparművészeti Múzeum in Budapest; see chap. 13, above.) An overview of Schlottheim in Augsburg and the reach of his influence is found in Maurice, *Die deutsche Räderuhr*, 1:118–22. A recent study of the politics and cultural milieu of such resplendent automata is Jessica Keating's *Animating Empire: Automata, the Holy Roman Empire, and the Early Modern World* (University Park: Pennsylvania State University Press, 2018).

12. Von Herz, "Androiden des 16. Jahrhunderts," 19, 30, 32, 46, 139–41.

13. Klaus Maurice, in Annette Beyer, *Faszinierende Welt der Automaten: Uhren, Puppen, Spielereien* (Munich: Callwey Verlag, 1983), 19, 36. He adds, the "automaton, probably from Augsburg, is not signed." (Both Maurice and von Herz refer to the instrument as a lute. The theorbo is a lute-like stringed instrument.)

14. Paulus Rainer, in *Making Marvels: Science and Splendor at the Courts of Europe*, exh. cat., ed. Wolfram Koeppe (New York: Metropolitan Museum of Art, 2019) 234–35: "These cups were designed in such a way as to not be immediately recognized as drinking vessels, or to hide the fact that drinking from them required great skill, as with so-called maiden cups. In virtually all such drinking games it was required that one completely empty the vessel in one gulp. In Dresden the 'fingernail test' was adopted as proof. After draining the cup, the drinker was required to rest the vessel upside down on his thumbnail. If more ran out than could fit on his nail, he was obliged to drink again."

15. Koeppe, *Making Marvels*, 240–41, 251.

16. A recent study of the relations between jest, drinking, and the history of robotics is Angela Vanhaelen's "Strange Things for Strangers: Transcultural Automata in Early Modern Amsterdam," *Art Bulletin* 103, no. 3 (September 2021): 42–68. By far the most elaborate and aggressive precedent of automata testing visitors at court was the fabulous grottoes and fountains of the château of Hesdin, in Artois, resplendent under the Dukes of Burgundy and particularly Philip the Good (1396–1467). Famed across Europe were the hydraulic and mechanical theaters at Hesdin, beguiling and entrapping visitors in every conceivable manner of "hospitable abuse," as Jessica Riskin has described it in *The Restless Clock*, 27. Not unrelated to the automata of jest in the Habsburg courts, the machines at Hesdin were more interactive and architectural. E. R. Truitt writes about them in delicious detail in *Medieval Robots: Mechanism, Magic, Nature, and Art* (Philadelphia: University of Pennsylvania Press, 2015), 122–37, and concludes, "The surprise, terror, dismay, or delight of the guests in the gallery were part of the spectacle for those who were more knowledgeable

or experienced. Networks of pleasure and privilege cohered around knowledge, even knowledge of jokes or tricks. Knowing the joke signaled inclusion into an elite group. Being the butt of the joke signified exclusion from the group, but the experience could also initiate courtiers into the elite court, by providing them with a way of gaining the knowledge that would lead to inclusion. And for courtiers, their surprise and humiliation came not only from physical pain or social embarrassment, but also from something deeper: the uncanny feeling of being confronted with something (the automata) that they could not anticipate and did not understand" (135).

17. Koeppe, *Making Marvels*, 256–57.

18. Klaus Maurice, *Von Uhren und Automaten, das Messen der Zeit* [Of clocks and automata, the measure of time] (Munich: Prestel, 1968), 57.

19. Schloss Ambras Innsbruck inv. no. PA 386: https://www.khm.at/de/object/33777f11a9/. The Ambras Castle collections are administered by the Kunsthistorisches Museum: https://www.schlossambras-innsbruck.at/. Literature on the Gondelautomat includes Elisabeth Scheicher, *Die Kunst- und Wunderkammern der Habsburger* (Vienna: Molden Edition, 1979), 99–102; and the exhibition catalog *Vienne à Versailles: Les grandes collections autrichiennes au Château de Versailles*, 48–49 (this catalog mentioned in chap. 15 n. 10, above). According to the latter, the automaton still functioned in 1964.

20. See our chap. 12 and n. 6 on the Deutsches Museum automaton, which arrived without a garment.

21. Adelheid Voskuhl offers a probing discussion of women musicians in the context of automata: *Androids of the Enlightenment*, 145–53.

22. Truitt, *Medieval Robots*, 142–45; Mayr, *Authority, Liberty, and Automatic Machinery*, 3–8. Both offer a history of the water clock, Truitt in depth. Mayr writes, "Water clocks were thus closely related to the highly developed ancient technology of automata: the steady sequence of hourly indications offered itself as a convenient way of programming automata performances" (4). See also the wonderful essay by Sigrid and Klaus Maurice, "Counting the Hours in Community Life of the 16th and 17th Centuries," in Maurice and Mayr, *Clockwork Universe*, 144–56.

23. Mayr, *Authority, Liberty, and Automatic Machinery*, 5, on the escapement: "The *verge-and-foliot* escapement was an invention of radical originality. The complexity of its functioning and the ingenuity with which it had been devised [in the late thirteenth century] far surpassed all previous mechanical inventions. Its characteristic feature, the carefully tuned dynamic interaction of several matching parts, is almost impossible to describe in words or to illustrate in a two-dimensional picture; to fully comprehend it, one should handle a working specimen." There are now good animation diagrams of escapements on the internet, starting with Wikipedia, https://en.wikipedia.org/wiki/Verge_escapement; and later developments involving the pendulum, https://en.wikipedia.org/wiki/Escapement. Smaller clocks mounted on the wall or installed in tall cabinets ("grandfather" or tall-case clocks) retained escapement technology. The spring and fusee, with regulating fly, replaced the escapement in later centuries, making possible small portable clocks and mobile automata.

24. On the history of the striking jack, see Chapuis and Gélis, *Le monde des automates*, chap. 7, "Les jaquemarts et autres automates des grandes horloges," 1:113–75; Bedini, "The Role of Automata," 29–30 n. 15; Mayr, *Authority, Liberty, and Automatic Machinery*, 3–8; Truitt, *Medieval Robots*, 145–47, 207 n. 32. See also Ernest L. Edwardes, *Weight-Driven Chamber Clocks of the Middle Ages and Renaissance* (Altrincham: John Sherratt and Son, 1965), 1–69; and Daniel J. Boorstin, *The Discoverers* (New York: Random House, 1983), 36–46.

25. Bedini, "The Role of Automata," 30.

26. Truitt, *Medieval Robots*, 149–50. As she declares, "The construction of massive clocks for public display dramatically widened the audience for automata" (144). See also Chapuis and Gélis, *Le monde des automates*, 1:120–27. The clock has survived the millennium, with major renovations in the sixteenth and nineteenth centuries.

27. Derek J. de Solla Price, "Automata and the Origins of Mechanism and Mechanistic Philosophy," *Technology and Culture* 5, no. 1 (1964): 9–23. This famous short essay is paired with Silvio Bedini's essay "The Role of Automata in the History of Technology" in the same issue of *Technology and Culture*. Both were papers delivered on January 19, 1963, for the symposium "Automata and Simulated Life as a Central Theme in the History of Science," organized by the William Andrews Clark Memorial Library and the Division of Medical History, University of California Los Angeles. The authors thank none other than Dr. C. Donald O'Malley for making the meeting possible. (For O'Malley, see our chap. 2 nn. 2, 4, and 5.) Both Price and Bedini use the phrase "biological automata." Price: "[astronomical clockwork and biological automata] go hand-in-hand and are indissolubly wedded in all their subsequent developments" (15). Bedini (citing Price): "It is consequently not surprising to discover that the greatest advances in the development of biological automata, of astronomical clocks, and of the fine mechanisms were probably simultaneous in point of time and derivation" (29).

28. Chapuis and Gélis, *Le monde des automates*, 1:101. For a spirited roundup of the role of the Catholic Church in the history of automata, see the opening chapter, "Machines in the Garden," in Jessica Riskin's *The Restless Clock*. Merriam Sherwood, in "Magic and Mechanics in Medieval Fiction," *Studies in Philology* 44, no. 4 (October 1947): 567–92, charges that we have underestimated the sophistication of medieval spectators of mechanical display: "It is not surprising, therefore, that statues that moved, images that spoke or sang and goblets from which the wine mysteriously vanished had their counterpart in actual machines of the time. It is, further, but reasonable to assume that the operation of such machines was familiar to a substantial portion of the people" (591–92).

CHAPTER 18

The Earliest Free-Walking Androids

1. Some of the more complex mobile automata mentioned in the previous chapter employ this innovation as well, to turn and roll in a sequence of changing directions. See Michael Wagner and Ulrike Weinhold, "Der Diana-Automat im Grünen Gewölbe: Technologische und künstlerische Aspekte einer Augsburger Figurenuhr" [The Diana automaton in the Green Vault: Technological and artistic aspects of an Augsburg figure clock], *Jahrbuch der Staatlichen Kunstsammlungen Dresden* (2002–3): 38. Our thanks to Paulus Rainer for bringing this excellent study to our attention.

2. Britten, *Britten's Old Clocks and Watches*, 29: "In early examples the movement is entirely of steel; and brass wheels do not become common until the seventeenth century." An ur-edition of this great book is F. J. Britten's *Old Clocks and Watches and Their Makers* (London: B. T. Batsford, 1899), where Britten writes, "During the first quarter of the [sixteenth] century the frames and wheels were of iron or steel; productions of the second quarter having brass plates and pillars are occasionally to be met with. But brass wheels before the middle of the century were quite exceptional" (69).

3. The sixteenth-century Swiss alchemist Paracelsus is credited among others for giving zinc its name, but its presence in calamine and its role in turning copper into brass was not understood in Europe until long after brass alloys were successfully produced and widely used. "It was not until the mid-17th century that the chemical structure of calamine was properly understood and it was discovered that it was the zinc constituent which fused with copper to give the alloy its character and colour. The production of zinc was established in the East well before it was carried on in Europe," writes Eric Turner, in *An Introduction to Brass* (Owings Mills, MD: Stemmer House, 1982), 6. The knowledge won by empirical craft practice prior to modern chemistry and physics is a fascinating history unto itself. Early descriptions of the role of calamine in the production of brass in Europe are found in the classic works of Theophilus (early twelfth century), Biringuccio (b. 1480), and Agricola (1494–1555), all available in Dover editions with invaluable editors' commentary and interpretation. See also Joan Day, "Brass and Zinc in Europe from the Middle Ages until the 19th Century," in *2000 Years of Zinc and Brass*, ed. P. T. Craddock (London: British Museum, 1990). Of note here is the difference between brass and bronze, the latter with a better-known history in the evolution of technology. Bronze is copper combined with tin (and other metals), whereas brass is copper cooked with zinc. The differences include brass's brighter yellow color but also its greater ductility under the hammer, hence its behavior was more similar to iron in sixteenth-century foundry processes for making the components for clocks and scientific instruments. Distinctions in name between bronze and brass in the sixteenth century were not always strict (Turner, *An Introduction to Brass*, 7). For brass vs. bronze in the history of clockmaking, see Edwardes, *Weight-Driven Chamber Clocks of the Middle Ages and Renaissance*, 23–24.

4. Britten, *Britten's Old Clocks and Watches*, 69: "The early German movements [of watches] were all of steel, but by 1575 steel was giving way to brass, especially for the plates which could be cast, whereas steel plates had to be laboriously hammered and filed to the correct thickness."

5. See chap. 10 for wheel and pinion definitions and ratios. Why wheels of brass and pinions of iron, and not the other way around? In clockmaking, in every wheel-pinion pair, the pinion is the driven element and the wheel the driver. Pinions by definition step up the rotational speed of the gear train: for every revolution of the wheel, the pinion revolves many more times, depending on wheel tooth and pinion leaf counts. Hence the pinion, turning again and again against one revolution of the wheel, receives the greater wear on each of its leaves, and iron is the stronger metal. To this day, standard practice is to make clockwheels from brass and pinions from iron or steel. The model of the monk with all-steel clockwork that David made in 1980 proves the point: it was in constant use seven days a week for more than fifteen years as an interactive display at the Smithsonian, and after this relatively short period, its ferrous wheel teeth and pinion leaves exhibited almost the same amount of wear as in a comparably sized and well-used brass and iron clock mechanism from the seventeenth century. Mention of brass-iron pairing in the history of clockwork in Edwardes, *Weight-Driven Chamber Clocks of the Middle Ages and Renaissance*, 119; and Britten, *Britten's Old Clocks and Watches*, 33, 69.

6. David Todd, "From the Ground Up: Making an Iron Clockwheel," unpublished essay.

7. Ambrosio de Morales records this, from direct witness. He presents Juanelo demonstrating its use in his shop, to make the clockwheels for his great astrarium for Charles V:

> In his imagination he conjured the fabrication of a clock showing all the celestial motions, so that it should be more than that of Archimedes of whom Plutarch writes.... It took him, as he has told me, twenty full years to imagine and comprehend its construction.... and having taken so long to imagine, not more than three and a half years to build it with his hands. This is much, for the clock contains in all one thousand eight hundred wheels, without counting many other necessary things made of iron and brass. Thus (not counting feast days) he had to produce more than three wheels each day, apart from all the rest, since the wheels are of different sizes, with varying numbers and shapes of teeth, in addition to the manner by which they are mounted and locked upon arbors. Though this diligence is so marvelous, what is even more amazing to us is the ingenious lathe that he invented, and now we see him cut iron wheels with its file to produce the size and equality of teeth that may be needed. And besides all of this, and considering that he made all this with his hands, we must not be surprised when Janelo states that no wheel was made twice, because it was always made the first time exactly as required. And if all that was said before was not enough, this would be held to be a strange wonder. (*Las antigüedades*, 92v–93, as transcribed in the 1792 edition reprint, 337–38; see our chap. 4, n. 3. Translation Alain Gelbman, with Elizabeth King.)

Morales's description is cited by Robert S. Woodbury, *History of the Gear-Cutting Machine* (Cambridge, MA: MIT Press, 1958), 45, as the earliest-known reference to a gear-cutting machine in horological history (acknowledging Derek J. de Solla Price for the reference to Morales). We should note that Morales's count of 1,800 clockwheels, much quoted by later writers, is quite impossible. José A. García-Diego may be the first to have noted this error (*Juanelo Turriano*, 63), speculating that Morales meant teeth, not wheels. Giuseppe Brusa also noticed the mistake ("Giovanni Torriano e i suoi favolosi orologi microcosmici," 14). Both authors discuss the wheel-cutting engine, García-Diego in detail (70–72). Ever eloquent, Morales himself concludes his descriptions of Juanelo's machines: "I've spoken of these marvels not because I can explain all that they are, but because I want to record that these noteworthy things were made in our time."

8. Returning to his 1979 conservation notes for the all-brass wheel train in the Smithsonian lute player, David encountered another memory shock. As he himself recorded it even then, the train terminates in the very inefficient four-leaf pinion for the fly, just like the Smithsonian monk and the IM saint (see chap. 13, above, and its n. 11). With its clockwork made of brass, this machine might have been made a half century later than the iron-clockwork figures. As David exclaimed out loud, "Surely the maker had plenty of time to correct the mistakes of his forebears!" Then he realized he himself had overlooked the wild card. Yet in the Smithsonian lute player, the pinion is of iron, and the wheel driving it (the third wheel in the gear train) is brass, hence the reduction of wear between mating gear teeth that had become standard practice by then. But the low leaf count of this pinion in all three figures is a detail that begs understanding.

9. Friess, "Restaurierung einer Automatenfigur," 48, 50 nn. 4 and 5.

10. Baxandall, *The Limewood Sculptors*, 27–31.

11. Imported ebony in particular required certification of its authenticity in Augsburg guild regulations beginning in 1625. See Georg Himmelheber, "The Clock and Its Base," in Maurice and Mayr, *The Clockwork Universe*, 124.

12. Murakami Kazuo, *Japanese Automata:* Karakuri Zui; *An Eighteenth Century Japanese Manual of Automatic Mechanical Devices* (Osaka: published by the author, 2012), reproduces a suite of extraordinary Edo-period diagrams first published in 1796 by Hosokawa Hanzō Yorinao, which include detailed diagrams for the tea-serving doll (102 ff.). This book, with Murakami's translations into English, is itself an extraordinary contribution to horological history. Chapuis and Droz mentioned the Japanese tea-serving doll: *Automata: A Historical and Technological Study*, 318–19.

13. George Kubler, *The Shape of Time: Remarks on the History of Things* (New Haven, CT: Yale University Press, 1962), 108.

14. Kubler, *The Shape of Time*, 49. In speaking of the appearance and development of an innovation, Kubler writes, "The linked series of solutions composing a sequence is not necessarily restricted to a single craft. On the contrary, it is more likely to appear when different crafts come into play at the same time."

15. The automaton is a Diana and centaur with four dogs and a clock, ca. 1606–10, in the collection of the Staatliche Kunstsammlungen Dresden, inv. no. IV 150, by the same Augsburg goldsmith who made a similar version now in the Kunsthistorisches Museum, Hans Jakob Bachmann. See n. 1, above. Wagner and Weinhold, "Der Diana-Automat im Grünen Gewölbe," 39, describe the probe.

CHAPTER 19

Craft Guild or Imperial Court?

1. See chap. 11 n. 3, above, for NGA conservators' comments on the monk's head. A well-aimed hammer blow on a chisel will split a piece of wood along its grain, to be later glued back together with barely the sign of a seam: this was standard period practice in sculpture requiring insertion of glass eyes or representing the open mouth. In 2006 Constancio del Álamo, curator of Sculpture, Archaeology, and Textiles, and Marcus Burke, senior curator of Paintings, Drawings, and Metalwork, at the Hispanic Society of America in New York kindly gave Elizabeth a tour of the Society's conservation lab where several large wooden figures were under restoration. Those whose heads were hollow, requiring the internal placement of eyes, revealed the technique. Once the two halves had been reassembled and glued back together, the seam all but hidden, the head would be primed and painted.

2. Gervase Rosser, "Crafts, Guilds and the Negotiation of Work in the Medieval Town," *Past and Present*, no. 154 (February 1997): 3–31, traces the history, origins, and economic conditions of artisanal labor in late medieval Europe. The organization of urban craft institutions, and their connections to religious confraternities and community life, are examined from the perspective of individual craftsmen and craftswomen in local settings. The roles of both men and women are touched on throughout the essay, one of its many virtues. Research on women in the clockmaking trades has only recently begun: the Antiquarian Horological Society in London launched "The AHS Women and Horology" project in 2017 to list women active in the watch- and clockmaking trades in the United Kingdom from the seventeenth century on, with plans to expand to Europe and the United States in the future; see https://www.ahsoc.org/resources/women-and-horology/.

3. Maurice, *Von Uhren und Automaten*, 56. See also Scheicher, *Die Kunst- und Wunderkammern der Habsburger*, 153, for her distinction "between the clockmaker who constructed the mechanism and the goldsmith who gave the casing and the ornaments their artistic form."

4. Maurice mentions that many of the bronze statuettes in Augsburg figure clocks originated in Nuremberg, where they were cast in multiples, though the makers of the original figures remain unnamed. Of these, some, he notes, were patterned after small Italian bronzes (Maurice, *Die deutsche Räderuhr*, 1:111–15). Wagner and Weinhold, "Der Diana-Automat im Grünen Gewölbe," 43, describing an automaton of Diana on a centaur now in Dresden (mentioned in chap. 18 nn. 1 and 15, above) speculates that the sculptor might have been Adriaen de Vries, who practiced in Augsburg and as imperial sculptor in Prague.

5. Maurice, *Die deutsche Räderuhr*, 1:116: "Another Nuremberg example of a drinking game [figure] was by Barthel Jamnitzer. He converted the statuette of an archery Cupid by Johan Gregor van der Schardt into a silver automaton figurine, probably executed in 1576, 'a child with a drawn bow, which moves and turns around, from the internal clockwork.'" (Maurice cites this quote from Hans Robert Weihrauch, "Italienische Bronzen als Vorbilder deutscher Goldschmiedekunst [Italian bronzes as models of German goldsmith art]," in *Studien zur Geschichte der europäischen Plastik: Festschrift Theodor Müller* [Munich, 1965], 264 n. 5.) Elizabeth spotted this reference late in our study, and we are now eager to learn more about the collaboration between Van der Schardt and the Jamnitzer goldsmith family. An excellent study in English on the work of Johan Gregor van der Schardt is Frits Scholten, "Johan Gregor van der Schardt in Nuremberg," in *Carvings, Casts & Collectors: The Art of Renaissance Sculpture*, ed. Peta

Motture, Emma Jones, and Dimitrios Zikos (London: Victoria and Albert Museum, 2013), 134–47. Certainly any clockmaker collaborated with woodworkers who made the virtuoso casings for clocks, though we continue to search for records of wood-carvers commissioned in the making of automata. Michael Baxandall finds examples of small piecework performed by wood-carvers for direct sale to merchant or patrician clientele, noting that Veit Stoss made small pieces for occasional sale (Baxandall, *The Limewood Sculptors*, 102–3). Further study to find records of sculptors producing such piecework would be invaluable.

6. See FIG. 17.3A, B. Regarding this nef, Clare Vincent writes, "The ship is not signed, but decorative objects made for the emperor are known to have been exempt from hallmarking" (Clare Vincent, "Diana, Goddess of the Showering Arrows: A Seventeenth-Century German Automaton Clock," *Yale University Art Gallery Bulletin* [2016]: 80). Schlottheim is a prime example of a clockmaker who moved between guild and court settings.

7. Morales, *Las antigüedades*, 93v.

CHAPTER 20

Two Masters, Two Cultures: Jakob Bulmann and Juanelo Turriano

1. Johann Neudörfer, *Nachrichten von Künstlern und Werkleuten . . . aus dem Jahre 1547* [Accounts of artists and craftsmen . . . in the year 1547], as transcribed by G. W. K. Lochner, *Quellenschriften für Kunstgeschichte und Kunsttechnik des Mittelalters und der Renaissance* (Vienna: Wilhelm Braumüller, 1875), 10:65–69. Neudörfer, friend of Albrecht Dürer, is famous for this treatise of the biographies of the best-known artists of midcentury Nuremberg, Dürer among them.

2. Neudörfer, *Nachrichten von Künstlern und Werkleuten*, 65–66:

> Dieser Bülmann, ob er gleich Schreibens und Lesens nicht geübt gewest, ist er doch in der Astronomie fast künstlich und gelehrt gewest, derhalben er auch mit 80 Pfd. Gewicht die Theoriam planetarum in Gang und Uhrwerk bracht hat, welches dann vor ihm Niemand hat thun mögen. Er macht auch für Uhrwerk Manns und Weibs Bilder die umgingen und schlugen ihre Mensur auf der Lauten und Pauken. König Ferdinandus, dem er viel Arbeit gemacht, liess ihn in seinem hohen Alter auf einer Sänften zu ihm gen Wien bringen, allein darum, dass er seiner Majestät ein Uhrwerk zeigte. Er hat aber sonsten von Schlosserwerk viel schönes Dings gemacht, sonderlich aber in der Wag die zwen schönen künstlichen Wagbalken, daran man ausserhalb der Wag die Güter auf der Fuhr wiegt.

Thanks to Henning Bauer for translation from the German for these and all following quotes in this chapter.

3. Lochner, in a long footnote examining trade and legal documents, corrects Neudörfer, who listed Bulmann's given name as Hans. Here is a glimpse: "Bullmann or Püllmann's name was not Hans but Jacob, and is more often mentioned as a locksmith who was, by way of an exception, granted a greater number of servants than permitted by law; e.g., on April 20th and October 12th, 1513, and on several more occasions, also on August 21st, 1528 . . ."

4. Johann Gabriel Doppelmayr, *Historische Nachricht von den nürnbergischen Mathematicis und Künstlern . . .* [Historical account of the mathematicians and artists in Nuremberg . . .] (Nuremberg: Peter Conrad Monath, 1730), 285: "Er stellte ferner unter der Gestalt Manns- und Weibspersonen verschiedene Figuren mit Uhrwercken dergestalten beweglich vor, daß sie hin und her giengen, und nach der Mensur auf der Paucken auch auf der Lauten schlugen." Bassermann-Jordan, Chapuis and Gélis, and Bedini all cite Bulmann via Doppelmayr.

5. "Dieses confirmiret der ältere Johann Neudörffer in einem MSto, von denen zu seiner Zeit berühmten Künstlern, da es dazumahl was neues und besonderes war. Zu unsern Zeiten sind dergleichen Kunst-Machinationes nicht mehr so rar, indeme sich bishero viele auf solche Wercke von allerhand Vorstellungen mit großen Fleiß geleget." Doppelmayr duly adds citations to this footnote, to Kaspar Schott, John Wilkins, and Johann Sturm. As von Herz did before us, we tracked these citations as follows: Kaspar Schott, *Magia universalis naturae et artis* (Würzburg, 1657), pt. 1, bk. 1, chap. 6, pp. 24–25, John Wilkins, *Mathematicall Magick, or, the wonders that may be performed by mechanicall geometry* (London, 1648), bk. 2, chap. 4, pp. 172–78; Johann Christoph Sturm, *Kurzer Begriff der Physic oder Naturlehre* (Hamburg, 1713), pt. 2, sec. 2, chap. 1, pt. 8, pp. 825–26. All three authors address automata in brief and general terms, without detailing sizes, mechanisms, or origins. Schott, however, in a section titled "Statua automata" describes what are clearly two tabletop walking maidens, one in the shape of a silver drinking cup and one a lute or cittern player. We return to Schott's report on these sightings in chap. 23.

6. Von Herz, "Androiden des 16. Jahrhunderts," 34–36. Add to this the question of whether the German phrase "schlugen ihre Mensur" means "striking their measure" or, in context, "striking their *hours*." If the latter, the phrasing is more suggestive of a clock, with automata as embellishments. (Thanks to Henning Bauer for this point.) The Männleinlaufen (little men walking) is the popular name of the famous church clock with automata on the facade of the Frauenkirche (Church of Our Lady) in Nuremberg. Still operating each day at noon, as the clock strikes, seven figures emerge from a hidden niche and march three

times around a seated emperor twice their size. The "little men" represent the prince electors who approve a new Holy Roman emperor. The Gothic Frauenkirche, initiated by Charles IV, was finished in 1362, the clock restored in 1509 (and many times since). The automata were made of hammered polychromed copper.

7. Lochner prints Neudörfer's sentence as, "Er macht auch *für* Uhrwerk Manns und Weibs Bilder die umgingen und schlugen ihre Mensur auf der Lauten und Pauken." Doppelmayr's version is, "Er stellte ferner unter der Gestalt Manns- und Weibspersonen verschiedene Figuren *mit* Uhrwercken dergestalten beweglich vor, daß sie hin und her giengen, und nach der Mensur auf der Paucken auch auf der Lauten schlugen." Von Herz's emphasis.

8. Lochner's note in Neudörfer, *Nachrichten von Künstlern und Werkleuten*, 66. See also von Herz, "Androiden des 16. Jahrhunderts," 36.

9. The story of this great astrarium is told by Bedini and Maddison, "Mechanical Universe." See chap. 4 n. 6, and chap. 16 n. 6, above. De' Dondi also produced a detailed codex on how he made the clock, the *Tractatus astrarii*. The clock did not survive, but copies of the codex did, thus preserving the reputation of one of the most famous planetary clocks in history. Modern reconstructions of it are based on precise details in the codex. The codex itself has been reproduced in *The Planetarium of Giovanni de Dondi Citizen of Padua*, trans. G. H. Baillie and H. Alan Lloyd (London: Antiquarian Horological Society, 1974). In addition to the Thwaites and Reed model made in 1960, shown in FIG. 20.1, a second model was built in Milan in 1963 by the master clockmaker Luigi Pippa and is now on display in the Museo Nazionale della Scienza e della Tecnologia Leonardo da Vinci, Milan. More recent scholarship on de' Dondi's astrarium is found in "Giovanni Dondi dall'orologio," in *Dizionario Biografico degli Italiani*, vol. 41 (1992). https://www.treccani.it/enciclopedia/giovanni-dondi-dall-orologio_(Dizionario-Biografico)/

10. See "Das Astrarium von Regiomontan" and "Jakob Bulmanns Theoria planetarum," in Maurice, *Die deutsche Räderuhr*, 1:56 and 58. Maurice examines three planetary clocks, among them the one Neudörfer cites for Bulmann, as possible appearances of the same clock in the generations following Regiomontanus. Neudörfer mentions the Theoriam planetarum again, under the name of the Nuremberg clockmaker Peter Henlein, contemporary to Bulmann, reporting that Henlein made further improvements to the clock using Bulmann's tables (Neudörfer, *Nachrichten von Künstlern und Werkleuten*, 71, 77). The supposition that it was Regiomontanus's astronomical clock that Bulmann restored, with Regiomontanus's tables, is mentioned by Ernst Zinner in his biography, *Regiomontanus: His Life and Work*, trans. Ezra Brown (Amsterdam: Elsevier Science, 1990), 135. Regiomontanus in fact saw the astrarium of Giovanni de' Dondi, wrote about it, and possibly consulted de' Dondi's calculations. See Bedini and Maddison, "Mechanical Universe," 26–27; Bedini and Maddison, 16–17, also helpfully comment on the connection between an instrument and a "theory."

11. Bassermann-Jordan, *Alte Uhren und ihre Meister*, 55–64.

12. Bassermann-Jordan, *Alte Uhren und ihre Meister*, 67–69. Henning Bauer's translation from the German for us, with spelling inconsistencies in Bulmann's name preserved. We omit from this quote only Bassermann-Jordan's full transcription of the Doppelmayr text, which we have already seen. As with the Morales report of Juanelo's dancing lady, one sees how texts travel through time from pen to pen, era to era.

13. Von Herz, "Androiden des 16. Jahrhunderts," 30–31.

14. Chapuis and Gélis, *Le monde des automates*, 2:181, where Bulmann and Turriano are paired in their chapter 21 in vol. 2, "Les constructeurs d'androïdes," directly following their introduction of the word *android* and its definition. They cite Doppelmayr for their mention of Bulmann (2:181). For Turriano (1:90–91 and 2:181–82), they cite Strada but have clearly used Basserman-Jordan's study, with its basis in Ambrosio de Morales's text, and William Stirling-Maxwell's *The Cloister Life of the Emperor Charles V*.

15. Bedini, "The Role of Automata in the History of Technology," 31–32. And see chap. 15 n. 5, above.

16. The catalog cites *Jakob* Bulmann, so the Lochner version of Neudörfer's manuscript was at last consulted. It took this long for Bulmann to be properly connected to a witness contemporary to him, for although freshly authenticated in the late nineteenth century along with Neudörfer himself, Jakob Bulmann's correct name and champion had to wait for Erwin Neumann's acknowledgment in 1966, in this entry in the KHM catalog. (See chap. 15 n. 10, above.)

17. By then, Klaus Maurice had published his 1976 history of the German wheel clock, *Die deutsche Räderuhr*, in which he quotes Neudörfer's praise of Bulmann, thus further strengthening the reference to Bulmann as a maker of automata (1:58).

18. See chap. 17 n. 27, above.

19. Bulmann had one published contemporary witness, Juanelo Turriano many more. Von Herz, "Androiden des 16. Jahrhunderts," 43, comments that García-Diego indirectly calls our attention to a problem of transmission, i.e., that historians may so readily choose Juanelo as the creator of automata because he, unlike many others of his time who were also constructing complex devices, was written about by so many contemporary scholars, making him more "available."

20. Marco Girolamo Vida, *Cremonensium orationes III: Adversus Papienses in controversia principatus*... (Cremona, 1550), 56r–56v. Transcribed and translated from the original Latin for us by William M. Barton, of the Ludwig Boltzmann Institute for Neo-Latin Studies, Innsbruck, Austria; with contributions from the historian Wenrui Zhao. "Bronte" and "Esterope," or Brontes and Steropes, were, with Arges, the three Cyclopes, children of Mother Earth, as Robert Graves tells us: "builders of gigantic walls and master-smiths, formerly of Thrace, afterwards of Crete and Lycia, whose sons Odysseus encountered in Sicily[,]... and their ghosts have dwelt in the caverns of the volcano Aetna since Apollo killed them." Robert Graves, *The Greek Myths* (Mt. Kisco, NY: Moyer Bell, 1955), 31.

21. García-Diego, *Juanelo Turriano*, 65.

22. Cardinal Federico Borromeo, *Il libro intitolato la gratia de'principi*... (Milan, 1632), 167–68:

> But it would be a highly commendable practice and worthy of being observed by wise men if they, in uttering such maxims, would avail themselves of any nice and opportune circumstance, as happened in the case of a famous Lombard craftsman, Master Giannelli, whose exquisite genius in the construction of the most diverse and artful devices was highly praised in his day. Among the many remarkable things that he used to say, he was in the habit of saying the following: that the three most beautiful and most excellent things, and most worthy of being valued, and held dear by all men, were withheld from Princes: namely, a person who will tell them the truth, the ability to behold the dawn, and to experience hunger.

Thanks to Evelyn Lincoln for this translation from the Italian original.

23. "[I]n memory, says tradition, of a puppet, of his making, which used to walk daily to the archiepiscopal palace, and return laden with an allowance of bread and meat, after doing ceremonious obeisance to the donor." Stirling-Maxwell, *The Cloister Life of the Emperor Charles V*, 444.

CHAPTER 21

A Nuremberg Clockmaker's Workshop

1. Penelope Gouk, *The Ivory Sundials of Nuremberg 1500–1700* (Cambridge: Whipple Museum of the History of Science, 1988), 46. Legend credits Regiomontanus with the making of two automata for the emperor: an iron fly and an iron and wood eagle, both said to rise and take flight. Johann Gabriel Doppelmayr (whose short biography of Bulmann is quoted in chap. 20, above) concluded his lengthy entry on Regiomontanus thus: "Finally the most remarkable thing, according to Petrus Ramus's first assertion, according to which Regiomontanus showed himself to be the most skillful mechanic in the world, [is] that he supposedly constructed an iron fly and wooden eagle that flew. We will leave the question open whether this is indeed true, since there is no sufficient basis to substantiate such a claim" (23). See Zinner, *Regiomontanus*, "The Nuremberg Master Years," 103–49.

2. Gouk, *The Ivory Sundials of Nuremberg*, 39–43, for trade routes, Nuremberg merchants in Venice, and the city's Protestant lean. For a close description of the clockmaking guild system but in Augsburg, see Eva Groiss, "The Augsburg Clockmakers' Craft," in Maurice and Mayr, *The Clockwork Universe*, 57–86. On clockmaking guilds in German-speaking central Europe, see Mayr, *Authority, Liberty, and Automatic Machinery*, 8–10; Maurice, *Von Uhren und Automaten*, 28–41; and Maurice, *Die deutsche Räderuhr*, 1:pt. 5: "Zentren des Uhrmacherhandwerks bis zur Mitte des 17. Jahrhunderts [Centers of the clockmaking trades to the middle of the seventeenth century]," where Nuremberg's unique case is addressed on pp. 141–46.

3. Benjamin A. Rifkin, in his 1971 introduction to Jost Amman and Hans Sachs, *Eygentliche Beschreibung aller Stände auff Erden [...]* (Nuremberg, 1568; in modern facsimile reprint as *The Book of Trades (Ständebuch)* [New York: Dover, 1973]), ix.

4. Britten, *Britten's Old Clocks and Watches*, 69–70: "The pioneer Nuremberg makers were not formed into a guild until 1656, and even as late as 1629 a locksmith or gunsmith could work as a clock maker, but not vice versa."

5. Amman and Sachs, *The Book of Trades (Ständebuch)*; see n. 3, above. The publisher of the *Ständebuch*, Sigmund Feyerabend, dedicated the book to the goldsmith Wenzel Jamnitzer. Precedent to the *Ständebuch* was the famous Mendel Housebook, an archive of illustrations memorializing fifteenth- and sixteenth-century Nuremberg artisans: *Das Hausbuch der Mendelschen Zwölfbrüder-stiftung zu Nürnberg* [The Housebook of the Mendel Twelve Brothers Foundation of Nuremberg] (Nuremberg, 1549; in modern reprint, Karlheinz Goldmann and Wilhelm Treue et al., eds., 2 vols., [Munich: F. Bruckmann, 1965]). The full series of historical images from the Mendel Housebook is now available online through the Nuremberg City Library and the German Research Foundation, https://www.nuernberger-hausbuecher.de/.

6. An alarm clock, in the age when clocks told time only by the hour, employed the wheel of the clock hand itself to set off a striking mechanism connected to a bell. Minutes came into fashion only in the next century, driven by clock technology itself.

7. Groiss, "The Augsburg Clockmakers' Craft," 69 n. 65.

8. Amman and Sachs, *The Book of Trades (Ständebuch)*, 75. Sachs's verses, each eight lines in rhyming couplets, are rendered in English by Theodore K. Rabb, trans., *A Sixteenth-Century Book of Trades: Das Ständebuch* (Palo Alto, CA: Society for the Promotion of Science and Scholarship, 2009). The clockmaker, p. 135.

9. Amman and Sachs, *The Book of Trades (Ständebuch)*, 69. Verse translation, the locksmith: Raab, *A Sixteenth-Century Book of Trades*, 123.

10. Amman and Sachs, *The Book of Trades (Ständebuch)*, 36. Verse translation, the sculptor: Raab, *A Sixteenth-Century Book of Trades*, 57.

11. Jan van der Straet, known as Johannes Stradanus, *Horologia ferrea* (Iron clocks), ca. 1591. A copperplate engraving from the historical series *Nova Reperta* (New discoveries), one of twenty prints based on drawings by Stradanus illustrating early modern technologies and discoveries, from windmills to silkworms to printing itself. Printed by Philips Galle in Antwerp. See Lia Markey, ed., *Renaissance Invention: Stradanus's "Nova Reperta,"* exh. cat. (Evanston, IL: Northwestern University Press, 2020). For a clockmaker's tour of the details depicted in both Amman's and Stradanus's clock shops, see John Robey, "Early Clockmakers' Workshops: Part 1, Jost Amman's Woodcut," and "Part 2, The Stradanus Engraving," in *Clocks Magazine*, October 2017 and November 2017. Our caption for the Stradanus print, FIG. 21.4, is informed by Robey's study.

12. Vannoccio Biringuccio, *De la pirotechnia* (Venice, 1540), as quoted in Cyril Stanley Smith, *A Search for Structure: Selected Essays on Science, Art, and History* (Cambridge, MA: MIT Press, 1981), 34–35. Smith and Martha Teach Gnudi translated and annotated Biringuccio's work as *The "Pirotechnia" of Vannoccio Biringuccio* (New York: American Institute of Mining and Metallurgical Engineers, 1942).

CHAPTER 22

The Art Cabinet of Rudolf II

1. Eliška Fučíková, "The Collection of Rudolf II at Prague: Cabinet of Curiosities or Scientific Museum?," in Impey and MacGregor, *The Origins of Museums*, 53.

2. R. J. W. Evans, *Rudolf II and His World: A Study in Intellectual History 1576–1612* (Oxford: Oxford University Press, 1973), 177.

3. Thomas DaCosta Kaufmann, "Remarks on the Collections of Rudolf II: The *Kunstkammer* as a Form of *Representatio*," *Art Journal* 38, no. 1 (Autumn 1978): 22.

4. Eliška Fučíková, "Prague Castle under Rudolf II, His Predecessors and Successors," in *Rudolf II and Prague: The Court and the City*, ed. Eliška Fučíková et al. (Prague: Prague Castle Administration; London: Thames & Hudson, 1997), 13: "Rudolf was born of a loveless marriage between Maximilian and his cousin, Maria of Spain, the daughter of Charles V. Maximilian's sympathies for Protestantism and Maria's bigoted Roman Catholicism both manifested themselves in their approach to the young prince's upbringing. Fearing for his faith, Maria of Spain, her brother King Philip II and their uncle Ferdinand I were able to convince Maximilian to entrust the education of the young pretender to the imperial throne to a responsible person at the Spanish court." See also Evans, *Rudolf II and His World*, 49–51.

5. Beket Bukovinská, in her essay, "The *Kunstkammer* of Rudolf II: Where It Was and What It Looked Like," in Fučíková et al., *Rudolf II and Prague*, 199 ff., describes the four rooms and their arrangement, based on the 1607–11 inventory.

6. Patrick Mauriès, *Cabinets of Curiosities* (London: Thames & Hudson, 2002), 171.

7. Bezoars were "gastric calculuses found in antelopes" believed to be antidotes to poison. (Bukovinská, "The *Kunstkammer* of Rudolf II," 203). Stirling-Maxwell is more specific in a footnote to the inventory of items belonging to Charles V in San Yuste, which included "a box of black leather, lined with crimson velvet, containing four bezuar stones, variously set in gold" (*The Cloister Life of the Emperor Charles V*, 477 n. 2): "The bezuar, bezoar, or bezar, was a stone found in the kidneys of the *cervicabra*, a wild animal of Arabia, partaking of the nature of the deer and the goat, and somewhat larger than the latter. The stone was supposed to be formed of the poison of serpents which had bitten the producer, combined with the counteracting matter with which nature had furnished it. It was a charm against plague and poison."

8. Fučíková, "The Collection of Rudolf II at Prague," 52.

9. "There can be no doubt that Rudolf's collection was never destined for a wider public, but was limited to those few initiates who knew how to use it." Fučíková, "The Collection of Rudolf II at Prague," 52. And see Bukovinská, "The *Kunstkammer* of Rudolf II," n. 13, below.

10. Eliška Fučíková, "The Fate of Rudolf II's Collections in Light of the History of the Thirty Years' War," in *1648: War and Peace in Europe*, ed. Klaus Bussmann and Heinz Schilling, vol. 2, *Art and Culture* (Münster: Westfälisches Landesmuseum für Kunst und Kulturgeschichte, 1998), 173–80. In the same volume, Thomas DaCosta Kaufmann, "War and Peace, Art and Destruction, Myth and Reality: Considerations on the Thirty Years' War in Relation to Art in (Central) Europe," 163–72. See also Rudolf Distelberger, "The Habsburg Collections in Vienna during the Seventeenth Century," in Impey and MacGregor, *The Origins of Museums*, 40; and Evans, *Rudolf II and His World*, 179–80.

11. Erwin Neumann, "Das Inventar der rudolfinischen Kunstkammer von 1607/11," in *Analecta Reginensia I, Queen Christina of Sweden, Documents and Studies*, (Stockholm, 1966), 262–65; quote from 262–63.

12. Rotraud Bauer and Herbert Haupt, eds., "Das Kunstkammerinventar Kaiser Rudolfs II., 1607–1611," *Jahrbuch der Kunsthistorischen Sammlungen in Wien*, 72 (1976). Neumann, "Das Inventar," 263: "We are looking at a bound parchment codex of 414 or 415 sheets." Fučíková, "The Collection of Rudolf II at Prague," 47–53, states that the 1607–11 inventory does not cover the entire collection. The work of comparing the principal extant inventories, what each lacks and what each offers, and what their sum allows us to know, is ongoing. See Beket Bukovinská, "The Known and Unknown *Kunstkammer* of Rudolf II," in *Collection—Laboratory—Theater: Scenes of Knowledge in the 17th Century*, ed. Helmar Schramm et al. (Berlin: Walter de Gruyter, 2005), originally published in German in 2003, for the series Theatrum Scientiarum.

13. "Today, it is universally presumed to have been drawn up by the emperor's miniature painter and court antiquarian Daniel Fröschl." Bukovinská, "The *Kunstkammer* of Rudolf II," 199. "If we can rely on marginal notes to reveal the true arrangement of the objects in the Prague *Kunstkammer* during the life of Rudolf II or immediately after his death, it is evident that a large number of objects were arranged in such a way that viewing them required time and a knowledge of the material. It is striking that if we use these pencilled notes to piece together the contents of the chests and cabinets, we discover that even ostentatious works we would expect to have been displayed on tables or shelves were shut away in chests and, often, additionally in small gilded leather cases" (207).

14. Bauer and Haupt, "Das Kunstkammer-inventar Kaiser Rudolfs II.," find it in the inventory as Nr. 1751 (their p. 93). Also see Scheicher, *Die Kunst- und Wunderkammern der Habsburger*, 157.

15. Bauer and Haupt, "Das Kunstkammer-inventar Kaiser Rudolfs II.," 110.

16. Bauer and Haupt, "Das Kunstkammer-inventar Kaiser Rudolfs II.," 113. The entry echoes in its phrasing Neudörfer's reference to Bulmann's "figures [for] clockworks, which would walk about and strike their measure on lutes and timbals." (A note on the double numbers for these and most other entries in the Bauer and Haupt transcription: the first number is assigned by Bauer and Haupt as their editorial index for all objects listed in the codex. In most cases a second number follows, from the codex itself, which had its own system of classification. Our thanks to Paulus Rainer for confirming this. The study of the original numbering system has helped historians shed light on how Rudolf identified and classified his collection.)

17. See chap. 15 n. 10, above.

18. A note on the translation of the German words *schlegt* and *schlecht* in the inventory entry: though the difference in spelling between the two words in the same passage gives pause, we conclude that they represent the same word, the New High German *schlägt*—in English, "strikes" or "strums." Given the *cittern*, a musical instrument that is strummed, and paired with the term *fantasia*, a translation reading, "Wound up like a clockwork, [it] strums a fantasia," is more credible than a supposition that "schlecht" represents another word altogether (such as, conceivably, New High German *schlicht* for English "simply"). "Fantasia" here must refer to the musical form. Our thanks to Henning Bauer for translation and interpretation.

19. Only Morales reports the *size* of an automaton: see chap. 4, above.

20. In fact, the clock contains two balls that take turns: one on view rolls the course while the other is lifted by a hidden mechanism to replace it at the right moment, for an illusion of a single ball running continuously. The device of the mirror and invisible glass plate is not unlike the phenomenon of "Pepper's ghost," a nineteenth-century stage illusion. The history of Margraf's clock and the many clocks by others made after his is presented in Hans von Bertele and Erwin Neumann, "Der kaiserliche Kammeruhrmacher Christoph Margraf und die Erfindung der Kugellaufuhr," *Jahrbuch der Kunsthistorischen Sammlungen in Wien* 59 (1963): 39–98. Further references for Margraf are in Manfred Leithe-Jasper and Rudolf Distelberger, *The Kunsthistorische Museum Vienna: The Treasury and the Collection of Sculpture and Decorative Arts*, trans. Elsie Callander and Ilse Seldon (London: Philip Wilson, 1982), 98, 112. See the KHM website: www.khm.at/de/object/415713fe57/. The museum collection has two signed and dated Rolling Ball Clocks by Margraf, and a third, unsigned, attributed to him. Our ongoing thanks to Paulus Rainer, curator of the *Kunstkammer* at the Kunsthistorisches Museum, who kindly read excerpts of this chapter and wrote to Elizabeth (December 20, 2021): "The first Rolling Ball Clock bears the inscription *Cum gratia et Priuilegio Sac: Caes: Maiest: ad Quindecim Annorum*. This is actually a kind of patent, meaning that Margraf had the exclusive right to make this kind of clock for fifteen years, a privilege received from the emperor." None of the Margraf rolling ball clocks are listed in the 1607–11 Rudolfian inventory, Rainer informs us, but the one we mention appears in a later 1619 inventory of the estate of Rudolf's brother Matthias.

21. Why did it take the discovery of the 1607–11 inventory to alert Neumann and Maurice to the entries that they brought forward as possible references for the extant automata? Later inventories, made in the decade after Rudolf's death, have long been known. Can we find entries in the later inventories that describe the same or similar figures, perhaps suggesting that they had not yet disappeared from the collection or the record? Some of the objects in Rudolf's collection can indeed be tracked from one inventory to another, and some have been tracked further in their diaspora to other lands and collections. Many more have simply been lost. Inventories posthumous to Rudolf include one drawn in 1619, and another in 1621, in the political struggle in Bohemia following Rudolf's death. The 1619 list was made at the time of the Protestant Bohemian estates' briefly successful revolt against the Catholic Habsburg reign and was intended to aid in selling off the collection to support the defense of the new kingdom. The 1621 inventory was made when the revolt was quelled, apparently on the orders of Habsburg Ferdinand II of Bohemia, returned to power but needing funds of his own as the Thirty Years' War escalated. (Fučíková, "The Fate of Rudolf II's Collections," 173–80; Bauer and Haupt, "Das Kunstkammerinventar Kaiser Rudolfs II.," xii–xiii; Neumann, "Das Inventar," 262). The 1621 inventory is transcribed in "Das inventar der Prager Schatz- und Kunstkammer vom 6. Dezember 1621," ed. Heinrich Zimmermann, *Jahrbuch der Kunsthistorischen Sammlungen des Allerhöchsten Kaiserhauses* 25 (1905): xiii–lxxv. We searched for any entries that could conceivably match entries singled out in the 1607–11 inventory. None offer sufficient detail for a sure connection, but for the record we offer our findings here, with thanks to Henning Bauer for help with both search and translation:

> ENTRY 349 ON PAGE XXVII:
> "One chest with a clockwork, [with] a drummer and a piper on top"
>
> ENTRY 721 ON PAGE XXXV:
> "A waxen child in a case, beating a drum"
>
> ENTRY 731 ON PAGE XXXVI:
> "A waxen bust of a woman, moving, with a clockwork"
>
> ENTRY 105 ON PAGE XXII:
> "A waxen child that moves"
>
> ENTRY 725 ON PAGE XXXV:
> "Two wooden chests with two waxen figures with clockwork, which dance"

22. See chap. 7, above, for the full catalog entry in *The Clockwork Universe*. As we mention in chap. 7 n. 3, Maurice and Mayr gave Rudolf inventory numbers 2193 and 2194 and do not include the Margraf entry 2145. A typo, as is the mistaken title of the inventory as "1607–1711"? While entries 2145 and 2194 describe solo figures, entry 2193 reads more like a stationary machine with an ensemble of moving figures, including a drummer. The entry appears under the heading "Clockworks in pictures and other diversions" (Bauer and Haupt, "Das Kunstkammerinventar Kaiser Rudolfs II.," 113) and reads, "A clockwork, wound up like a clock, in a green box, a piper, a drummer, and a soldier, dressed appropriately, each performs his office."

23. We will assume Neudörfer (1497–1563) saw the works of Bulmann (d. ca. 1535) in person.

24. Findlen, *Possessing Nature*, 34. See chap. 16 n. 5, above. The Ambrosiana website offers a brief history and a portrait of Settala: https://www.ambrosiana.it/en/opere/portrait-of-manfredo-settala/

CHAPTER 23

Searching the Lost and Found

1. For this image, and a discussion of automata as features of decorative table settings, see Maurice and Mayr, *The Clockwork Universe*, 235; and Maurice, *Von Urhen und Automaten*, 54. Daston and Park, *Wonders and the Order of Nature*, 100–108 ("Wonders at Court"), help us envision the will to capture something more than dramatic effect in the coming theater of the mechanical age, with automata appearing in the extravagant wedding banquets and court spectacles of dukes and princes.

2. Von Herz, "Androiden des 16. Jahrhunderts," 58–74.

3. A good summary of the "talking heads" of Gerbert of Aurillac, Albertus Magnus, Robert Grosseteste, and Roger Bacon can be found in Minsoo Kang's excellent book, *Sublime Dreams of Living Machines: The Automaton in the European Imagination* (Cambridge, MA: Harvard University Press, 2011), 68–79, "The Talking Oracular Head." Don Quixote is tricked by such a head in part 2 of Cervantes's epic, chap. 62: "Which relates the adventure of the enchanted head, as well as other foolishness that must be recounted." Miguel de Cervantes, *Don Quixote*, trans. Edith Grossman (New York: HarperCollins, 2003), 864–72. Cervantes (1547–1616), a native of Alcalá de Henares, was born just two years after Don Carlos. Part 1 of *Don Quixote* was published in 1605; part 2, in 1615.

4. Von Herz, "Androiden des 16. Jahrhunderts," 60.

5. Von Herz, "Androiden des 16. Jahrhunderts," 43.

6. Von Herz, "Androiden des 16. Jahrhunderts," 43.

7. But one panel of the Brueghel and Rubens series (in the collection of the Museo Nacional del Prado in Madrid), *The Sense of Hearing*, may show a clock by Juanelo. García-Diego (*Juanelo Turriano*, 66–69) offers a fascinating discussion about whether a certain octagonal clock in the background at far right, in a wooden case standing on the floor, could be Juanelo's astrarium.

8. Translating from the French (and losing its rhyme), Elizabeth wonders, are "Les Grands" gods or kings? Chapuis and Droz, *Automata: A Historical and Technological Study*, 317, translate the lines thus:

> *Without joking, what we are*
> *Is often our fate.*
> *The gods when it pleases them*
> *make very much the same thing of men as a hidden spring does of Catin.*

9. Horst Bredekamp, *The Lure of Antiquity and the Cult of the Machine: The Kunstkammer and the Evolution of Nature, Art and Technology*, trans. Allison Brown (Princeton, NJ: Markus Wiener, 1995), 2–4.

10. Kaspar Schott, *Magia universalis naturae et artis* (Würzburg: Pigrin, 1657), Prolegomena, Partis I. Lib. I, pp. 24–25. Our thanks to William M. Barton for transcription into modern Latin from the seventeenth-century original and to Wenrui Zhao for her translation into English. The Latin word *citharam* for the instrument the automaton plays could refer to a cittern as well as a lute or a lyre, among stringed instruments of the period. We note that in the Latin original, the phrase "per mensam discurrendo," translated here as "roamed across the table," hinges on the phrase "per mensam," which literally means "through the table." Elizabeth asked the translator Wenrui Zhao if the phrase might mean "around" the table rather than "on" the table—for "on the table" lets us infer a small tabletop figure, but "around the table" may mean that Schott is describing a larger, even life-size android. Zhao responded, "[The Latin] '*per*' as a proposition is often associated with a linear sense of 'through' or 'throughout'—spatially as well as temporally. Sometimes it can also mean 'all over'—but that would still imply it was *on* the table. If 'around' was meant, a proposition like 'circum' would probably have been used" (email correspondence, December 28, 2021).

11. Stirling-Maxwell, *The Cloister Life of the Emperor Charles V*, appendix, 477–85: "A Selection from the Extracts made by Don Tomas Gonzales from the inventory of the jewels, wardrobe, and furniture of the Emperor Charles V., at Yuste, drawn up after his death, by Fray Juan de Regla, Martin de Gaztelu, and Luis Quixada." The inventory lists "Portraits of the Emperor and Empress, on canvas, by Titian" and three pictures of the Virgin, among them, "A picture of Our Lady, on wood, by Titian" (483).

12. In Jadraque's depiction of the portrait, the sitter's head and collar are close to those in Titian's painting, but the hands are in a different pose. Titian's canvas *El caballero del reloj*, featuring a Maltese cross sewn on the figure's doublet, may not represent Juanelo, though the likeness is compelling when compared to other portraits of the clockmaker.

13. Stirling-Maxwell, *The Cloister Life of the Emperor Charles V*, 483–84; García-Diego, *Juanelo Turriano*, 123–28; Bassermann-Jordan, *Alte Uhren und ihre Meister*, on clocks in the San Yuste inventory, 56; and Brusa, "Giovanni Torriano e i suoi favolosi orologi microcosmici," 16–18.

14. "...about two feet in diameter, and somewhat less in height" (Stirling-Maxwell, *The Cloister Life of the Emperor Charles V*, 179). In Stirling-Maxwell's appendix inventoryof items at San Yuste "the great clock" appears on p 483. Giuseppe Brusa gives 60 cm (23⅝ in.) for its diameter and particularly notes its octagonal shape (12). See also n. 7, above, on García-Diego's proposal of an image of this clock in the Brueghel and Rubens panel *The Sense of Hearing*.

15. Peter Friess and Reinhard Steiner, "Frömmigkeits-Maschinen in der Frühen Neuzeit" [Machines of piety in the early modern period], in *Automaten in Kunst und Literatur des Mittelalters und der Frühen Neuzeit*, ed. Klaus Grubmüller and Markus Stock, Wolfenbütteler Mittelalter-Studien, vol. 17 (Wiesbaden: Harrassowitz Verlag, 2003), 223–45. The authors note their discovery of the painting (239 n. 37):

> This was a chance finding in: Luisa Cuesta, "Lo que conocemos de Carlos V" in: *Revista de archivos, bibliotecas y museos*, 68 (1960), 29–79, Abb. XVII; also: Museo del Prado, Inventario General de Pinturas, III, Nuevas adquisiciones Museo Iconocráfico Tapices (Madrid, 1990), 130, no. 476 (Inv. actualizado No. 4826). The painting *Carlos V en Yuste*, oil on canvas, 146 × 193 cm, was executed by Miguel Jadraque y Sánchez Ocaña, a historical and genre painter born in Valladolid. His vita is not precisely known (ca. 1840–1912). The painting was exhibited in Paris in 1878 by Jadraque.

(Our thanks to Henning Bauer for translation from the German of Friess and Steiner's text.) We were grateful and fascinated to receive a reprint of the essay from Peter Friess in 2003.

The painting has been on long-term loan from the Prado to the Cámara Oficial de Comercio e Industria in Salamanca. Elizabeth saw it there in 2006, with the kind permission and accompaniment of Gabriel Hortal Castaño, secretary general. The Prado restored the painting in the 1990s, Hortal Castaño said, and only then did the small figures become visible. It is Hortal Castaño's opinion that the tiny standing figure is shown making the sign of the cross over the kneeling "Moor," converting him. The assembled onlookers are startled, he said, even a little frightened at what they are seeing. See also http://cvc.cervantes.es/el_rinconete/anteriores/octubre_00/31102000_02.htm. The Prado website itself now carries a good reproduction of the Jadraque painting, allowing close-in zoom inspection: Carlos V en Yuste-Colección-Museo Nacional del Prado (museodelprado.es)

16. Friess and Steiner, "Frömmigkeits-Maschinen in der Frühen Neuzeit," 239:

> Even if at present no definitive evidence can be proffered (yet), there is a historical painting that would seem to suggest that up to the late nineteenth century, it was believed that not only had Charles V automatic table soldiers do battle, but that he possessed small monk figurines. It is more than a fading trace in the sand when, in the background of Miguel Jadraque y Sánchez Ocaña's painting showing Charles V in Yuste, we see the figure of a monk in a white habit next to an astronomical clock atop a piece of furniture. Judging by the proportions, it would appear to have roughly the same measurements of the known monk automata. Since it appears unlikely that an individual pleurant would have served as an interior ornament, one may surmise that this monk figurine was part of an ensemble of six—perhaps even eight—automata. These may not have held an actual wake during the melancholic emperor's lifetime, but their playful gravity would nevertheless have afforded him a memento mori without subjecting him to the condemnation of the church.

CHAPTER 24

A Monk's Opinion, Continued

1. Stirling-Maxwell, "Author's Preface," xx–xxiv. Stirling-Maxwell died in 1878; the first edition of his book, with this preface, was published in 1852. The fourth edition was published posthumously in 1891 by his son John Stirling-Maxwell. (González's work was essential in identifying objects in the inventory at San Yuste; see chap. 23 n. 11, above.)

2. Stirling-Maxwell's story of the González manuscript is a tale all its own, as he related it in the preface to his own book in 1852. Prescott reprised the story in 1855 in his *History of the Reign of Philip the Second, King of Spain* (vol. 1, 325 ff., of the 1874 ed.); see above, chap. 2 n. 3 and chap. 3 n. 5.

3. Friess and Steiner, "Frömmigkeits-Maschinen in der Frühen Neuzeit," 239 n 37.

4. Bassermann-Jordan, *Alte Uhren und ihre Meister*, summarizes his sources (59 n. 4) and singles out Stirling-Maxwell's book as most useful.

5. The letter is given in full in chap. 6, above.

6. A recent and remarkable set of essays on this tradition, worldwide, is Ittai Weinryb, ed., *Agents of Faith: Votive Objects in Time and Place*, exh. cat. (New York: Bard Graduate Center, 2018).

7. The Istituto Storico dei Cappuccini was formally established in Assisi in 1930 and moved to Rome during World War II. The Museo Francescano predates the Institute and was founded in 1889 in Marseilles, moved to Rome in 1912, and in 1927 was associated with the then-nascent Istituto in Assisi. The Institute's mandate is the pursuit of scholarly research on the history of the Franciscan order, and in particular its Capuchin family, formed during the sixteenth-century Counter-Reformation. Servus Gieben, associated with the Institute from 1953 to 2013, was president in 1971–77 and again in 1998–2004 (http://www.ofmcap.org/en/istituzioni-centrali/museo-francescano). Our warm thanks to Yohannes Teklemariam Bache, president of the Institute from 2014 to 2017 and current director of the Museo Francescano, for providing citations for the history of the institution.

8. Quoted from the tribute by Cecilia Pantu (University of Rome) and Pietro B. Rossi (University of Turin), which appeared as "In memoriam Servus Gieben, OFM Cap.," *Bulletin de philosophie médiévale* 56 (2014): 554–58, and is reproduced on the website of the International Robert Grosseteste Society, https://grossetestesociety.org/2015/12/05/in-memoriam-servus-gieben-ofm-cap/. Our thanks to Evelyn Lincoln for translation from the Italian.

9. A bibliography of Servus Gieben's publications is found in "Bibliografia de Servus Gieben," in *Verum, pulchrum et bonum: Miscellanea di studi offerti a Servus Gieben in occasione del suo 80th compleanno*, ed. Yohannes Teklemariam (Rome: Istituto Storico dei Cappuccini, 2006), 15–30.

10. Lydia Beerkens, managing director and senior conservator of Modern and Contemporary Art at Stichting Restauratie Atelier Limburg in Maastricht, helped us with this modest biography of Servus Gieben, her uncle. She added that Servus Gieben's father was a blacksmith in the village of Sint Anthonis. We have also depended on Vincenzo Criscuolo, ed., *Quarant'anni di servizio nell'Istituto Storico dei Cappuccini (1953–1993): Isidoro da Villapadierna, Mariano D'Alatri, e Servus Gieben* (Rome: Istituto Storico dei Cappuccini, 1993), 69–84.

11. Servus Gieben, "Christelijke iconografie: Problemen en oplossingen," *Desipientia: Zin & waan* 2, no. 3 (1995): 8–9; quote from p. 8. Our thanks to Daniëlle Slootjes, of Radboud University, for her translation from the Dutch and to historian Michael Maas for introducing us to Professor Slootjes. Much hangs on these sandaled feet, for this was the moment the automaton was identified as a Franciscan icon.

12. Gieben, "Christelijke iconografie," 8.

13. Gieben, "Christelijke iconografie," 8–9.

14. Our still-astonished thanks to Teresita Fernández, friend and colleague, who initiated the contact with Servus Gieben for Elizabeth in July 1999, through associates at the American Academy in Rome. See chap. 7, above. The feast day of Diego is November 13 on the liturgical calendar, but Franciscans have traditionally observed it on November 12, the anniversary of Diego's death.

15. We are grateful to Servus Gieben for his permission to quote or paraphrase excerpts from this and subsequent letters. His references to sources were important to our work, in particular, the materials in the Franciscan index *Archivo Ibero-Americano*. He also sent us excellent photographs of the two engravings of San Diego that he had forwarded to Sedlmajer. A note on the engravings: Gieben's letter to Sedlmajer had given a date of 1588 for one of them, the other a little later (see chap. 6, above). To Elizabeth's request for confirmation of dates he wrote on December 31, 1999:

> John Sadeler's engraving of Diego d'Alcala (II-C-2/9) must be dated between 1595 and 1600. [Between] the first year he settled in Venice and [his death] during the pestilence of 1600. The engraving by Cornelis Galle (IV-G-9) first appeared in the picture-book by Hendrik Sedulius, "Imagines seraphicae santissimi P. Francisci Assisiatis, illustriumque virorum et feminarum, qui ex tribus Ordinibus relati sunt inter sanctos…, Antverpiae 1614." The edition of our Print Collection was issued from the same copperplates, by John Galle, Antwerp 1653.

Cornelius (or Cornelis) Galle (1576–1650) belonged to a renowned printing and publishing dynasty in Antwerp with his brother Theodore (1571–1633), sons of Philips Galle (1537–1612), whose engraving of Stradanus's clock shop we show in FIG. 21.4. John (or Joannes) Galle was Theodore's son.

16. It is unknown if Sedlmajer himself had yet made any connection to the Vienna automaton or to Bassermann-Jordan's work when he wrote to Gieben in summer 1975. The Zurich auction—the monk's first public appearance (see chap. 5 n. 5, above)—came a little later that year, on September 29, 1975, before Gieben's return letter. In that auction catalog (*Uto Auktionen: 2.Auktion, 29. September bis 3. October 1975*, Uto Auktions AG, Zurich, item no. 351, p. 40), the monk is described only as follows:

> Android Figure of a Monk
> South Germany, 16th century
> Driven by an iron mechanism. Walks in a triangle. Moves both hands, eyes, head, and opens its mouth. A similar piece from the same period is in the Historisches Museum in Vienna. Described throughout as the oldest such android in the literature of von Bassaman-Jordan [*sic*], *Alte Uhren und Automaten.*

(We assume the cited Bassermann-Jordan book title refers to his *Alte Uhren und ihre Meister*.) The Geneva auction the next year, June 21, 1976, cites Servus Gieben's work, as we've seen, transcribing verbatim Gieben's citation of "Lucio M. Núñez in Archivo Ibero-Americano 2" on Don Carlos's illness and cure. José A. García-Diego wrote in his book on Juanelo that he saw the monk for the first time in summer 1975. We do not know at what point Georges Sedlmajer was brought in to handle the transaction. He was certainly connected to Edgar Mannheimer by the time of the Geneva auction and may have been involved in the Zurich auction as well (see chap. 14 n. 1, above). When Elizabeth asked Father Servus if she might publish excerpts from his letters to her, he wrote (on January 16, 2000), "I have no problem that you quote from my letters. Just remember that when I first told my story of the mechanical monk to Mr. Sedlmajer, I had no idea at all of… the consequences of that letter. Nor did he ever tell me which use he actually made of my findings."

17. Bassermann-Jordan, *Alte Uhren und ihre Meister*, 59.

18. Stirling-Maxwell, *The Cloister Life of the Emperor Charles V*, 391.

19. Chapuis and Gélis, *Le monde des automates*, 2:182. Brusa, "Giovanni Torriano e i suoi favolosi orologi microcosmici," 6, also mentions the highly skilled Jean Valin (1530/35–1598), a Flanders-born clockmaker and close associate of Juanelo. Distinguished historian of clocks Luis Montañés Fontenla (1926–2011) discusses Juanelo Turriano and Jorge de Diano in *Capítulos de la relojería en España* (Madrid: Roberto Carbonell Blasco, 1954), 103–6, and also names Hans de Evalo and Juan Cocart (or Cocquart) of Brussels as clockmakers for Philip II. A clock by Hans de Evalo survives and is now in the Escorial: Montañés Fontenla tells the story of its discovery in *El escape y el péndulo* [The escapement and the pendulum] (Madrid: Ediciones Antiqvaria, 1991), 11–22. It was Montañés Fontenla who supported and arranged for the publication of José A. García-Diego's first book on Juanelo, and wrote its forward.

20. Personal correspondence from Otto Mayr to Elizabeth King, March 13, 2000.

21. Personal correspondence from Servus Gieben to Elizabeth King, January 16, 2000.

22. FIG. 24.1, identified as a death mask of San Giacomo della Marca (1393–1476), is kept in the Convento S. Maria La Nova, Naples. Servus Gieben's example of such a mask. The photograph is from P. Daniele Capone, OFM, *Iconografia di S. Giacomo della Marca nell' ambiente napoletano lungo i secoli* (Naples: 1976), 38–39.

23. Personal correspondence from Servus Gieben to Elizabeth King, January 22, 2000.

CHAPTER 25

Miracles and Walking Sculptures

1. Freud's definition of the uncanny (*unheimlich*)—uncertainty as to whether a thing is, or is not, alive—is taken from Ernst Jentsch, "Zur Psychologie des Unheimlichen," *Psychiatrisch-Neurologische Wochenschrift* 8, no. 22 (August 25, 1906): 195–98; and 8, no. 23 (September 1, 1906): 203–5. Available online in English translation by Roy Sellars, "On the Psychology of the Uncanny." Federico García Lorca's immortal 1933–34 essay, "Play and Theory of the Duende," is found in *Deep Song and Other Prose*, ed. and trans. Christopher Maurer (New York: New Directions, 1980).

2. Burke, *The Historical Anthropology of Early Modern Italy*, 218. The quote, from chap. 15, "Rituals of Healing in Early Modern Italy," addresses conflicting medical practices that reveal beliefs about illness and the body. This chapter and Burke's chap. 5, "How to Be a Counter-Reformation Saint," have been essential to our study.

3. David Freedberg, *The Power of Images: Studies in the History and Theory of Response* (Chicago: University of Chicago Press, 1989), 431: "[When our study] is too strictly based on reclamation of context…[it] resolutely refuses to allow the integration of ourselves into the past, or—to put it less sentimentally—to allow the lessons we learn from our own responses to inform our judgments about the past."

4. Freedberg, *The Power of Images*, 28; original emphasis.

5. Freedberg, *The Power of Images*, 32.

6. Second Commandment, Exodus 20, KJV.

7. English translation from Władysław Tatarkiewicz, *History of Aesthetics*, vol. 2: *Medieval Aesthetics* (The Hague: Mouton, 1970), 105.

8. The quote is from Baxandall, *The Limewood Sculptors*, 53, summarizing this passage attributed to Nicholas of Lyra: "If it is said that under the Old Law God had no image and that this was a sign for the New Law, the answer is that, if God had not afterwards become Man, then indeed he should not have had images."

9. For this distinction, and an analysis of the effect of the Counter-Reformation on art in Spain, see Marjorie Trusted, *The Arts of Spain: Iberia and Latin America 1450–1700* (University Park: Pennsylvania State University Press, 2007), 25 ff. Trusted quotes the prolix language of the 1563 Council of Trent:

> The images of Christ, of the Virgin Mother of God, and of the other saints are to be placed and retained especially in the churches, and that due honour and veneration is to be given them; not, however, that any divinity or virtue is believed to be in them by reason of which they are to be venerated, or that something is to be asked of them, or that trust is to be placed in images, as was done of old by the Gentiles who placed their hope in idols; but because the honour which is shown them is referred to the prototypes which they represent, so that by means of the images which we kiss and before which we uncover the head and prostrate ourselves, we adore Christ and venerate the saints whose likeness they bear.

(Her source: 25th Session of the Council of Trent, from H. J. Schroeder, trans., *Canons and Decrees of the Council of Trent* [Rockford, IL: Tan Books, 1978], 215–16.)

10. Pedro de Ribadeneira, *Flos sanctorum, libro de las vidas de los Santos* (Madrid: Luis Sanchez, [1599–1601] 1616), 708–9. An English translation was published in 1669: *The Lives of Saints with Other Feasts of the Year According to the Roman Calendar*, trans. W. Petre (St Omers: Ioachim Carlier, 1669), 769. Pedro de Ribadeneira (or Ribadeneyra) (1527–1611) was a Spanish chronicler, a Jesuit priest, and the first biographer of Ignatius of Loyola.

11. Our thanks to Susan Verdi Webster, who advises caution about current-day interpretations of reports of apparitions in the sixteenth century (phone conversation, March 8, 2022). Her warning echoes that of Peter Burke: we are not to trust our modern distinctions between natural and supernatural. Webster speaks of "the lived, real, sensory experience" of those who witnessed visions. The very odor of the crypt is mentioned by Ribadeneira's source, Marcos de Lisboa: "And when the doors were opened, an odor of such sweetness came out that we could not bear its fragrance."

12. Ford, quoted in *The Sacred Made Real: Spanish Painting and Sculpture 1600–1700*, exh. cat., ed. Xavier Bray (London: National Gallery, 2009), 173. For Gregorio Fernández and copies of his and Mena's sculptures of Saint Francis, see Trusted, *The Arts of Spain*, 54; and Marjorie Trusted, "Art for the Masses: Spanish Sculpture in the Sixteenth and Seventeenth Centuries," in *Sculpture and Its Reproductions*, ed. Anthony Hughes and Erich Ranfft (London: Reaktion Books, 1997), 57–60. For Antonio de Montúfar, see http://collections.lacma.org/node/215693. Montúfar's painting, *Saint Francis of Assisi Appearing before Pope Nicholas V, with Donors* (La aparición de san Francisco de Asís al Papa Nicolás V, con donantes), dated 1628, is in the collection of the Los Angeles County Museum of Art.

13. See note 12, above. Curated by Xavier Bray at the National Gallery in London, the show opened in London in 2009 and traveled the next year to the National Gallery of Art in Washington, DC.

14. Elizabeth King, "Notes from the Field: Anthropomorphism," *Art Bulletin* 94, no.1 (March 2012): 9–12. Later published as "Inhale, Exhale, Pause: Breath and the Open Mouth in Sculpture," in *Field Notes on the Visual Arts*, ed. Karen Lang (Bristol: Intellect, 2019), 2–7.

15. For the Mena: José Luis Romero Torres and Gonzalo Otalecu Guerrero, eds., *Pedro de Mena: Granatensis Malacae*, exh. cat. (Málaga: Palacio Episcopal, 2019), 282–84. For the Zurbarán (painted in 1640): Joan Yeguas, in *The Spanish Golden Age: Painting and Sculpture in the Time of Velázquez*, exh. cat., ed. Roberto Contini and María López-Fanjul y Díez del Corral (Munich: Hirmer Verlag, 2016), 214.

16. Marcos de Lisboa, *Primera parte de las crónicas de la orden de los frayles menores* (Alcalá: Casa de Andres de Angulo, 1562), bk. 10, chap. 1 (fol. CCXXIX). Fray Marcos wrote that the cardinal related the story only on his deathbed. Grateful thanks to Susan Verdi Webster for her translation of Fray Marcos's narrative for us from the original sixteenth-century Spanish text. Pedro de Ribadeneira cites Fray Marcos's *crónicas* as the source for the cardinal's report, but in his own narration Ribadeneira refers not to Pope Nicholas V (r. 1447–55) but to Nicholas IV (a Franciscan, r. 1288–92). Trusted has also noted this discrepancy but cites modern sources naming Nicholas V ("Art for the Masses," 57 and n. 44). The *crónicas* text gives only the name *Nicolao* for the pope. For more on the legend, see Émile Mâle, *L'art religieux de la fin du XVIe siècle, du XVIIe siècle et du XVIIIe siècle, étude sur l'iconographie après le Concile de Trente: Italie, France, Espagne, Flandres* (Paris: Librairie Armand Colin, 1972), 480–83.

17. Trusted, "Art for the Masses," 57–58: "The first sculpted interpretations of the subject seem to have been produced by Gregorio Fernández; one, which had been dated to not later than 1620, is in the Royal Convent of the Discalced Nuns (Monasterio de las Descalzas Reales) in Valladolid." See Luis Luna Moreno, ed., *Pedro de Mena y Castilla*, exh. cat. (Valladolid: Museo Nacional de Escultura, 1989), 22–23; and Juan José Martín González, *El escultor Gregorio Fernández* (Madrid: Ministerio de Cultura, 1980), 249.

18. José Luis Romero Torres in Andreas Pampoulides, ed., *Pedro de Mena: The Spanish Bernini*, exh. cat. (London: Coll & Cortés, 2014), 165–66. Cano and Mena together made *San Diego de Alcalá* (1657–58), commissioned by the convent of Santo Ángel Custodio in Granada, among a group of four saints "larger than life size," now in the Museo de Bellas Artes, Granada. Other *Diegos* known or attributed to Mena, according to Romero Torres, include figures in the Iglesia de Capuchinas de San Antonio Abad, Granada; the Museo Diocesano de Arte Sacro de Vitoria; and the San Diego Museum of Art (see FIG. 25.2). Here, too, as with the *Saint Francis*, Gregorio Fernández may have been the first sculptor to depict the saint: a 1605 *San Diego de Alcalá* now in the Museo Nacional de Escultura, Valladolid, is attributed to Fernández (Martín González, *El escultor Gregorio Fernández*, 251).

19. Antonio Rojo, *Historia de San Diego de Alcalá: Fundación y frutos de santidad que ha produzido su convento de Santa Maria de Iesús [...]* (Madrid: Imprenta Real, 1663), 130–31. Andrew Villalon mentions that witnesses in the monastery reported that Diego had never until then been heard to speak Latin (Villalon, "The Miracle Book of Diego de Alcalá," 22). An early image of Diego embracing the cross is the 1594 Assumption retablo by Alonso Vázquez in the Cathedral of Seville, depicting the saint as a witness (Romero Torres, "Iconografía y devoción regia," 103–4, 106).

20. John Marciari, *Italian, Spanish, and French Paintings before 1850 in the San Diego Museum of Art* (San Diego: San Diego Museum of Art, 2015), 325: "Although Diego was in his late fifties or sixties when the Miracle of the Roses occurred, he is usually shown in Spanish art as a younger man, perhaps to emphasize the sweet simplicity of his charitable act."

21. Romero Torres, in *Pedro de Mena: The Spanish Bernini* (2014), 158–69. We wish to thank José Luis Romero Torres, Conservador del Patrimonio Histórico, Real Academia de Bellas Artes de San Telmo de Málaga y Real Academia de Nobles Artes de Antequera, for graciously responding to our questions about this sculpture, and more broadly about early depictions of Diego de Alcalá in works of art, in email correspondence between 2020 and 2022.

22. Romero Torres, in *Pedro de Mena: The Spanish Bernini* (2014), 166.

23. Romero Torres, "Iconografía y devoción regia," 101–2.

24. The earliest image we have found of Diego, an anonymous engraving, appears in a copy of the documents Philip II submitted to Sixtus V in Rome in pursuit of the canonization. It is seen in Francisco de la Peña's *De vita miraculis et actis canonizationis sancti Didaci libri tres* (Rome: In Aedibus Populi Romani, 1589), published the year following canonization. On death masks and death portraits of saints, see Fernando Quiles, "Between Being, Seeming and Saying: The *Vera Effigies* in Spain and Hispanic America during the Baroque," trans. Anna Moorby, in *Fiction sacrée: Spiritualité et esthétique durant le premier âge moderne*, ed. Ralph Dekoninck, Agnès Guiderdoni, and Emilie Granjon (Leuven: Peeters, 2013), 181–200.

25. Susan Verdi Webster provides this quote in *Art and Ritual in Golden-Age Spain: Sevillian Confraternities and the Processional Sculpture of Holy Week* (Princeton, NJ: Princeton University Press, 1998), 181. Her source is Juan de la Cruz, *Subida del Monte Carmelo* (1579), in *Biblioteca de autores españoles* (Madrid: M. Rivadeneyra, 1853), 27:92.

26. Trusted, *The Arts of Spain*, shows how the Catholic processional tradition itself originated in part in the carrying of the Host through the streets on the annual feast of Corpus Christi in the thirteenth century, as confirmation of the body and blood of Christ (41–47). See also Marjorie Trusted, "Moving Church Monuments: Processional Images in Spain in the Seventeenth Century," *Church Monuments: Journal of the Church Monuments Society* 10 (1995): 56–58.

27. See Patrick Lenaghan, *Images in Procession: Testimonies to Spanish Faith* (New York: American Bible Society and Hispanic Society of America, 2000).

28. Trusted, "Moving Church Monuments," 62–64, writes of one confraternity, "Its procession on Maundy Thursday in the seventeenth century was a solemn procession of discipline, comprising 650 brothers of light (holding torches), dressed in black, and 1,500 brothers of blood, who took part in the procession 'in the trance of flagellation', beating themselves with knotted cords until the blood flowed."

29. Webster, *Art and Ritual in Golden-Age Spain*, 167. Our description of Spanish processional sculpture is drawn from Webster's portrait of the processional culture in Seville, with its focus on the traditions and practices connected to Holy Week. See also Webster's essay "Shameless Beauty and Worldly Splendor: On the Spanish Practice of Adorning the Virgin," in *The Miraculous Image in the Late Middle Ages and Renaissance*, ed. Erik Thunø and Gerhard Wolf (Rome: L'Erma di Bretschneider, 2004), 249–71.

30. Gervase Rosser gives a vivid portrait of the origins and social structure of confraternities in his essay "Crafts, Guilds and the Negotiation of Work in the Medieval Town." See chap. 19 n. 2, above.

31. Webster, *Art and Ritual in Golden-Age Spain*, 133.

32. Lenaghan, *Images in Procession*, 19–20; and Suzanne L. Stratton, ed., *Spanish Polychrome Sculpture 1500–1800 in United States Collections*, exh. cat. (New York: Spanish Institute, 1993), 175.

33. Webster, *Art and Ritual in Golden-Age Spain*, 64.

34. Trusted, *The Arts of Spain*, 47–48.

35. Trusted, *The Arts of Spain*, 59 n. 172. She quotes Carl Justi in his introduction to K. Baedeker, *Spain and Portugal: Handbook for Travellers* (Leipzig, 1913), lxxi. Richard Ford, quoted above on Mena's *Saint Francis*, nonetheless joined other foreigners in their objections to putting paint on sculpture: "From being clothed and painted they are failures as works of art, strictly speaking, for they attempt too much. The essence of statuary is *form*, and to clothe a statue, said Byron, is like translating Dante: a *marble* statue never deceives; it is the colouring it that does, and is a trick beneath the severity of sculpture. The imitation of life may surprise, but... it can only please the ignorant and children" (R. Ford, *A Handbook for Travellers in Spain* [1845], ed. I. Robertson, 3 vols. [London, 1966], 1:169). Cited by Marjorie Trusted, *Spanish Sculpture: Catalogue of the Post-Medieval Spanish Sculpture in Wood, Terracotta, Alabaster, Marble, Stone, Lead and Jet in the Victoria and Albert Museum* (London: Victoria and Albert Museum, 1996), 1, in her chapter "'Failures as Works of Art'? British Attitudes to Spanish Sculpture and the Formation of the Collection." Trusted's revealing discussion of the battle between religious and aesthetic responses to the art of Spain refers to the "almost moral disapproval" of the aesthete.

36. Marcel Dieulafoy (1844–1920) from his 1908 *La statuaire polychrome en Espagne* (Paris), as quoted by Gridley McKim-Smith, "Spanish Polychrome Sculpture and Its Critical Misfortunes," in Stratton, *Spanish Polychrome Sculpture*, 13. McKim-Smith's essay is a fierce cross-examination of art history's repeated failures to recognize the genre.

37. Baxandall, *The Limewood Sculptors*, 151 ff.

38. Baxandall, *The Limewood Sculptors*, chap. 3; for this characterization of Riemenschneider, see 92, 172–90; and for Stoss, 191–202. But see Johannes Taubert, *Polychrome Sculpture: Meaning, Form, Conservation* (Los Angeles: Getty Conservation Institute, 2015), a superb new English translation of Taubert's now-classic 1978 study, *Farbige Skulpturen*. And see chap. 19 and n. 5, above, for the work of Johan Gregor van der Schardt and a glimpse of polychrome portrait commissions in Nuremberg. A keen analysis of the centuries-old prejudice against putting color on sculpture is Luke Syson, "Polychrome and Its Discontents: A History," in *Like Life: Sculpture, Color, and the Body*, exh. cat., ed. Luke Syson et al. (New York: Metropolitan Museum of Art, 2018), 14–41.

39. "San Ignacio Loyola advised active participation of all the senses in meditating on the torments of Hell": Webster, *Art and Ritual in Golden-Age Spain*, 177, quoting Palma Martínez-Burgos García, *Ídolos e imágenes: La controversia del arte religioso en el siglo XVI español* (Valladolid: Universidad de Valladolid, 1990), 166–67. And see n. 11, above.

40. Webster, *Art and Ritual in Golden-Age Spain*, 163 and chap. 5, "Public Response to the Sculptures," 164 ff.

41. Freedberg, *The Power of Images*, 30. A very different Passion cycle, in which the viewer walks and the sculpture "watches," is the famous *Sacro Monte* in Varallo in Italy's Piedmont, with its trompe l'oeil frescoes merging with sculptures in wood and terracotta by Gaudenzio Ferrari (ca. 1470–1546). Now a UNESCO World Heritage Site, it was finished in the first half of the sixteenth century: a series of forty-five small chapels replicate holy sites from Eden to Calvary for the pilgrim to discover by walking a winding path up the mountain toward its summit. See Fredrika H. Jacobs, *The Living Image in Renaissance Art* (Cambridge: Cambridge University Press, 2005), 16–20.

42. Webster, *Art and Ritual in Golden-Age Spain*, 79.

43. Koeppe, *Making Marvels*, 256–57.

44. For the film, see https://www.metmuseum.org/exhibitions/listings/2019/making-marvels-science-splendor/art-in-motion

CHAPTER 26

Perpetual Devotion

1. Dutch-born sculptors Jean de Marville (active 1366–1389) and Claus Sluter (ca. 1360–1406) began the massive tomb monument for Philip the Bold: Marville in 1381, Sluter in 1389. Sluter's nephew Claus de Werve (? -1439) took over and completed the monument in 1410, six years after the duke's death. By that time, the duke's son John the Fearless had commissioned a second monument for himself and his wife, Duchess Margaret of Bavaria, finished half a century after John's own death. The Spanish sculptor Juan de la Huerta (active 1431–1462) began the work in 1443, and the French sculptor Antoine Le Moiturier (1425–1495) finished it in 1470.

2. The "eye of northern Europe," Hugh Trevor-Roper's phrase, *Princes and Artists*, 127. Sophie Jugie writes, "For the Hapsburgs, the descendants of the Dukes of Burgundy, the dynastic value of the Chartreuse de Champmol was undiminished: in his first will in 1522, Holy Roman Emperor Charles V stated his intention to be buried in Champmol, but ultimately his remains were transferred to El Escorial near Madrid." Sophie Jugie, *The Mourners: Tomb Sculptures from the Court of Burgundy*, exh. cat. (Dijon: Musée des Beaux-Arts and FRAME, French/Regional/American Museum Exchange, in association with Yale University Press, 2010), 35.

3. The tombs were dismantled and partly destroyed in 1793 in the French Revolution, the effigies of dukes and duchess, hated symbols of the ancien régime, obliterated (Jugie, *The Mourners*, 35). The mourners fared better: "Most of the mourners have never been outside of Dijon, but in the chaos following their dismantling in 1793, three of them vanished, while eight others went into private hands. Of these latter, three later returned to Dijon through national museum loans, four belong to the Cleveland Museum of Art, and one is still in a private collection" (47). Although the monuments are called tombs, they are in fact cenotaphs: memorial tombs. The bodies of the dukes were interred not within the monuments but in coffins beneath the church chancel. As with the monuments, these too suffered destruction. The Cathedral of St Bénigne in Dijon today preserves what are considered only partial remains.

4. *The Mourners: Tomb Sculptures from the Court of Burgundy*, Sophie Jugie, curator and director, Musée des Beaux-Arts, Dijon. The exhibition opened at the Metropolitan Museum of Art in New York in March 2010 and traveled to the Saint Louis Art Museum, Dallas Museum of Art, Minneapolis Institute of Art, Los Angeles County Museum of Art, Fine Arts Museums of San Francisco, Virginia Museum of Fine Arts, and Musée national du Moyen Âge-Musée de Cluny in Paris.

5. Anthony Grafton, "In a Fantastic, Lost World," *New York Review of Books*, May 13, 2010, 8–10.

6. William A. Christian Jr., *Local Religion in Sixteenth-Century Spain* (Princeton, NJ: Princeton University Press, 1981). This work is a panoramic survey of the religious practices of lay communities in New Castile under the reign of Philip II.

7. Philip the Good (1396–1467), who made certain that the tomb monument for his father, John the Fearless, was brought to completion, in his own reign renovated the theatrical grottoes and automata of Hesdin, resplendent under earlier counts and dukes. See chap. 17 n. 16, above.

8. Webster's work is invaluable for its portrait of popular responses to images of faith outside direct church rule, for the processions brought every class of citizen to the street. She has especially acknowledged the study of the practices of local lay communities across sixteenth-century Spain in the work of William A. Christian Jr. Peter Burke's *Historical Anthropology of Early Modern Italy*, with its intimate portraits of social subgroups from the full "urban mosaic" of a given population in the past (not just its rulers, ambassadors, and bishops), gives us a chance to imagine less documented kinds of settings for an encounter with the monk.

CHAPTER 27

The Machine and the Corpse

1. Vesalius at the prince's bedside: Villalon, "Putting Don Carlos Together Again," 352 ff.; Skaarup, *Anatomy and Anatomists in Early Modern Spain*, 129–37; and O'Malley, *Andreas Vesalius*, 296–302. Vesalius performed the postmortem on Henry II in France, injured in a jousting accident (O'Mally, *Andreas Vesalius*, 283–88).

2. Villalon, "Putting Don Carlos Together Again," 353, drawn from Daza Chacón.

3. The words of the witness Alonso de Mendoza, theologian and professor at the University in Alcalá, transcribed and reproduced in Núñez, "Documentos," 1, no. 4 (July-August 1914): 431–32.

4. Núñez, "Documentos," 1, no. 4 (July-August 1914): 439–40.

5. *De humani corporis fabrica*, published in Basel in 1543, as groundbreaking for the history of woodblock printing as it was for the history of medicine, corrected the ancient Galenic theory of the humors with direct observation of the body in dissection. Vesalius published an abbreviated version the same year, the *Epitome*, and dedicated it to the young Philip II, then a boy of sixteen.

6. Núñez, "Documentos," 3, no. 13 (January-February 1916): 119–22. Philip amassed holy relics at the Escorial in Spain as passionately as his nephew Rudolf amassed artworks and machines in his Prague *Kunstkammer*. Marjorie Trusted counts them: "Philip II collected 7,422 relics between 1571 and his death in 1598, all of them authenticated and labelled" (*The Arts of Spain*, 50).

In 2005 Elizabeth visited the Iglesia Catedral Magistral in Alcalá de Henares to view the tomb of San Diego. A young docent, a student at the university, mentioned that on November 13 of each year, the current feast day of the saint in Spain, the tomb is opened, to display the body. Elizabeth commented on the very small size of the saint's tomb. "Yes," the docent said, "it is missing its lower legs."

Selected Bibliography

Bassermann-Jordan, Ernst von. *Alte Uhren und ihre Meister* [Historical clocks and their masters]. Leipzig: Wilhelm Diebener, 1926.

Battisti, Eugenio. "For Some Time Now, in Order to Write This Wretched Article, I Have Been Thinking About Automatons." *Artforum* 21, no. 9 (May 1983): 54-61.

Bauer, Rotraud, and Herbert Haupt, eds. "Das Kunstkammerinventar Kaiser Rudolfs II., 1607–1611." *Jahrbuch der Kunsthistorischen Sammlungen in Wien* 72 (1976): XI-XLV, 1–191.

Baxandall,Michael. *The Limewood Sculptors of Renaissance Germany*. New Haven, CT: Yale University Press, 1980.

Beaune, Jean-Claude. "The Classical Age of Automata: An Impressionistic Survey from the Sixteenth to the Nineteenth Century." In *Fragments for a History of the Human Body, Part One*, edited by Michel Feher, with Ramona Naddaff and Nadia Tazi, 430–80. New York: Zone, 1989.

Bedini, Silvio A. "The Role of Automata in the History of Technology." *Technology and Culture* 5, no. 1 (1964): 24-42.

Bedini, Silvio A., and Francis R. Maddison. "Mechanical Universe: The Astrarium of Giovanni de' Dondi." *Transactions of the American Philosophical Society*, n.s., 56, pt. 5 (October 1966): 1–69.

Bray, Xavier, ed. *The Sacred Made Real: Spanish Painting and Sculpture 1600–1700*. Exh. cat. London: National Gallery, 2009.

Bray, Xavier, and José Luis Romero Torres. *Pedro de Mena: The Spanish Bernini*. Exh. cat. Luxembourg City: Musée national d'histoire et d'art Luxembourg, Coll & Cortés, Hannibal, 2019. Revised edition of Andreas Pampoulides, ed., *Pedro de Mena: The Spanish Bernini*. Exh. cat. London: Coll & Cortés, 2014.

Bredekamp, Horst. *The Lure of Antiquity and the Cult of the Machine: The Kunstkammer and the Evolution of Nature, Art and Technology*. Translated by Allison Brown. Princeton, NJ: Markus Wiener, 1995.

Brown, Jonathan. *Kings and Connoisseurs: Collecting Art in Seventeenth-Century Europe*. A. W. Mellon Lectures in the Fine Arts, 1994, National Gallery of Art, Washington, DC. Princeton, NJ: Princeton University Press, 1995.

Brown, Peter. *The Cult of the Saints: Its Rise and Function in Latin Christianity*. Chicago: University of Chicago Press, 1981.

Brusa, Giuseppe. "Giovanni Torriano e i suoi favolosi orologi microcosmici." *La voce di hora: Pubblicazione dell'Associazione Italiana Cultori di Orologeria Antica*, no. 15 (December 2003): 3–22.

Bukovinská, Beket. "The Known and Unknown *Kunstkammer* of Rudolf II." In *Collection—Laboratory—Theater: Scenes of Knowledge in the 17th Century*, edited by Helmar Schramm et al. Berlin: Walter de Gruyter, 2005. (Originally published in German, 2003, for the series Theatrum Scientiarum.)

Burke, Peter. *The Historical Anthropology of Early Modern Italy: Essays on Perception and Communication*. Cambridge: Cambridge University Press, 1987.

Bussmann, Klaus, and Heinz Schilling, eds. *1648: War and Peace in Europe*. Exh. cat., 3 vols. Münster: Westfälisches Landesmuseum für Kunst und Kulturgeschichte, 1998.

Case, Thomas E. *La historia de San Diego de Alcalá: Su vida, su canonización y su legado / The Story of San Diego de Alcalá: His Life, His Canonization and His Legacy*. Alcalá de Henares: Universidad de Alcalá, 1998. Bilingual edition.

Cervantes, Miguel de. *Don Quixote*. Translated by Edith Grossman. New York: HarperCollins, 2003.

Chapuis, Alfred, and Édouard Gélis. *Le monde des automates: Étude historique et technique*. 2 vols. Paris, 1928. Reprint Geneva: Éditions Slatkine, 1984.

Christian, William A., Jr. *Apparitions in Late Medieval and Renaissance Spain*. Princeton, NJ: Princeton University Press, 1981.

———. *Local Religion in Sixteenth-Century Spain*. Princeton, NJ: Princeton University Press, 1981.

Cohen, John. *Human Robots in Myth and Science*. London: George Allen & Unwin, 1966.

Daston, Lorraine, and Katharine Park. *Wonders and the Order of Nature 1150–1750*. New York: Zone, 1998.

Evans, R. J. W. *Rudolf II and His World: A Study in Intellectual History 1576–1612*. Oxford: Oxford University Press, 1973.

Evans, R. J. W., and Alexander Marr, eds. *Curiosity and Wonder from the Renaissance to the Enlightenment*. Aldershot: Ashgate, 2006.

Fane, Lawrence. *M.T./L.F.: A Sculptor's Dialogue with Mariano Taccola, Fifteenth-Century Italian Artist-Engineer*. Lake Placid, NY: Pont La Vue Press, 2006.

Findlen, Paula. *Possessing Nature: Museums, Collecting, and Scientific Culture in Early Modern Italy*. Berkeley: University of California Press, 1994.

Freedberg, David. *The Power of Images: Studies in the History and Theory of Response*. Chicago: University of Chicago Press, 1989.

Fritsch, Julia, ed. *Ces curieux navires: Trois automates de la Renaissance* [Ships of curiosity: Three Renaissance automata]. Paris: Éditions de la Réunion des musées nationaux, 1999.

Fučíková, Eliška, James M. Bradburne, Beket Bukovinská, Jaroslava Hausenblasová, Lumomír Konečný, Ivan Muchka, and Michal Šroněk, eds. *Rudolf II and Prague: The Court and the City*. Prague: Prague Castle Administration, London: Thames & Hudson, 1997.

Gachard, Louis-Prosper. *Don Carlos et Philippe II*. 2 vols. Brussels: E. Devroye, imprimeur du roi, 1863.

García-Diego, José A. *Juanelo Turriano, Charles V's Clockmaker: The Man and His Legend*. Translated by Charles David Ley. Madrid and Sussex: Editorial Castalia and Antiquarian Horological Society, 1986.

———. *Los relojes y autómatas de Juanelo Turriano*. Tempvs Fvgit, Monografías Españolas de Relojería, vol. 6. Madrid: Albatros Ediciones, 1982.

García Lorca, Federico. *Deep Song and Other Prose*. Edited and translated by Christopher Maurer. New York: New Directions, 1980.

Garnett, Jane, and Gervase Rosser. *Spectacular Miracles: Transforming Images in Italy from the Renaissance to the Present*. London: Reaktion Books, 2013.

Gieben, Servus. "Christelijke iconografie: Problemen en oplossingen" [Christian iconography: Problems and solutions]. *Desipientia: Zin & waan* 2, no. 3 (December 1995): 4–11.

Gila Medina, Lázaro. *Pedro de Mena, escultor 1628–1688*. Madrid: Editorial Arco/Libros, 2007.

Gouk, Penelope. *The Ivory Sundials of Nuremberg 1500–1700*. Cambridge: Whipple Museum of the History of Science, 1988.

Gross, Kenneth, ed. *On Dolls*. London: Notting Hill, 2018.

———. *Puppet: An Essay on Uncanny Life*. Chicago: University of Chicago Press, 2011.

Grubmüller, Klaus, and Markus Stock, eds. *Automaten in Kunst und Literatur des Mittelalters und der Frühen Neuzeit* [Automata in medieval and early modern art and literature]. Wolfenbütteler Mittelalter-Studien, vol. 17. Wiesbaden: Harrassowitz Verlag, 2003.

Hall-van den Elsen, Catherine. *Luisa Roldán*. Los Angeles: Getty Publications, 2021.

Haspels, J. J. L., ed. *Royal Music Machines*. Exh. cat. Zutphen, the Netherlands: Walburg Pers, 2006.

Hawthorne, Nathaniel. "The Artist of the Beautiful." In *Mosses from an Old Manse*, 164–91. London: Wiley & Putnam, 1846.

Hoffmann, E. T. A. "The Sandman." In *Tales of Hoffmann*, 85–125. Translated by R. J. Hollingdale. London: Penguin, 1982.

Impey, Oliver, and Arthur MacGregor, eds. *The Origins of Museums: The Cabinet of Curiosities in Sixteenth- and Seventeenth-Century Europe*. Oxford: Clarendon Press, 1985.

Jacobs, Fredrika H. *The Living Image in Renaissance Art*. Cambridge: Cambridge University Press, 2005.

Jugie, Sophie. *The Mourners: Tomb Sculptures from the Court of Burgundy*. Exh. cat. Dijon: Musée des Beaux-Arts and FRAME, French/Regional/American Museum Exchange, in association with Yale University Press, 2010.

Kamen, Henry. *Philip of Spain*. New Haven, CT: Yale University Press, 1997.

Kang, Minsoo. *Sublime Dreams of Living Machines: The Automaton in the European Imagination*. Cambridge, MA: Harvard University Press, 2011.

Kaufmann, Thomas DaCosta. "Remarks on the Collections of Rudolf II: The *Kunstkammer* as a Form of *Representatio*." *Art Journal* 38, no. 1 (Autumn 1978): 22–28.

Keating, Jessica. *Animating Empire: Automata, the Holy Roman Empire, and the Early Modern World*. University Park: Pennsylvania State University Press, 2018.

King, Elizabeth. "Clockwork Prayer: A Sixteenth-Century Mechanical Monk." *Blackbird: An Online Journal of Literature and the Arts* 1, no. 1 (Spring 2002). https://blackbird.vcu.edu/v1n1/nonfiction/king_e/prayer_introduction.htm.

———. "Clockwork Prayer: A Sixteenth-Century Mechanical Monk." In *Breaking the Disciplines: Reconceptions in Knowledge, Art and Culture*, edited by Martin L. Davies and Marsha Meskimmon, 84–128. London: I. B. Tauris, 2003.

———. "Inhale, Exhale, Pause: Breath and the Open Mouth in Sculpture." In *Field Notes on the Visual Arts*, edited by Karen Lang, 2–7. Bristol: Intellect, 2019.

———. "Perpetual Devotion: A Sixteenth-Century Machine That Prays." In *Genesis Redux: Essays in the History and Philosophy of Artificial Life*, edited by Jessica Riskin, 263–92. Chicago: University of Chicago Press, 2007.

Kleist, Heinrich von. "On the Marionette Theater," translated by Roman Paska, and "The Inanimate Incarnate." In *Fragments for a History of the Human Body, Part One*, edited by Michel Feher, with Ramona Naddaff and Nadia Tazi, 410–29. New York: Zone, 1989.

Koeppe, Wolfram, ed. *Making Marvels: Science and Splendor at the Courts of Europe*. Exh. cat. New York: Metropolitan Museum of Art, 2019.

Kowar, Helmut. "'Wir machen Musik!': Trompeterautomat, Schiff und Bacchuswagen des Hans Schlotheim" ["We make music!": The Trumpeter automaton, the ship, and the Bacchus Chariot by Hans Schlotheim]. *Jahrbuch des Kunsthistorischen Museums Wien* 13–14 (2011–12): 118–59.

Kubler, George. *The Shape of Time: Remarks on the History of Things*. New Haven, CT: Yale University Press, 1962.

Leithe-Jasper, Manfred, and Rudolf Distelberger. *The Kunsthistorische Museum Vienna: The Treasury and the Collection of Sculpture and Decorative Arts*. Translated by Elsie Callander and Ilse Seldon. London: Philip Wilson, 1982.

Lenaghan, Patrick. *Images in Procession: Testimonies to Spanish Faith*. New York: American Bible Society and Hispanic Society of America, 2000.

Long, Pamela O. *Artisan/Practitioners and the Rise of the New Sciences, 1400–1600*. Corvallis: Oregon State University Press, 2011.

Luna Moreno, Luis, ed. *Pedro de Mena y Castilla*. Exh. cat. Valladolid: Museo Nacional de Escultura, 1989.

Mâle, Émile. *L'art religieux de la fin du XVIe siècle, du XVIIe siècle et du XVIIIe siècle, étude sur l'iconographie après le Concile de Trent: Italie, France, Espagne, Flandres*. Paris: Librairie Armand Colin, 1972.

Martín González, Juan José. *El escultor Gregorio Fernández*. Madrid: Ministerio de Cultura, 1980.

Maurice, Klaus. *Die deutsche Räderuhr: Zur Kunst und Technik des mechanischen Zeitmessers im deutschen Sprachraum* [The German wheel clock: On the art and technology of the mechanical timepiece in the German-speaking world]. 2 vols. Munich: C. H. Beck, 1976.

———. *Von Uhren und Automaten, das Messen der Zeit* [Of clocks and automata, the measure of time]. Munich: Prestel, 1968.

Maurice, Klaus, and Otto Mayr, eds. *The Clockwork Universe: German Clocks and Automata, 1550–1650*. Exh. cat. New York: Neale Watson Academic Publications, for the Smithsonian Institution, 1980. German edition: *Die Welt als Uhr: Deutsche Uhren und Automaten, 1550–1650*. Munich: Deutscher Kunstverlag, for the Bayerisches Nationalmuseum, 1980.

Mayr, Otto. *Authority, Liberty, and Automatic Machinery in Early Modern Europe*. Baltimore: Johns Hopkins University Press, 1986.

———. "Automatenlegenden in der Spätrenaissance" [Legends of automata in the late Renaissance]. *Technikgeschichte: Beiträge zur Geschichte der Technik und Industrie* 41, no. 1 (1974): 20–32.

Melville, Herman. "The Belltower." *Putnam's Monthly Magazine*, August 1855. Reprinted in *The Piazza Tales*. New York: Dix, Edwards, and Co., 1856.

Montañés Fontenla, Luis. *Capítulos de la relojería en España* [Chapters on clockmaking in Spain]. Madrid: Roberto Carbonell Blasco, 1954.

——, ed. *Cuadernos de relojería: Una antología imprescindible* [Notes on clockmaking: An essential anthology]. Tempvs Fvgit, Monografías Españolas de Relojería, vol. 5. Madrid: Albatros Ediciones, 1982.

——. *Relojes españoles*. Madrid: Editorial Prensa Española, 1968.

Motture, Peta, Emma Jones, and Dimitrios Zikos, eds. *Carvings, Casts and Collectors: The Art of Renaissance Sculpture*. London: Victoria and Albert Museum, 2013.

Murakami Kazuo. *Japanese Automata:* Karakuri Zui; *An Eighteenth Century Japanese Manual of Automatic Mechanical Devices*. Osaka: Published by the author, 2012.

Nelson, Victoria. *The Secret Life of Puppets*. Cambridge, MA: Harvard University Press, 2001.

Neumann, Erwin. "Das Inventar der rudolfinischen Kunstkammer von 1607/11 [The inventory of the Rudolfian Kunstkammer from 1607/11]." In *Analecta Reginensia I, Queen Christina of Sweden, Documents and Studies*, 262–65. Stockholm, 1966.

Newman, William R. *Promethean Ambitions: Alchemy and the Quest to Perfect Nature*. Chicago: University of Chicago Press, 2004.

Núñez, Fr. Lucio María, OFM, ed. "Documentos sobre la curación del príncipe D. Carlos y la canonización de San Diego de Alcalá." *Archivo Ibero-Americano* 1, no. 4 (July–August 1914); 2, no. 9 (July–August 1915); 3, no. 13 (January–February 1916); 4, no. 19 (January–February 1917).

O'Malley, Charles Donald. *Andreas Vesalius of Brussels, 1514–1564*. Berkeley: University of California Press, 1965.

——. *Don Carlos of Spain: A Medical Portrait*. Los Angeles: University of California, 1970.

Parker, Geoffrey. *Imprudent King: A New Life of Philip II*. New Haven, CT: Yale University Press, 2014.

Plank, Robert. "The Golem and the Robot." *Literature and Psychology* 15 (1965): 12–28.

Prescott, William H. *History of the Reign of Philip the Second, King of Spain*. Edited by John Foster Kirk. 3 vols. Philadelphia: J. B. Lippincott, 1874.

Price, Derek J. de Solla. "Automata and the Origins of Mechanism and Mechanistic Philosophy." *Technology and Culture* 5, no. 1 (1964): 9–23.

Proske, Beatrice Gilman. *Juan Martínez Montañés, Sevillian Sculptor*. New York: Hispanic Society of America, 1967.

Rabb, Theodore K., trans. *A Sixteenth-Century Book of Trades: Das Ständebuch*. Palo Alto, CA: Society for the Promotion of Science and Scholarship, 2009.

Rihouet, Pascale. *Art Moves: The Material Culture of Processions in Renaissance Perugia*. London: Harvey Miller, 2017.

Riskin, Jessica, ed. *Genesis Redux: Essays in the History and Philosophy of Artificial Life*. Chicago: University of Chicago Press, 2007.

——. *The Restless Clock: A History of the Centuries-Long Argument over What Makes Living Things Tick*. Chicago: University of Chicago Press, 2016.

Romero Torres, José Luis. "Iconografía y devoción regia de un santo franciscano andaluz: San Diego de Alcalá" [Iconography and royal devotion of an Andalusian Franciscan Saint: San Diego de Alcalá]. In *Actas del Simposio Cuatro siglos de presencia de los franciscanos en Estepa, celebrado el 11–13 de septiembre de 2003*, 94–113. Estepa: Ayuntamiento, 2007.

Romero Torres, José Luis, and Gonzalo Otalecu Guerrero, eds. *Pedro de Mena: Granatensis Malacae*. Exh. cat. Málaga: Palacio Episcopal, 2019.

Sánchez de Mayendia, José Cristóbal. "Los retratos de Juanelo." In *Cuadernos de relojería: Una antología imprescindible*, edited by Luis Montañés Fontenla, 123–26. Tempvs Fvgit, Monografías Españolas de Relojería, vol. 5. Madrid: Albatros Ediciones, 1982.

Sawday, Jonathan. *Engines of the Imagination: Renaissance Culture and the Rise of the Machine*. London: Routledge, 2007.

Scheicher, Elisabeth. *Die Kunst- und Wunderkammern der Habsburger*. Vienna: Molden, 1979.

Selznick, Brian. *The Invention of Hugo Cabret*. New York: Scholastic, 2007.

Sherwood, Merriam. "Magic and Mechanics in Medieval Fiction." *Studies in Philology* 44, no. 4 (October 1947): 567–92.

Smith, Cyril Stanley. *A History of Metallography: The Development of Ideas on the Structure of Metals before 1890*. Chicago: University of Chicago Press, 1960.

——. *A Search for Structure: Selected Essays on Science, Art, and History*. Cambridge, MA: MIT Press, 1981.

Smith, Pamela H. *The Body of the Artisan: Art and Experience in the Scientific Revolution*. Chicago: University of Chicago Press, 2004.

Stafford, Barbara Maria, and Frances Terpak. *Devices of Wonder: From the World in a Box to Images on a Screen*. Los Angeles: Getty Research Institute, 2001.

Stirling-Maxwell, William. *The Cloister Life of the Emperor Charles V*. 4th ed. London: John C. Nimmo, 1891.

Stoichita, Victor I. *The Pygmalion Effect: From Ovid to Hitchcock*. Translated by Alison Anderson. Chicago: University of Chicago Press, 2008.

———. *Visionary Experience in the Golden Age of Spanish Art*. Translated by Anne-Marie Glasheen. London: Reaktion Books, 1995.

Stratton, Suzanne L., ed. *Spanish Polychrome Sculpture 1500–1800 in United States Collections*. Exh. cat. New York: Spanish Institute, 1993.

Syson, Luke, et al. *Like Life: Sculpture, Color, and the Body*. Exh. cat. New York: Metropolitan Museum of Art, 2018.

Taubert, Johannes. *Polychrome Sculpture: Meaning, Form, Conservation*. Edited by Michele D. Marincola. Translation from the German by Carola Schulman. Los Angeles: Getty Conservation Institute, 2015.

Teklemariam Bache, Yohannes. *Verum, pulchrum et bonum: Miscellanea di studi offerti a Servus Gieben in occasione del suo 80 compleanno*. Rome: Istituto Storico dei Cappuccini, 2006.

Thomas, Keith. *Religion and the Decline of Magic*. London: Penguin Books, [1971] 1991.

Trevor-Roper, Hugh. *Princes and Artists: Patronage and Ideology at Four Habsburg Courts 1517–1633*. London: Thames & Hudson, 1976.

Truitt, E. R. *Medieval Robots: Mechanism, Magic, Nature, and Art*. Philadelphia: University of Pennsylvania Press, 2015.

Trusted, Marjorie. "Art for the Masses: Spanish Sculpture in the Sixteenth and Seventeenth Centuries." In *Sculpture and Its Reproductions*, edited by Anthony Hughes and Erich Ranfft, 46–60. London: Reaktion Books, 1997.

———. *The Arts of Spain: Iberia and Latin America 1450–1700*. University Park: Pennsylvania State University Press, 2007.

———. "Moving Church Monuments: Processional Images in Spain in the Seventeenth Century." *Church Monuments: Journal of the Church Monuments Society* 10 (1995): 55–69.

———. *Spanish Sculpture: Catalogue of the Post-Medieval Spanish Sculpture in Wood, Terracotta, Alabaster, Marble, Stone, Lead and Jet in the Victoria and Albert Museum*. London: Victoria and Albert Museum, 1996.

Vanhaelen, Angela. "Strange Things for Strangers: Transcultural Automata in Early Modern Amsterdam." *Art Bulletin* 103, no. 3 (September 2021): 42–68.

Villalon, L. J. Andrew. "The 1562 Head Injury of Don Carlos: A Conflict of Medicine and Religion in Sixteenth-Century Spain." *Mediterranean Studies* 22, no. 2 (2014): 95–134.

———. "Putting Don Carlos Together Again: Treatment of a Head Injury in Sixteenth-Century Spain." *Sixteenth Century Journal* 26, no. 2 (1995): 347–65.

Vincent, Clare. "Diana, Goddess of the Showering Arrows: A Seventeenth-Century German Automaton Clock." *Yale University Art Gallery Bulletin* [European Art] (2016): 75–87.

Von Herz, Adelheid. "Androiden des 16. Jahrhunderts." Master's thesis, University of Hamburg, 1990.

Voskuhl, Adelheid. *Androids of the Enlightenment: Mechanics, Artisans, and Cultures of the Self*. Chicago: University of Chicago Press, 2013.

Wagner, Michael, and Ulrike Weinhold. "Der Diana-Automat im Grünen Gewölbe: Technologische und künstlerische Aspekte einer Augsburger Figurenuhr" [The Diana automaton in the Green Vault: Technological and artistic aspects of an Augsburg figure clock]. *Jahrbuch der Staatlichen Kunstsammlungen Dresden* (2002–3): 35–46.

Webster, Susan Verdi. *Art and Ritual in Golden-Age Spain: Sevillian Confraternities and the Processional Sculpture of Holy Week*. Princeton, NJ: Princeton University Press, 1998.

———. "Shameless Beauty and Worldly Splendor: On the Spanish Practice of Adorning the Virgin." In *The Miraculous Image in the Late Middle Ages and Renaissance*, edited by Erik Thunø and Gerhard Wolf, 249–71. Rome: L'Erma di Bretschneider, 2004.

Weinryb, Ittai, ed. *Agents of Faith: Votive Objects in Time and Place*. Exh. cat. New York: Bard Graduate Center, 2018.

Wethey, Harold E. *Alonso Cano: Painter, Sculptor, Architect*. Princeton, NJ: Princeton University Press, 1955.

Wood, Gaby. *Edison's Eve: A Magical History of the Quest for Mechanical Life*. New York: Knopf, 2002.

Zanetti, Cristiano. *Janello Torriani and the Spanish Empire: A Vitruvian Artisan at the Dawn of the Scientific Revolution*. Leiden: Brill, 2017.

Acknowledgments

We began this book more than twenty years ago, our progress always interrupted by workshop and studio. From the start, colleagues who believed in our project helped us: Carlton Newton, Evelyn Lincoln, Carlene Stephens, Rosamond Purcell, Fredrika Jacobs, Clifford Edwards, and Ashley Kistler. In the past six years we asked Lory Frankel, Erin Hollaway Palmer, Jeanne Schinto, and David Morris to read our full manuscript and act as our editors, each at a key moment as we kept working. During this time, Judith Hudson, Scott Nolley, Brittany Nelson, Pat Keck, Michael B. Arthur, Bob Frishman, and Cody Whitby continually contributed their knowledge and skills. For advice and counsel on points across the horizon of our study, we thank Susan Bielstein, Pamela H. Smith, Susan Verdi Webster, Jessica Riskin, Alison Luchs, Mary Levkoff, Steve and Jeremie Ryder, Andrew Villalon, Karen Lang, Nancy Princenthal, Paulus Rainer, Marina Belozerskaya, Hanoch Ben-Yami, Michael Schreffler, William R. Newman, Jane Kamensky, Daniel Dennett, Garry Wills, Frank Fehrenbach, Peter Friess, Murakami Kazuo, Ian Greaves, José Luis Romero Torres, and Wolfram Koeppe. Our editors at Getty Publications, Karen Levine, Tevvy Ball, Nola Butler, and consultant Sheila Berg, as well as Danielle Brink, Michelle Deemer, Jeffrey Cohen, and consultant Amy McFarland, together with the outside readers the Getty engaged, shepherded the book to completion.

The monk walks out of Rosamond Purcell's pictures into the reader's hands. We saw things in her photographs that we missed looking at the monk with our own eyes. Her camera captures signals, as from a distant star, marking the centuries it took the monk to reach us. Dennis Purcell, magus in his own right, scanned and color corrected the 35mm transparencies Rosamond made at the museum.

Curators and conservators in museums in France, Germany, Austria, Italy, and Hungary contributed essential data on the automata in their collections and welcomed Elizabeth on visits to study them in 2019 and 2020. We thank Cyrille Foasso and Rémi Catillon at the Musée des arts et métiers, Conservatoire national des arts et métiers in Paris; Thomas Rebényi at the Deutsches Museum in Munich; Raphael Beuing at the Bayerisches Nationalmuseum in Munich; Paulus Rainer at the Kunsthistorisches Museum in Vienna; and Giorgio Gregato at the Giorgio Gregato Orologeria in Milan. From Budapest, although the Iparművészeti Múzeum was closed for restoration, Anna Ecsedy, curator, sent photographs, videos, bibliography, and details of the automaton saint; the former curator, Ágnes Prékopa, responded to Elizabeth's first queries in 1999. Rudolf Distelberger, from 1971 to 2002 the distinguished curator and director of the *Kunstkammer* in Vienna's Kunsthistorisches Museum, answered our first questions in 1999-2000 on the mechanism and history of the cittern player.

The automaton monk appeared in public for the first time (in our day) in the landmark 1980 exhibition, *The Clockwork Universe*, curated by Klaus Maurice and Otto Mayr. Mayr, a distinguished curator at the Smithsonian's National Museum of History and Technology, had pursued the automaton for the museum collection and acquired it in time for the show. Silvio Bedini, Mayr's colleague at the museum, contributed one of the essays to the exhibition catalog. Bedini, curator at the founding of the National Museum of History and Technology in 1964, later rose to deputy director of the museum, then keeper of rare books at the museum's Dibner Library. Both Mayr (1992) and Bedini (2000) were awarded the Leonardo da Vinci Medal by the Society for the History of Technology. With the scholars Lynn T. White Jr., Cyril Stanley Smith, Bern Dibner, Derek J. de Solla Price, Eugene Ferguson, and others, they helped establish and shape the discipline of the history of technology as we know it. Both also wrote about the history of automata. In 1978 Otto and Silvio invited David Todd to become specialist in the physical care of the timekeeping collection. Silvio's books on Thomas Jefferson and on the great African American clockmaker and surveyor Benjamin Banneker were an inspiration to David, who was newly arrived from England. Otto returned to his native Germany in 1982 to become general director of the Deutsches Museum in Munich, but he imparted his level wisdom in correspondence with us over many years. Silvio had agreed to write our preface. He was just completing a book of his own, his twenty-third—a biography of the seventeenth-century Italian clockmaker and astronomer Giuseppe Campani—at the time of his death in 2007 at the age of ninety. We yet dream of how he would have launched our story. Carlene Stephens, curator in the Division of Work and Industry at the National Museum of American History (NMAH; renamed in 1980), played a major role behind the scenes in preparations for *The Clockwork Universe* exhibition. She was there when the monk arrived in 1977, and she was there when David himself arrived the next year. Her own exhibitions on the history of timekeeping, navigation, sound recording, robots, and GPS technology in American culture have documented the shift from mechanical to electronic and then atomic timekeeping in the twentieth century. Robotics appears everywhere in her world. She knows more of the monk's history than anyone alive. Through Carlene, many members of the NMAH staff helped us: Beth Richwine, senior objects conservator; Scott Nolley, former conservator of exhibits and loans (now head conservator at the Hirshhorn Museum); Dawn P. Wallace and her staff in the Conservation Department; and the archivists Stephanie Kurasz and Kay Peterson of the NMAH Digital Programs Office.

We thank the staff in libraries at Harvard, Brown, Dartmouth, the Library of Congress, and the Smithsonian's Dibner Library of the History of Science and Technology for granting access to rare books. For the far greater number of books still in circulation that we sought, the longtime librarian Jean Scott and her interlibrary loan staff at James Branch Cabell Library, Virginia Commonwealth University, never failed us.

Between 1999 and 2002, the historian Alex Keller of the University of Leicester guided Elizabeth through the then-current scholarship on the life and work of Juanelo Turriano and the groundbreaking research on Juanelo by his close friend José A. García-Diego. García-Diego, much quoted in this book, established the Fundación Juanelo Turriano in Madrid in 1987, which has since published significant new research on Juanelo, including the work of Cristiano Zanetti. Equal to such pioneers, from a separate sphere, Adelheid von Herz's thesis will remain foundational to all future study of this family of walking androids.

We thank Leslee Asch and Cheryl Henson, who invited us to present our work on the monk for the symposium "Forms in Motion" at the 2000 Henson International Festival of Puppet Theater in New York. Andrew Periale, editor of the journal *Puppetry International*, published Elizabeth's first essay on the monk that year. Expanded essays were published between 2002 and 2007 by Mary Flinn and Michael A. Keller, editors of *Blackbird: An Online Journal of Literature and the Arts*, hosted by the Virginia Commonwealth University Department of English; Marsha Meskimmon and Martin L. Davies in their anthology, *Breaking the Disciplines: Reconceptions in Knowledge, Art, and Culture*; and Jessica Riskin in *Genesis Redux: Essays on the History and Philosophy of Artificial Life*. In 2002 we cochaired a panel, "The Automaton, the Homunculus, and Other Artificial Beings: Towards an Interdisciplinary History," for the annual College Art Association in the United States. These opportunities, and diverse venues where we have given talks since, pushed us to consider the monk from many points of view.

Henning Bauer's translations of German sources from across centuries carried us through the full span of our research and writing. Thanks also to Susanne Böer and, for eleventh-hour wisdom, Joe Hannan. Wenrui Zhao, William M. Barton, and Anthony T. Padovano translated Latin texts. Daniëlle Slootjes translated Servus Gieben's Dutch. For help with Spanish: María Lourdes De Panbehchi, Alain Gelbman, Adam Meuse, Susan Verdi Webster, and Carlton Newton. Lory Frankel edited Elizabeth's French. Italian sources were translated for us by Silvio Bedini himself and by Evelyn Lincoln, Cinzia Corubolo, Sally Hill, Miriam Hurley, and Mario Pereira. In Milan, when the clockmaker Giorgio Gregato set the *nobil matrona* in motion for Elizabeth, his son-in-law, Alessandro Sperlari, translated our animated conversation.

Robert and Karen Duncan and Marc and Kathy LeBaron granted Elizabeth a 2016 residency at PVArt Studio in Puerto Vallarta, Mexico, giving her time to write. In 2021 Ellen Cassilly and Frank Konhaus gave her support and time at the Cassilhaus artist residency in Durham, North Carolina. Through Frank and Ellen, Elizabeth met Wallace K. Lightsey, who provided counsel on rights and publication.

Friends supported our work—some directly, some by example, many by grace: Joseph Seipel, Larry Fane, Gail Mazur, Michael Mazur, Wendy Ewald, Michael Maas, Chris Taggart, Heji Kim, John Powers, Karen Rader, Joan Gaustad, Stephen Vitiello, Dan Hurlin, Mike Belzer, Bill Barrette, George Ferrandi, Kathleen Graber, Kenneth Gross, Denise Markonish, Mark Sloan, Paul Jacobs, Sara Monroe, Rob Carter, Lynton Gardiner, Leslie Wayne, Teresita Fernández, Carol Corey, Charlotte Potter Kasic, Allan M. Heyward, and Olympia Stone. Elizabeth's father, John S. King, read drafts. Matthew Charboneau made the first digital scans of David's drawings. Brittany Nelson scanned more drawings and David's photographs, cleaning and correcting the files to David's specifications. Carlton Newton made 3-D CAD renderings of parts of the monk's clockwork for David. Eric Stepp assisted on the day Rosamond photographed the monk at the museum. Kate Scherer and Rachael Starbuck aided Elizabeth's research. Pamela H. Smith gave us confidence, as artisans ourselves, to add our voices to the conversation about the history of artificial life. Inspiration came always from the fabled David Beck.

Cyril Boniface taught David Todd the art of clockmaking. The last in a venerable line of English clockmakers, Cyril learned the craft from his father, who scorned the use of electricity and lit his workshop with oil lamps. David rose from apprentice under Cyril to senior clockmaker at Meyrick Neilson of Tetbury, in Gloucestershire, England, a firm run by Brigadier James Meyrick Neilson and his wife, Moyra, specializing in the restoration of seventeenth- and eighteenth-century English clocks. David credits the ten years he worked with Cyril and "the General" for his lifelong dedication to the history and practice of the clockmaker's craft. David's father, William Joseph Todd, whose sketches in letters from North Africa during World War II became family treasures, was his first inspiration.

Of all the people to whom we wrote who knew about this automaton in the early years of its recovery, Servus Gieben wrote us back. Our correspondence with him brought us closest, looking at the monk, to thinking back in time. He pointed us to sources essential to our search and trusted us to quote from his letters. We are equally indebted to two people close to him. Yohannes Teklemariam Bache, president of the Istituto Storico dei Cappuccini in Rome from 2014 to 2017 and current director of its Museo Francescano, edited the comprehensive tribute to Servus Gieben, *Verum, pulchrum et bonum: Miscellanea di studi offerti a Servus Gieben in occasione del suo 80th compleanno*, which includes a bibliography of Gieben's published works. For milestones in the life of Father Servus, we thank his niece Lydia Beerkens, managing director and senior conservator of modern and contemporary art at Stichting Restauratie Atelier Limburg in Maastricht, the Netherlands.

Over twenty years we have been graced with august help. All mistakes will be our own, for time will shine more light on these mysterious clockwork automata. Whatever survives of what we have written, let it include our deepest thanks to Sheila Todd and Carlton Newton.

About the Authors

Elizabeth King, professor emerita of sculpture and extended media at Virginia Commonwealth University, is a sculptor and writer interested in the ways artists have crafted the human form, from sculptural traditions across cultures, to medical models, puppets, and automata. She works in wood, porcelain, and bronze, and often animates her precisely movable sculptures on stop-motion film. Museums that own her work include the Hirshhorn Museum, in Washington, DC; the Minneapolis Institute of Art; the Virginia Museum of Fine Arts, in Richmond; the Hood Museum of Art at Dartmouth; and the Metropolitan Museum of Art, in New York. She is the author of *Attention's Loop: A Sculptor's Reverie on the Coexistence of Substance and Spirit* (1999).

From 1978 to 2005 **W. David Todd** was conservator of timekeeping at the Smithsonian Institution's National Museum of American History, where he is now associate curator emeritus. Trained in his native England, he has been a practicing clockmaker since the late 1960s. He was accepted as a Craft Member of the British Horological Institute in 1973. An expert on sixteenth- to eighteenth-century English, European, and American clockwork, David has conserved and restored some of the world's finest early clocks and automata, among them the clocks Thomas Jefferson commissioned for Monticello. He loves best, while working on a clock, to listen to music from the same period, the sixteenth century his favorite.

Rosamond Purcell is known for her photography in natural history museums. She has collaborated with scientists as well as scholars of Shakespeare and magic on a range of publications and projects. Among the many books featuring her photography is *A Glorious Enterprise: The Academy of Natural Sciences of Philadelphia and the Making of American Science*, by Robert M. Peck and Patricia Tyson Stroud; and her reconstruction of an iconic curiosity cabinet by Ole Worm, a seventeenth-century Danish linguist, physician, and philosopher, is now in the permanent collection of the Natural History Museum, in Copenhagen. A retrospective exhibition of her work, *Rosamond Purcell: Nature Stands Aside*, was held at the Addison Gallery of American Art, in Andover, MA, from September 2022 through January 2023.

Illustration Credits

Front cover: photograph © Rosamond W. Purcell
Color plates on pp. ii, iv, vi, 5, 6, 8–9, 36, 38, 39, 72, 74–75, 126, 128, 129, 166, 168, 169, 197: photographs © Rosamond W. Purcell. **Figs. I.1–I.2, 10.20, 20.1:** Image courtesy of the National Museum of American History, Smithsonian Institution. **Fig. I.3:** Image courtesy of the National Museum of American History/ Photograph by Margery Albertini. **Fig. 1.1:** Historic Collection/Alamy Stock Photo. **Fig. 1.2:** Niday Picture Library/Alamy Stock Photo. **Fig. 2.1:** © Museu Nacional de Soares dos Reis Direção-Geral do Património Cultural/Arquivo de Documentação Fotográfica (DGPC/ADF). **Fig. 3.1:** Fundación Pública Andaluza Rodríguez-Acosta. Photograph by Javier Algarra. **Figs. 3.2, 12.2, 12.6a–12.6b, 14.2a, 16.2–16.3, 17.1, 20.2:** Photographs by Elizabeth King. **Figs. 4.1–4.2:** Span 165.1.7. Houghton Library, Harvard University. **Fig. 4.3:** Fundación Juanelo Turriano of Madrid. **Fig. 4.4:** © Archivo Fotográfico Museo Nacional del Prado. **Figs. 5.1, 15.1–15.2, 17.3a–17.3b, 17.5a–17.5b, 22.1:** KHM-Museumsverband. **Fig. 6.1:** Museo Francescano dell'Istituto Storico dei Cappuccini di Roma. **Figs. 8.1–8.6, 10.3a–10.3d, 10.4–10.5, 10.6a–10.6b, 10.7, 10.8a–10.8b, 10.9a–10.9b, 10.10–10.12, 10.13a–10.13c, 10.14, 10.16–10.17, 10.18a–10.18b, 10.19:** Courtesy of David Todd, Smithsonian Associate Curator Emeritus, from his Smithsonian project files. **Fig. 8.7:** Photograph © Rosamond W. Purcell. **Fig. 9.1:** Photograph by Margery Albertini. **Fig. 10.1:** National Museum of American History & Smithsonian Museum Conservation Institute (Radiograph taken by staff at the Museum Conservation Institute, formerly the Conservation Analytical Laboratory). **Fig. 10.2:** Diagram by Carlton Newton (based on NMAH video, shot from above, of the monk in performance, by Scott Nolley, Conservator, Exhibits and Loans; Dawn P. Wallace, Objects Conservator; and the Conservation Department of the National Museum of American History). **Fig. 10.15:** National Museum of American History & Smithsonian Museum Conservation Institute (radiograph taken by staff at the Museum Conservation Institute, formerly the Conservation Analytical Laboratory). **Figs. 12.1, 12.3–12.4:** Photo: Deutsches Museum/Konrad Rainer. **Fig. 12.5:** Courtesy of Bob Frishman. **Figs. 13.1–13.3, 13.5:** © Iparművészeti Múzeum/Photo by Ágnes Soltész-Haranghy. **Fig. 13.4:** © Iparművészeti Múzeum Archive. **Fig. 14.1a:** Division of Work and Industry, National Museum of American History, Smithsonian Institution. **Figs. 14.1b, 14.2b, 18.2:** Courtesy of David Todd, Smithsonian Associate Curator Emeritus, from his Smithsonian project files/Photo by David Todd. **Fig. 14.1c:** National Museum of American History & Smithsonian Museum Conservation Institute (Radiograph taken by staff at the Museum Conservation Institute, formerly the Conservation Analytical Laboratory). **Figs. 14.3a–14.3b:** © Musée des arts et métiers-Cnam, Paris/photo J-C Wetzel. **Fig. 14.3c:** © 1960 Conservatoire national des arts et métiers. **Fig. 16.1:** private collection, Milan, Italy/photo by Mario Liguigli ©. **Fig. 17.2:** © Musée des arts et métiers-Cnam, Paris/photo P. Faligot. **Fig. 17.4:** Musée d'horlogerie du Locle-Château des Monts, Switzerland/Photo R. Sterchi. **Fig. 18.1:** David Todd. **Fig. 18.3:** Photo by Murakami Kazuo. **Fig. 19.1:** Photo credit: bpk Bildagentur/Skulpturensammlung und Museum für Byzantinische Kunst, Berlin, Germany/Photo: Antje Voigt/Art Resource, NY. **Figs. 21.1–21.3:** SLUB Dresden/Digital Collections/Lit.Germ.rec.B.2039. **Fig. 21.4:** Getty Research Institute, Los Angeles (2020.PR.2). **Fig. 22.2:** © LIECHTENSTEIN. The Princely Collections, Vaduz-Vienna. **Fig. 23.1:** Getty Research Institute, Los Angeles (86-B22668). **Fig. 23.2:** Print Collection, University of San Diego, purchased through the Legler Benbough Student Acquisition fund by Katelyn Allen (Class of 2018). **Fig. 23.3:** Getty Research Institute, Los Angeles (2015.PR.2). **Fig. 23.4:** © Museo Nacional del Prado. **Fig. 23.5:** Photo Credit: Copyright of the image Museo Nacional del Prado/Art Resource, NY. **Fig. 24.1:** Collection of Convento S. Maria La Nova, Naples. **Fig. 24.2:** Museo Francescano dell'Istituto Storico dei Cappuccini di Roma/ Photograph by Lydia Beerkens, Roma 2005. **Figs. 25.1, 25.5:** Instituto del Patrimonio Cultural de España, Ministerio de Cultura y Deporte. **Fig. 25.2:** The San Diego Museum of Art. **Fig. 25.3:** Photo by José Luis Gutiérrez-jlgfotografo.com. **Fig. 25.4:** © Photography: Image M.A.S./photographers. **Fig. 25.6:** Photographer: Jose Luis Gutierrez. Bishopric of Malaga©. **Fig. 25.7:** © Photo José María Pérez Concellón. **Fig. 25.8a:** © Institut Amatller d'Art Hispànic (foto Mas C-85682 Foto Pelai Mas 1936) Image courtesy of Susan Verdi Webster. **Fig. 25.8b:** Courtesy of The Hispanic Society of America, New York/Image courtesy of Susan Verdi Webster. **Figs. 25.9–25.11:** Courtesy of The Hispanic Society of America, New York. **Fig. 25.12:** Photograph by Scott W. Nolley, then Conservator of Exhibits and Loans, National Museum of American History, Smithsonian Institution. **Fig. 26.1:** From the collection of the Minneapolis Institute of Art/ Photographer Charles Walbridge. **Fig. 26.2:** Photo Credit: © National Portrait Gallery, London/Art Resource, NY. **Figs. 26.3, 26.4:** © Musée des Beaux-Arts de Dijon/François Jay.

Index

NOTE: *Italic* page numbers indicate illustrations.

Published by Getty Publications, Los Angeles
1200 Getty Center Drive, Suite 500
Los Angeles, California 90049-1682
getty.edu/publications

Tevvy Ball
PROJECT EDITOR

Sheila Berg
MANUSCRIPT EDITOR

Amy McFarland
DESIGNER

Michelle Deemer
PRODUCTION

Danielle Brink
IMAGE AND RIGHTS ACQUISITION

Distributed in the United States and Canada by the University of Chicago Press

Distributed outside the United States and Canada by Yale University Press, London

Printed in China

FRONT COVER: Automaton of a Friar, mid-sixteenth century. Spain or South Germany. Wood, fabric, iron clockwork (detail). National Museum of American History, Smithsonian Institution, Washington, DC

BACK COVER: X-ray of the automaton's head (see **FIG. 10.15**)

PAGE II: Head and torso of the automaton, seen from the back

PAGE IV: David Todd's model of the automaton

PAGE VI: Detail of the mainspring barrel clickwork

LIBRARY OF CONGRESS CATALOGING-IN-PUBLICATION DATA

Names: King, Elizabeth, 1950- author. | Todd, W. David, author. | Purcell, Rosamond Wolff, photographer. | Getty Publications, publisher.
Title: Miracles and machines : a sixteenth-century automaton and its legend / Elizabeth King and W. David Todd ; with photographs by Rosamond Purcell.
Description: Los Angeles : Getty Publications, [2023] | Includes bibliographical references and index. | Summary: "This richly illustrated volume tells the uncanny story of a sixteenth-century automaton and the legend that has grown up around it"— Provided by publisher.
Identifiers: LCCN 2023001741 (print) | LCCN 2023001742 (ebook) | ISBN 9781606068397 (hardcover) | ISBN 9781606068403 (ebook)
Subjects: LCSH: Robots—Europe—History—16th century. | Sculpture, European—16th century. | Art and technology—Europe—History—16th century. | Monks in art.
Classification: LCC NK3649 .K56 2023 (print) | LCC NK3649 (ebook) | DDC 629.8/92—dc23/eng/20230315
LC record available at https://lccn.loc.gov/2023001741
LC ebook record available at https://lccn.loc.gov/2023001742